AF719183

70 Structure and Bonding

Editors:
M. J. Clarke, Chestnut Hill · J. B. Goodenough, Oxford
J. A. Ibers, Evanston · C. K. Jørgensen, Genève
D. M. P. Mingos, Oxford · J. B. Neilands, Berkeley
G. A. Palmer, Houston · D. Reinen, Marburg
P. J. Sadler, London · R. Weiss, Strasbourg
R. J. P. Williams, Oxford

Bioinorganic Chemistry

With Contributions by
P. Aisen B. C. Antanaitis S. J. Berners-Price K. Doi H. Köpf
P. Köpf-Maier P. J. Sadler

With 96 Figures and 31 Tables

QD Structure and bonding ;
461
.S92 ct/t
v.

Springer-Verlag
Berlin Heidelberg NewYork
London Paris Tokyo

Editorial Board

Professor *Michael J. Clarke,* Boston College, Department of Chemistry, Chestnut Hill, Massachusetts 02167, U.S.A.

Professor *John B. Goodenough,* Inorganic Chemistry Laboratory, University of Oxford, South Parks Road, Oxford OX1 3QR, Great Britain

Professor *James A. Ibers,* Department of Chemistry, Northwestern University, Evanston, Illinois 60201, U.S.A.

Professor *Christian K. Jørgensen,* Dépt. de Chimie Minérale de l'Université, 30 quai Ernest Ansermet, CH-1211 Genève 4

Professor *David Michael P. Mingos,* University of Oxford, Inorganic Chemistry Laboratory, South Parks Road, Oxford OX1 3QR, Great Britain

Professor *Joe B. Neilands,* Biochemistry Department, University of California, Berkeley, California 94720, U.S.A.

Professor *Graham A. Palmer,* Rice University, Department of Biochemistry, Wiess School of Natural Sciences, P. O. Box 1892, Houston, Texas 77251, U.S.A.

Professor *Dirk Reinen,* Fachbereich Chemie der Philipps-Universität Marburg, Hans-Meerwein-Straße, D-3550 Marburg

Professor *Peter J. Sadler,* Birkbeck College, Department of Chemistry, University of London, London WC1E 7HX, Great Britain

Professor *Raymond Weiss,* Institut Le Bel, Laboratoire de Cristallochimie et de Chimie Structurale, 4, rue Blaise Pascal, F-67070 Strasbourg Cedex

Professor *Robert Joseph P. Williams,* Wadham College, Inorganic Chemistry Laboratory, Oxford OX1 3QR, Great Britain

QD
461
.S92
v.70

ISBN 3-540-50130-4 Springer-Verlag Berlin Heidelberg New York
ISBN 0-387-50130-4 Springer-Verlag New York Berlin Heidelberg

This work is subject to copyright. All rights are reserved, whether the whole or part of the material is concerned, specifically the rights of translation, reprinting, re-use of illustrations, recitation, broadcasting, reproduction on microfilms or in other ways, and storage in data banks. Duplication of this publication or parts thereof is only permitted under the provisions of the German Copyright Law of September 9, 1965, in its version of June 24, 1985, and a copyright fee must always be paid. Violations fall under the prosecution act of the German Copyright Law.

© Springer-Verlag Berlin Heidelberg 1988
Printed in Germany

The use of general descriptive names, trade marks, etc. in this publication, even if the former are not especially identified, is not to be taken as a sign that such names, as understood by the Trade Marks and Merchandise Marks Act, may accordingly be used freely by anyone.

Typesetting: Mitterweger Werksatz GmbH, 6831 Plankstadt, Germany
Printing: Druckhaus Beltz, 6944 Hemsbach, Germany
Binding: J. Schäffer GmbH & Co.KG., 6718 Grünstadt, Germany
2152/3140-543210 – Gedruckt auf säurefreiem Papier

Table of Contents

The Binuclear Iron Centers of Uteroferrin and the Purple Acid Phosphatases
K. Doi, B. C. Antanaitis, P. Aisen 1

Phosphines and Metal Phosphine Complexes: Relationship of Chemistry to Anticancer and Other Biological Activity
S. J. Berners-Price, P. J. Sadler 27

Transition and Main-Group Metal Cyclopentadienyl Complexes: Preclinical Studies on a Series of Antitumor Agents of Different Structural Type
P. Köpf-Maier, H. Köpf 103

Author Index Volumes 1–70 187

The Binuclear Iron Centers of Uteroferrin and the Purple Acid Phosphatases*

Kei Doi[1], Bradley C. Antanaitis[2], and Philip Aisen[1,3]

[1]Department of Physiology and Biophysics, and
[2]Department of Physics, Lafayette College, Easton, Pennsylvania 18042, U.S.A.
[3]Department of Medicine, Albert Einstein College of Medicine, Bronx, New York 10461, U.S.A.

The purple acid phosphatases comprise a group of proteins distinguished by their enzymic activity, distinctive color and, most importantly, the presence of a spin-coupled binuclear iron center. This center exists in two stable interconvertible states: pink, reduced, EPR-visible and enzymically active, with an antiferromagnetically exchanged Fe(II)–Fe(III) cluster, and purple, oxidized, EPR-silent and inert, with the binuclear pair as Fe(III)–Fe(III). In uteroferrin, the purple acid phosphatase of uterine secretions, a transient intermediate species has also been identified. Engendered by the interaction of phosphate with the pink form of uteroferrin, this transient intermediate is purple, EPR-silent and devoid of the contact-shifted proton resonances seen in its pink parent. Nevertheless, it is paramagnetic, with an Fe(II)–Fe(III) couple demonstrable by Mössbauer spectroscopy. Although considerable progress toward characterizing the properties of uteroferrin and the purple acid phosphatases has been achieved in recent years, enigmas persist. The identity of the ligands of each iron atom and how these change in nature or arrangement during redox transitions is still unknown. Perhaps most interestingly, the mechanism of enzymic activity, and the relation of the redox and enzymatic properties of the purple acid phosphatases to their yet uncertain physiological roles, remain to be established.

I. Introduction: History and Perspective 3

II. Molecular Properties 3
  A. Molecular Weight and Carbohydrate Content 3
  B. Primary Structure 4

III. Physical and Chemical Properties 5
  A. Iron Content 5
  B. Spectroscopic Properties 6
    1. Absorption Spectra 6
    2. Resonance Raman Spectra 6
    3. Circular Dichroism Spectra 8
    4. EPR and Magnetic Susceptibility 9
    5. Vector Coupling Model 11
    6. Mössbauer Spectra 12

* This work was supported, in part, by grant DK 15056 from the National Institutes of Health.

© Springer-Verlag Berlin Heidelberg 1988

7. ^{1}H NMR Spectra . . . 13
8. LEFE, ESEEM, and ENDOR Spectra . . . 15
9. EXAFS Spectra . . . 16
10. Interaction With Phosphate and Other Perturbants . . . 17

IV. Enzymatic Properties . . . 19
A. Substrate Specificities . . . 19
B. Activators and Inhibitors . . . 20

V. Biological Roles of the Purple Acid Phosphatases . . . 22
A. Intracellular Mammalian Phosphatases . . . 22
B. Uteroferrin . . . 22
C. Purple Acid Phosphatases in Plants . . . 23

VI. Problems and Perspectives . . . 23

VII. References . . . 24

I. Introduction: History and Perspective

Acid phosphatases are ubiquitously distributed enzymes defined by the pH optimum (usually 4.9–6.0) of their hydrolytic activity toward orthophosphate monoesters. A metal-dependent subclass of these enzymes was first recognized in 1973, when iron-bearing acid phosphatases were isolated from porcine uterine fluid[1] and bovine spleen[2]. Because of their intense colors they have become known as the purple acid phosphatases. Its source, content of iron, and presumed role in iron transport from pregnant sow to fetal pig have earned the porcine protein the euphonious name of uteroferrin.

The presence of iron in these hydrolases came as something of a surprise since, with few exceptions[3–8], iron-requiring enzymes function primarily in electron-transfer reactions. Even more curious was the demonstration that each molecule of the purple acid phosphatases binds two iron atoms in a binuclear cluster[9–12]. As is now appreciated, the redox activity of iron in the purple acid phosphatases plays a central role in regulating their enzymic activities, while the binuclear configuration confers their unique spectroscopic properties. In this review we will consider what has been learned of the chemistry, spectroscopic features, and enzymic properties of the purple acid phosphatases, emphasizing the problems and paradoxes still evading explanation. An historical perspective is provided in Ref. 13.

In addition to its archetypical members, uteroferrin and bovine spleen acid phosphatase, the class of purple acid phosphatases includes proteins isolated from: rat bone and spleen[14,15], spleens of patients with Gaucher's disease[16] or leukemic reticuloendotheliosis[17], equine uterine flushings[18], bovine cortical bone[19], giant cell tumors[20], human placenta[21], and microorganisms[13]. The plant enzymes include an Fe–Zn phosphatase from red kidney beans[22] and an Fe–Fe or Mn(III) protein from sweet potato tubers[23,24]. Although less well-defined and more heterogeneous than their mammalian counterparts, the color and iron content of the plant enzymes warrant their designation as purple acid phosphatases.

II. Molecular Properties

A. Molecular Weight and Carbohydrate Content

Uteroferrin, probably the most studied of the purple acid phosphatases because of its relative abundance, has a molecular weight near 36,000 as calculated from its carbohydrate structure[3] and what is known of its amino acid sequence[25]. This agrees well with estimates of its molecular weight based on SDS-PAGE and sedimentation equilibrium measurements[26]. As evidenced by gel electrophoresis[26] and glycosidase digestion[27], the protein consists of a single polypeptide chain bearing one branched oligosaccharide chain. The carbohydrate chain is composed of five or six mannose and two N-acetylglucosamine residues in a structure recently elucidated by ^{1}H NMR (Sect. III.B.7.)[3]. A small proportion of the oligosaccharide on uteroferrin isolated from allantoic fluid may be phosphorylated, presumably as mannose-6-phosphate since it is in this form that the phosphate is found on uteroferrin released by cultures of endometrium[3]. Only the six-mannose glycan is capable of undergoing enzymic phosphorylation *in vitro,* indicating that an α1,2-linked mannose residue is required[28]. The relationship of the carbohydrate-bound phosphate to the tightly bound inorganic phosphate discussed in Sect. III.B.10. is not known.

Recently, a pink high molecular weight ($M_r \approx 80,000$) form of uteroferrin has been isolated from the uterine secretions of pigs[29]. This form appears to be a heterodimer comprised of the "usual" species of uteroferrin ($M_r \approx 36,000$) noncovalently complexed to an antigenically unrelated polypeptide of $M_r \approx 50,000$. Its optical, EPR, and enzymatic properties resemble closely those of its low molecular weight counterpart, except that the protein remains pink in the presence of inorganic phosphate.

B. Primary Structure

Sequence analyses of uteroferrin and bovine spleen acid phosphatase show greater than 90% homology between the two proteins[25]. This remarkable finding, together with the strong immunochemical cross-reactivity of the proteins[30, 31], suggests they may be products of the same gene. As yet, however, the molecular biology of these proteins is essentially unexplored.

Unlike uteroferrin, which produces a single band of expected mobility for its molecular size by SDS-PAGE, bovine spleen acid phosphatase yields two bands, even when isolated in the continuous presence of proteolytic inhibitors[32]. However, the extensive sequence homology between the spleen and uterine enzyme, and the fact that isolation of the spleen enzyme is accomplished by prolonged acid extraction of spleen homogenates, make it likely that the purple acid phosphatase of spleen is also a single-chain protein[25].

An early analysis of the spleen enzyme showed the presence of hexose and hexosamine[33]. A more recent study has demonstrated an enzyme from human hairy leukemic cells which appears identical to the purple spleen phosphatase, and which binds to ConA-Sepharose and is eluted with methylmannoside[30]. Very likely, therefore, the spleen enzyme, like uteroferrin, is a high mannose glycoprotein, but the structure of its carbohydrate has yet to be determined. In each protein the oligosaccharide chain appears to be attached to an asparagine residue (at position 97 in uteroferrin), since this residue was observable only after treatment with N-glycanase which removes N-asparagine-linked oligosaccharides[25].

Human acid phosphatases (designated Type 5 acid phosphatases on the basis of electrophoretic mobility) have been isolated from the spleens of patients with Gaucher's disease[34] and hairy cell leukemia[30]. These are almost certainly human analogs of bovine spleen acid phosphatase, with similar size, iron content, and enzymic properties.

A purple acid phosphatase, substantially similar to uteroferrin and bovine spleen acid phosphatase in molecular weight, iron content, and binding to ConA-Sepharose, has been isolated from rat spleen[15]. However, the protein is heterogeneous by isoelectric focusing, perhaps because of variations in carbohydrate content. This protein appears identical to a purple acid phosphatase found in developing rat bone[35], and also bears some similarities to a less well characterized acid phosphatase with protein phosphotyrosyl phosphatase activity recently purified from bovine bone[19]. What evidence there is now available therefore indicates that the mammalian purple acid phosphatases constitute a class of proteins diverse in origin, but similar in structure and enzymatic properties.

The purple acid phosphatases of plant origin, in contrast, comprise a more heterogeneous class of enzymes[13]. A phosphatase from sweet potato, originally thought to depend upon Mn(III) for activity[24], has recently been shown to consist of two apparently identical 53 kDa subunits, the holoenzyme bearing two atoms of iron[23]. However, discordance in the metal content may reflect differences in the species of sweet potato used in the isolation procedure, the Mn(III) protein being isolated from Japanese "Kintoki"[24] and the two-iron phosphatase from "local" American tubers[23]. In either case, at least one of these metal atoms is likely to be coordinated to tyrosyl residue(s) to account for the intense violet color and resonance Raman spectrum of the protein[24]. No direct evidence that the iron atoms in the two-iron protein are magnetically coupled is yet available, but spectroscopic similarities between the sweet potato and animal enzymes make this a reasonable conjecture.

Another plant phosphatase, from kidney bean, is also a dimer of approximately 130 kDa and further resembles the sweet potato enzyme in amino acid composition and visible absorption spectrum[22]. Surprisingly, however, it is a zinc-iron enzyme and in some ways resembles the reconstituted zinc-iron derivative of uteroferrin[36]. Whether the zinc and iron atoms are neighbors is not yet known, but the possibility is appealing.

III. Physical and Chemical Properties

A. Iron Content

Until recently the iron-binding capacity of uteroferrin was controversial, with claims of either one or two irons bound per molecule[26, 37–39]. Since similar preparative and analytic procedures had been used by all groups studying the protein, the sources of the discrepancies were not apparent[13]. While disparities in molecular weight and extinction coefficients used to estimate protein concentration existed, these were too small to account for the differences in iron content. Recent work comparing iron analyses on protein samples that were either phosphate-free or had one phosphate bound per molecule indicate that earlier assays were confounded by the presence of tightly bound phosphate which interfered with colorimetric detection of iron by acid-release methods[37]. It is now clear that uteroferrin is isolated only as a two iron protein which, depending upon its history, may have up to one tightly bound phosphate per molecule. This intimate association of iron and phosphate has also been confirmed for the bovine spleen enzyme since the high molecular weight fragment ($M_r \approx 24{,}000$) of TCA-precipitated protein retains one iron atom and one phosphate group[40].

Preparations of uteroferrin-like proteins have also been reported with less than two irons bound per molecule[29]. However, present work on the stable mixed-metal (Fe–Zn, Fe–Cu and Fe–Hg) forms of uteroferrin, as well as the isolation of a naturally occurring

Fe–Zn acid phosphatase, raise the possibility that in these "low iron" preparations zinc or another metal ion had replaced a portion of the protein's iron, thus stabilizing it with fewer than two iron atoms bound per molecule[22, 36)].

B. Spectroscopic Properties

1. Absorption Spectra

When treated with oxidants such as ferricyanide or hydrogen peroxide, uteroferrin and other two-iron acid phosphatases, freed of orthophosphate or other strongly interacting anionic inhibitors, are driven to their purple forms characterized by a broad intense absorption maximum between 550–570 nm and a prominent near-UV shoulder between 315–320 nm[12, 13, 37)]. The pink form of these proteins, generated by mild reductants such as 2-mercaptoethanol, ascorbate, or ferrous ion, show absorption maxima shifted to 505–510 nm and their near-UV shoulders, now less conspicuous, shifted to 310 nm[13, 32, 41)]. As expected, both redox forms of these proteins have a sharp, protein-dominated peak at 280 nm. In the case of uteroferrin, this promontory is flanked by several more or less well-defined shoulders[13)]. Further, the intensity of the 280 nm peak in uteroferrin is sensitive to the binding of anionic inhibitors, most notably orthophosphate[37)]. Quite remarkable is the preservation of the integrated intensity (oscillator strength) of the protein's primary visible absorption band following these redox-initiated color conversions[13, 42–44)]. This observation provided the first clue that only one of uteroferrin's iron atoms is chromophoric. Studies of several mixed-metal (Fe–Zn, Fe–Cu and Fe–Hg) forms of the protein, showing little loss of visible absorption per molecule, confirm this suggestion[36)]. The recent discovery of a naturally occurring Fe–Zn purple acid phosphatase from the red kidney bean, *Phaseolus vulgaris,* having essentially the same absorption characteristics as two-iron uteroferrin, further substantiates this inference[22)]. Short exposure of uteroferrin or the bovine spleen enzyme to dithionite in the presence of suitable chelators allows the selective removal of only one of the protein's iron atoms and preparation of stable mixed-metal forms of either enzyme[9, 36)]. In contrast, prolonged exposure to the same reductant produces a colorless, iron-free protein which possesses no enzymic activity[13, 33, 41, 42, 45)]. This demonstrates that iron is not only essential for enzymic activity, but is also an integral constituent of the redox-sensitive chromophores of the purple acid phosphatases.

2. Resonance Raman Spectra

Two iron-bearing purple acid phosphatases, uteroferrin and the bovine spleen enzyme, have been extensively studied by laser-Raman spectroscopy (Fig. 1)[13, 42–44)]. The detection of resonance-enhanced internal tyrosyl vibrations indicates that the deep purple color of these proteins arises primarily from tyrosine-to-Fe(III) charge-transfer transitions, thus identifying purple acid phosphatases as members of a burgeoning class of iron-tyrosinate proteins[38, 42, 46)]. The preservation of the high-frequency (1150–1600 cm^{-1}) tyrosyl quartet in the pink forms (Fig. 1B) of both proteins further indicates that tyrosine coordination to the chromophoric iron is maintained following reduction[13, 42, 43)]. This

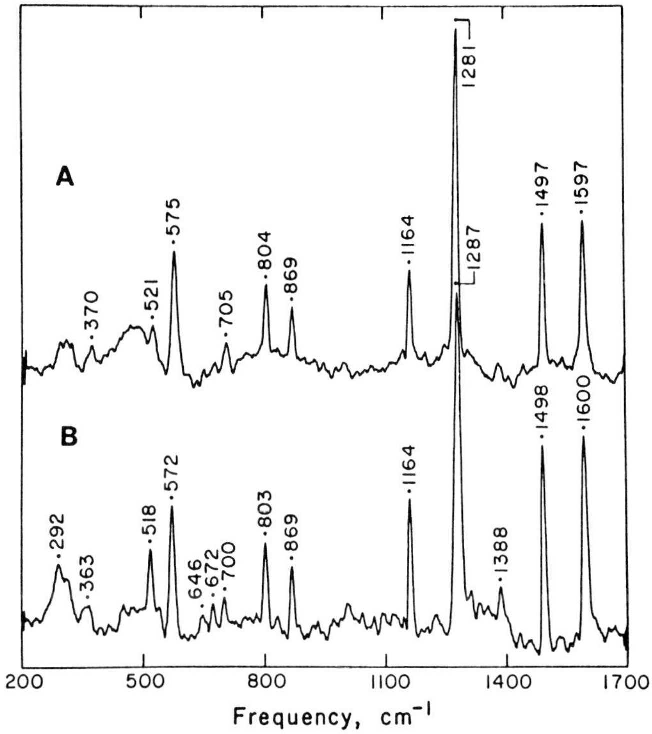

Fig. 1A, B. Resonance Raman spectra of 5 mM purple **(A)** and 2.7 mM pink **(B)** bovine spleen acid phosphatase at pH 5.0 and 5 °C. Data were collected with use of 514.5 nm excitation, 100 mW incident power, and 140° back-scattering geometry. The spectra, representing an average of 3 **(A)** and 8 **(A)** scans were subjected to a 13-point smoothing procedure. (Adapted from Ref. 42)

observation leaves little doubt that one or more tyrosyl residues are coordinated to an iron that remains ferric after reduction, since Fe(II)-phenolate complexes are not expected to give rise to ligand-metal charge-transfer transitions in the visible region[43, 47]. Thus, laser-Raman and optical studies of native acid phosphatases, as well as optical studies of mixed-metal hybrids, convincingly demonstrate that these phosphatases bind two types of iron, a chromophoric tyrosine-coordinated species that remains ferric and a colorless species that reversibly cycles between the ferric and ferrous states during redox-induced color changes[13, 42, 43, 47].

Bands other than those associated with pure internal tyrosyl vibrations have also been detected in both phosphatases[12, 42, 46]. Most notable among these are intense, polarized bands at 872, 805, 575, and 521 cm^{-1} [46]. The pair of bands at 872 and 805 cm^{-1} have been ascribed to a Fermi doublet, which are bands of coupled vibrations arising when an overtone of one vibration interacts with the fundamental of another vibration having the same symmetry[46]. Most probably, this Fermi resonance is between a tyrosyl ring breathing mode ($\nu = 830$ cm^{-1} [42, 46]) and an overtone of an out-of-plane bending mode of the same symmetry ($\nu_{16a} = 410$ cm^{-1} (nomenclature of Ref. 48[42, 46])). This assignment also implies resonance enhancement of the ν_{16a} fundamental (the mode near 400 cm^{-1}) and recent low temperature Raman studies of both acid phosphatases have detected a band at

417 cm^{-1}, making this assignment all the more plausible[42, 46]. The line at 575 cm^{-1} has been attributed to a combination mode with substantial Fe–O character[42, 46]. The 521 cm^{-1} feature is particularly interesting because its intensity doubles upon reduction. Its sensitivity to the oxidation state of the protein, strong resonance enhancement, and insensitivity to ^{18}O substitution suggest that it may represent a coupled Fe-tyrosinate and Fe-(bridging ligand) vibration, perhaps involving the oxygen of a proposed bridging carboxylate group[42]. In analogy to hemerythrin and ribonucleotide reductase, it is tempting to suppose that the strong antiferromagnetic coupling between the irons of these phosphatases is due to a μ-oxo bridge[49–51]. The earmark of such a bridge is a resonance-enhanced vibration in the vicinity of 500 cm^{-1} which is sensitive to ^{18}O-substitution[42]. However, isotopic substitution experiments on the splenic phosphatase have failed to detect any such band[42]. As Fig. 1 shows, the low frequency spectrum of the splenic enzyme is rich in additional structure, but no credible assignment of these peaks is yet possible. We note that the 292 cm^{-1} peak is in the right region for Fe–N vibrations[52] and, indeed, histidine ligation is supported by both NMR and electron spin-echo studies[47, 53, 54].

3. *Circular Dichroism Spectra*

So far, CD spectra have been reported only for uteroferrin among the iron-containing acid phosphatases[12, 46]. These studies show that all absorption bands in the visible and near-UV regions of both forms of the protein have low optical activity, an observation consistent with their assignment as tyrosinate-to-Fe(III) charge-transfer transitions[12, 13, 46]. CD spectra of synthetic phenolate-ferric complexes typically reveal a pair of widely split transiticns attributable to phenolate-to-Fe(III) charge-transfer transitions, one of which lies in the visible and the other in the near-UV region of the spectrum[55]. Moreover, as illustrated by the model compound EDDHA (ethylenediamine di(o-hydroxy-phenylacetate)), the difference in energy between this pair of transitions provides a rough measure of the crystal field strength (10 Dq) about the ferric ion[55]. It is tempting to assume, therefore, that the splitting of uteroferrin's primary visible absorption band and prominent shoulder, each into two optically active transitions, signifies Fe(III) coordination by two inequivalent tyrosyl residues[13, 43, 56]. In keeping with this interpretation, the pair of lower energy transitions (i.e., those at 530 and 345 nm) would be assigned to one tyrosine, as this pair is more strongly influenced by the reductive purple-to-pink conversion, while the remaining pair (at 475 and 305 nm) which appear more sensitive to the binding of phosphate would be assigned to the second tyrosine[12, 56]. The high extinction coefficient of the protein's visible absorption band (4000 M^{-1} cm^{-1} at 550 nm) is consistent with the binding of two tyrosines to the protein's chromophore[12]. The corresponding energy differences would then lead to an average value of approximately 11,000 cm^{-1} for 10 Dq. This value falls in the range expected for ionic octahedrally coordinated high-spin ferric complexes and, perhaps more pertinently, is very close to the value estimated for hemerythrin, a protein to which the acid phosphatases are often compared[55, 57].

The striking similarity between the CD spectra of pink and purple uteroferrin reinforces the conclusion drawn from both optical and laser-Raman studies that mercaptoethanol reduction fails to disrupt the purple protein's iron-tyrosine coordination. Small shifts in peak positions and intensities, however, are consistent with a redox-induced rearrangement of these residues. Further, the virtual coincidence of the pink and purple protein's far-UV and aromatic spectral regions argues against any substantial change in the secondary structure of the protein[12,13]. Preservation of a shoulder at 255 nm, a wavelength at which disulfide bonds may make a sizable contribution, suggests that such bonds, if present, are also unaffected by reduction. In contrast, the binding of phosphate and presumably other tetrahedral anionic inhibitors apparently increases the polypeptide's unordered structure at the expence of its α-helical and β-pleated sheet structures[12,13]. Even more significant, perhaps, is that the binding of phosphate induces a negative Cotton effect at 380 nm, a feature which has no obvious counterpart in either the pink or purple phosphate-free forms of the protein[12]. The identity of this new band is unknown, but comparison with corresponding features in the absorption spectra of hemerythrin and other model compounds suggests that it may represent a bridging ligand-to-Fe(III) charge-transfer transition[51].

4. EPR and Magnetic Susceptibility

EPR and magnetic susceptibility data have been reported for only two acid phosphatases, uteroferrin (both low and high molecular weight forms)[13,29] and the bovine splenic enzyme[9,42,58]. The fully oxidized ($\lambda_{max} \geq 550$ nm) forms of these enzymes are EPR-silent and nearly diamagnetic[13,42,56]. Reversible one-electron reduction, however, elicits an intense rhombic EPR signal with principal g-values of 1.93, 1.75 and 1.59 for uteroferrin (Fig. 2) and a similar signal for the splenic enzyme. It is now clear that this distinctive $g_{ave} = 1.74$ EPR signal, much like the $g_{ave} = 1.94$ EPR signal of two-iron ferredoxins, is

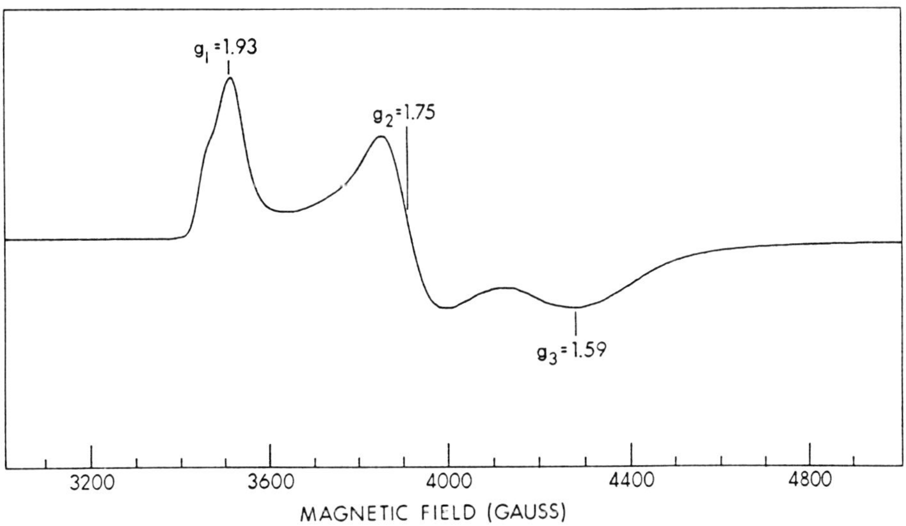

Fig. 2. X-band EPR spectrum of 2.4 mM native pink uteroferrin in 0.1 M sodium acetate buffer, pH 4.9, at 9 K. Approximate principal g-values are indicated. (From Ref. 54)

the spectral signature of a new class of iron-binding proteins[13, 42, 43, 56]. The $g_{ave} = 1.74$ signal, observable only at temperatures below 35 K, accounts for up to one unpaired spin per two iron atoms and never more, a result suggesting that the irons of these proteins are juxtaposed in a binuclear spin-coupled $S = 1/2$ paramagnetic center[13, 42]. This hypothesis is supported by magnetic susceptibility measurements of both uteroferrin and bovine spleen purple acid phosphatase and perhaps more convincingly, by ^{57}Fe-Mössbauer studies (Sect. III.B.6.). Similar spin-coupled paramagnetic centers exist in other iron-binding proteins, specifically, hemerythrin[59], ribonucleotide reductase[60], and the two-iron ferredoxins[61]. In particular, semimethemerythrin yields EPR spectra[59] and LEFE which are remarkably similar to those of uteroferrin, indicating an underlying structural similarity between the paramagnetic centers of these proteins[53].

From the temperature dependence of EPR spectral intensities over the range 9–18 K it is possible to estimate J, the isotropic exchange coupling constant for the irons of their spin-coupled centers[12, 42]. This procedure yields $J = 14 \pm 1\ cm^{-1}$ for uteroferrin and $J = 11 \pm 1\ cm^{-1}$ for the splenic enzyme (taking the exchange Hamiltonian, $H_{ex} = +JS_1 \cdot S_2$). This low value for J has been corroborated by ^{1}H NMR studies at room temperature for uteroferrin[47]. Recent studies of mixed-valence, spin-coupled Fe(III)–Fe(III) model compounds for uteroferrin indicate that such low values for J are consistent with iron atoms joined by a μ-hydroxo or μ-phenoxo bridge[62].

Magnetic susceptibility measurements on oxidized, purple splenic enzyme show that the antiferromagnetic coupling in this state is far greater than in the reduced state (cf. $J_{ox} \geq 300\ cm^{-1}$ to $J_{red} \geq 11\ cm^{-1}$). The dramatic increase in coupling constant strength accompanying oxidation is far greater than that found, for instance, in the two-iron ferredoxins[61]. It seems conceivable that oxidation is accompanied by an alteration in bridge structure, so that, for example, a μ-hydroxo bridge might become deprotonated to form a μ-oxo bridge[42, 47]. Evans susceptibility measurements at room temperature indicate that uteroferrin also shows a marked increase in coupling strength upon oxidation, with $J \geq 80\ cm^{-1}$ [47]. A coupling of this magnitude, in the absence of any sulfur bridging groups, is consistent with a μ-oxo bridged structure[10, 42, 47, 63]. At this point, a binuclear center, bridged by multiple ligands as in hemerythrin, is also an attractive possibility.

Heterogeneity is evident in the EPR spectra of virtually all forms of iron-containing acid phosphatases[9, 12, 42, 58, 64, 65]. More specifically, the EPR spectrum of the bovine splenic enzyme is a composite of two overlapping rhombic signals with g = 1.94, 1.78 and 1.65 for one signal, and g = 1.85, 1.73 and 1.58 for the other[42]. Further, the proportion of each signal in the spectrum is a sensitive function of pH and varies with the nature of the buffer as well[42]. Quantitative studies of these signals as a function of pH suggest that the transition from the low-pH to the high-pH form depends upon the state of ionization of a single group of the protein[42]. Uteroferrin also exhibits heterogeneous EPR spectra, but their dependence on pH is not so obvious. Moreover, the spectral heterogeneity of uteroferrin's signal can be virtually eliminated by freezing the protein in a 1:1 (v/v) mixture of aqueous buffer and methanol[13]. Taken together, these observations on the two proteins suggest that the heterogeneity problem is both interesting and not completely understood, and therefore deserving of further study.

Mixed-metal, Fe–Zn forms of both uteroferrin and the splenic enzyme possessing full enzymic activity have been prepared[9, 36]. Susceptibility and EPR measurements of the splenic Fe–Zn enzyme indicate that its single iron is high-spin ferric in a state of rhombic symmetry[9]. Its $g_{ave} = 4.3$ signal, which accounts for the protein's full complement of iron, exhibits unusual spin relaxation properties which are sensitive to the binding of phosphate[9].

5. Vector Coupling Model

As detailed further below, the evidence for a spin-coupled binuclear iron center in the purple acid phosphatases is overwhelming. The magnetic properties of this center can be understood, at least in part, by invoking the vector coupling model for spin angular momentum. This model has been used successfully in explaining the properties of similar spin-coupled paramagnetic centers in other binuclear iron proteins, specifically, the two-iron ferredoxins[61] and hemerythrin[57, 59]. For such centers, the coupling of the ferric ion spin (S_1) to that of the ferrous ion (S_2) yields a spin-coupled manifold of states characterized by the resultant spin values S = 1/2, 3/2, 5/2, 7/2, and 9/2[61]. If the coupling is antiferromagnetic, as is the case for the purple acid phosphatases, the S = 1/2 state lies lowest. In this state the effective g-tensors for the paramagnetic complex can be related to the g-tensor of its individual constituent ions:

$$g_{eff} = (7/3)g_1 - (4/3)g_2 ,$$

where g_1 is the g-tensor of the high-spin ferric ion and g_2 is the g-tensor of the high-spin ferrous ion. The coefficients, 7/3 and − 4/3, represent the projections of the subsite spin vectors (divided by the magnitude of the resultant spin) along the resultant spin direction. In a first approximation, one may take the g-tensor of the ferric ion (a nominal S-state ion) as isotropic and equal to 2.0[57, 61]. To describe the ligand field states of the ferrous ion, we may adapt the theory of Bertrand and Gayda[57], originally formulated for the 2-Fe iron-sulfur clusters, to the present situation. In this theory the five ligand-field states of the ferrous ion are:

$$\begin{aligned} |\phi_0\rangle &= \cos\theta|x^2 - y^2\rangle + \sin\theta|z^2\rangle , \\ |\phi_1\rangle &= |xz\rangle , \\ |\phi_2\rangle &= |yz\rangle , \\ |\phi_3\rangle &= |xy\rangle , \\ |\phi_4\rangle &= \cos\theta|z^2\rangle - \sin\theta|x^2 - y^2\rangle , \end{aligned} \tag{1}$$

where θ is the mixing parameter for the rhombic component of the ligand field and x, y and z have their usual meaning for a field of this symmetry. Arguments similar to those used for hemerythrin indicate that $|\phi_0\rangle$ must be the ground state for a site of distorted octahedral symmetry[57]. The g-tensor for the ferrous ion can now be calculated using standard second-order perturbational techniques[66]. When the components of the ferrous g-tensor calculated this way are combined with the expression for g_{eff}, Eq. (1) above, the components for the g-tensor of the paramagnetic complex become in agreement with[57]:

$$g_x = \frac{7\,g_1 - 8}{3} + \frac{32\,\lambda_2 \sin^2(\theta + \Pi/6)}{3\,\Delta_{yz}},$$

$$g_y = \frac{7\,g_1 - 8}{3} + \frac{32\,\lambda_2 \sin^2(\theta - \Pi/6)}{3\,\Delta_{xz}},$$

$$g_z = \frac{7\,g_1 - 8}{3} + \frac{32\,\lambda_2 \cos^2\theta}{3\,\Delta_{xy}}.$$

Using $\lambda = -80\ \mathrm{cm}^{-1}$, the same value assumed for the spin-orbit coupling constant in hemerythrin and $\Delta_{xy} = 11{,}000\ \mathrm{cm}^{-1}$, the value of the ligand field strength estimated from CD data, the following approximate energy level diagram for uteroferrin is obtained:

$$|\phi_4\rangle \ \text{———————} = \ ?\ \mathrm{cm}^{-1},$$
$$|\phi_3\rangle \ \text{———————} = 11{,}000\ \mathrm{cm}^{-1},$$

$$|\phi_2\rangle \ \text{———————} = 1{,}190\ \mathrm{cm}^{-1},$$
$$|\phi_1\rangle \ \text{———————} = 137\ \mathrm{cm}^{-1},$$
$$|\phi_0\rangle \ \text{———————} = 0\ \mathrm{cm}^{-1}.$$

It should be emphasized that the spacing of the energy levels depends critically upon choices for several physical parameters, as well as implicit assumptions in the model itself[57]. Accordingly, the energy level diagram will have to be modified as additional pertinent information becomes available. All in all, however, one can expect the present diagram to be semiquantitatively correct. The nature of the low-lying orbitals, as well as their relatively small energy gaps, provides a simple and satisfying explanation for the large anisotropy and unusually low values of g_{ave} of the EPR signals of the pink acid phosphatases. The existence of such low-lying levels can be tested by measuring the temperature dependence of the quadrupole splittings in the Mössbauer spectrum over a wide range of temperatures.

6. *Mössbauer Spectra*

Mössbauer spectroscopy of ^{57}Fe-enriched uteroferrin (Fig. 3)[11] and nonenriched bovine spleen acid phosphatase[42] demonstrates that the pair of irons in each of these proteins are sequestered in spin-coupled binuclear iron clusters. In the oxidized, purple protein (Fig. 3B), both iron atoms are high-spin ferric with their spins coupled antiferromagnetically to produce an $S = 0$ diamagnetic ground state[11, 42]. One-electron reduction to the pink form (Fig. 3A) yields an Fe(III)–Fe(II) pair whose spins are now coupled to give an EPR-active $S = 1/2$ ground state[11, 42]. The quadrupole splittings of the ferric sites of phosphate-free uteroferrin are unusually large for a nominal S-state ion, but no larger than those found for oxidized hemerythrin and related model compounds. Current evidence suggests that these large splittings are the result of a highly asymmetric ligand environment about the irons[11, 42, 67–71]. It is also noteworthy that the ferric ions of purple uteroferrin are inequivalent, an observation consistent with the notion that the coordina-

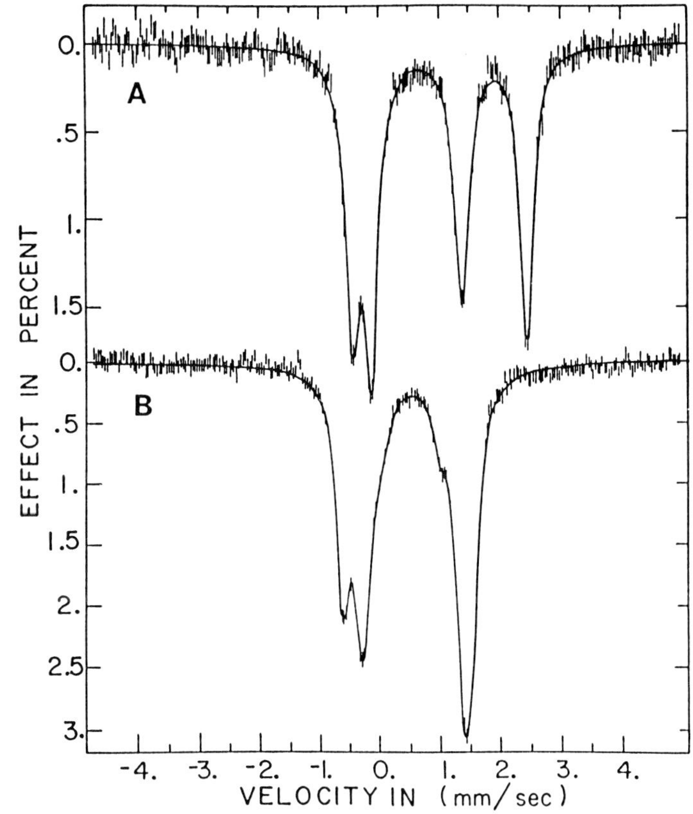

Fig. 3A, B. Mössbauer spectra of the ^{57}Fe-enriched pink form of uteroferrin at 185 K (**A**) and purple form at 10 K (**B**). The *solid lines* are Lorentzian fits obtained with the parameters cited in Ref. 11. (Adapted from Ref. 11)

tion spheres of the protein's iron atoms are significantly different. The noticeable broadening observed as the temperature of the purple enzyme is increased indicates a population of thermally accessible magnetic states, presumably states of the spin-coupled manifold with $S \geq 1$[11, 42)]. The large isomer shift ($\delta_{Fe} \approx 1.2$ mm/s) of the ferrous ion in the reduced form of uteroferrin and the large internal field at the Fe(III) nucleus (saturation field, $H_{sat} \approx -55$ T) are both consistent with oxygen and nitrogen, not sulfur, ligation to the protein's iron atoms[11, 67, 69)]. The value of the saturation field, in particular, which is very close to that found in transferrin, enterobactin, and the model compound (Fe(EDDHA)H_2O), suggests an octahedral oxygen-nitrogen coordination environment[55)]. Thus, sulfur, if present at all in the protein's active site, cannot be bound directly to either iron atom, a result in accord with recent EXAFS work[42, 72)].

7. ^1H NMR Spectra

Proton NMR studies of uteroferrin have provided: insights into the nature of the two oxidation states of the protein; an estimate for the strength of the antiferromagnetic interaction (– J) in both pink (reduced) and purple (oxidized) forms; evidence for tyrosine and histidine ligation to the binuclear iron center[47]; and identification of uteroferrin's oligosaccharide structure[3].

The room temperature ^{1}H NMR spectrum of pink uteroferrin (Fig. 4) reveals a pattern of well-resolved paramagnetically shifted resonances spanning 90 ppm downfield to 70 ppm upfield from the water signal. The intensities of these lines are proportional to the degree of reduction, the oxidized form of the protein exhibiting no contact-shifted resonances (Fig. 4). Together with EPR experiments performed at liquid helium temperatures, these results confirm the presence of two oxidation states of uteroferrin: an EPR-active, reduced species which gives rise to an NMR spectrum and an oxidized form which is both EPR and NMR-silent.

An estimate of the antiferromagnetic coupling constant, using the Hamiltonian $H = -2\,JS_1 \cdot S_2$, was made for reduced uteroferrin based on the temperature dependence of isotropically shifted proton resonances. The data indicate that the Fe(III)–Fe(II) cluster of the pink protein has weak antiferromagnetic coupling ($-J = 10\ cm^{-1}$), similar to estimates derived from EPR studies. Evans susceptibility measurements on pink and purple uteroferrin, comparing relative shifts in the DSS probe, show a strong exchange coupling ($-J > 40\ cm^{-1}$) for the oxidized, Fe(III)–Fe(III) form of the protein. The results are in agreement with low temperature magnetic susceptibility experiments and suggest the presence of an oxo bridge.

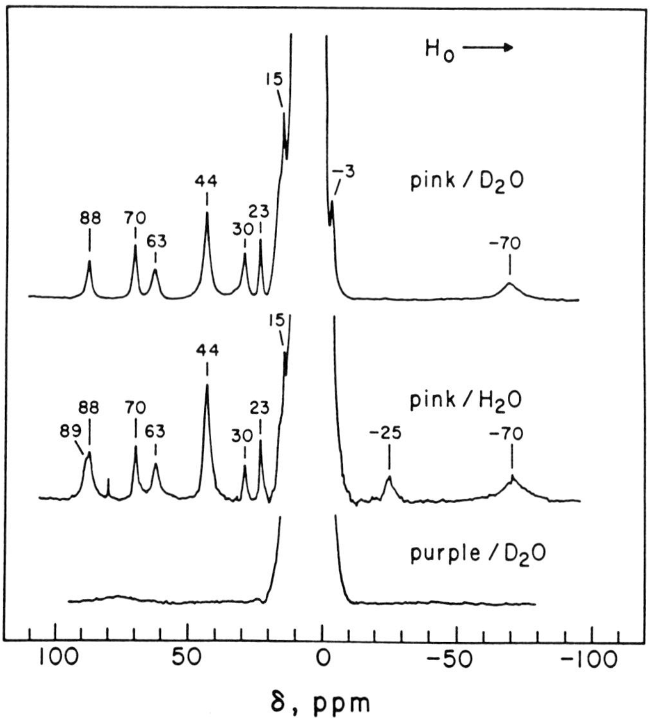

Fig. 4. 300-MHz ^{1}H NMR spectra of pink and purple uteroferrin (1 mM) in 0.1 M sodium acetate buffer, pH 4.9, at 30 °C. Spectra were recorded using a 10 μs 90° pulse, a 125 kHz band width, 8 K data points, and 30,000–60,000 transients. (Adapted from Ref. 47)

Tyrosine and histidine ligation to the binuclear iron center of uteroferrin is established by comparing the spectrum of the pink protein to those of synthetic high spin ferric and ferrous complexes. ^1H NMR resonances are observed at 89, 88, 70, 63, 44, 30, 23, and 15 ppm downfield, and 25 and 70 ppm upfield from the water signal (Fig. 4). The 70 ppm (upfield) line is assigned to the ortho protons, and the 63 and 70 ppm (downfield) lines to the meta protons of a tyrosine coordinated to the ferric iron of the Fe(II)–Fe(III) pair. This is consistent with earlier results of Raman spectroscopy in which resonance-enhanced phenolate vibrations are detected for both pink and purple forms of the protein. Further assignments are made by comparisons of NMR spectra in H_2O to spectra in D_2O (Fig. 4). Solvent exchangeable resonances at 89 and 44 ppm (downfield) are attributed to histidine coordination to the ferric and ferrous sites, respectively. The remaining features in the NMR spectrum are yet to be assigned.

The identification of uteroferrin's oligosaccharide structure was also made possible by ^1H NMR[3]. Uteroferrin, isolated in the purple form from either uterine secretions or allantoic fluid, is a single-chain glycoprotein with 4.8% carbohydrate content by weight[3]. This carbohydrate exists as a chain, consisting mainly of mannose residues, which is released by endoglycosidase II. The structures of the $Man_5GlcNAc$ and $Man_6GlcNAc$ oligosaccharides, separated and purified from uteroferrin, were determined using ^1H NMR by comparing the chemical shifts of the resonances to corresponding features in the spectra of model compounds. The data are consistent with the following structure for the Man_6 species (Man_5 lacks the terminal α1,2-linkage):

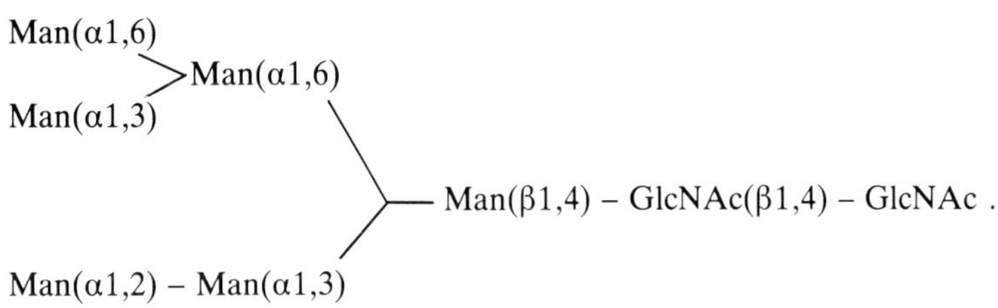

8. *LEFE[1], ESEEM, and ENDOR Spectra*

The large linear electric field effect (LEFE) exhibited by pink uteroferrin clearly demonstrates that its paramagnetic center is noncentrosymmetric and, in accord with the Mössbauer studies, suggests that the reducing electron resides primarily on one of the protein's iron atoms[53]. Though subtle differences exist, the overall similarity of uteroferrin's LEFE to that of semimethemerythrin again indicates a close structural similarity between the active sites of these two proteins, which clearly have different functions[53]. The marked deviation from centrosymmetry is also consistent with other evidence indicating that the tyrosyl residues coordinated to the pink protein's paramagnetic center are unequally partitioned between its two iron atoms[13, 56]. The magnetic field dependence of LEFE singles out the g_{min} (i.e., the high-field axis) as the direction of most facile electron

1 The abbreviations used are: LEFE, linear electric field effect; ESEEM, electron spin echo envelope modulation; ENDOR, electron nuclear double resonance; XANES, x-ray absorption near edge structure; EXAFS, x-ray absorption fine structure.

polarization. This makes it tempting to argue, as for the two-iron ferredoxins, that the g_{min} axis points away from the ferrous ion and toward an electron-accepting center, most likely to the other iron atom. It is also noteworthy that the g_{min} axis for both types of proteins is the axis showing the greatest g-strain effects, suggesting that the paramagnetic center is least rigid along that axis[53, 61)].

Electron spin-echo studies indicate that pink uteroferrin's unpaired electron interacts with at least one and possibly two classes of nitrogen nuclei, one of which is the imidazole nitrogen of an iron-coordinated histidine[53)]. These results support similar conclusions drawn from ^{1}H NMR studies showing histidine ligation to both ferrous and ferric ions of the binuclear cluster[47)]. Recent ENDOR experiments have also detected ^{14}N hyperfine interactions with the $S = 1/2$ center of uteroferrin, presumably due to the remote nitrogen(s) of an iron coordinated histidine(s)[54)].

Proton ENDOR spectra of reduced uteroferrin reveal at least 6 sets of lines mirrored about the ^{1}H Larmor frequency[54)]. Two pairs of these lines become reduced in intensity upon deuteration of the protein. In addition, ESEEM and ^{2}H ENDOR display resonances at the ^{2}H Larmor frequency. The spin-coupled cluster of uteroferrin is therefore accessible to solvent. Moreover, deuterons which replace a population of strongly coupled and readily exchangeable protons are observable by ESEEM[54)]. The hyperfine couplings for these deuterons are orientation dependent and it is intriguing to consider if the exchangeable proton is from a bridging hydroxy group between the two iron atoms[54)].

9. *EXAFS Spectra*

To date, X-ray absorption studies have been performed only on the purple acid phosphatase from bovine spleen[72)]. Iron K-edge near-edge (XANES) and extended X-ray absorption fine structure (EXAFS) results, comparing the splenic enzyme with oxo-bridged model complexes, support the presence of a binuclear site in which the iron atoms are multiply bridged[72)].

The XANES show that upon reduction of the purple phosphatase to the pink form, the absorption edge is shifted to lower energy by 2.0 ± 0.5 eV. This is consistent with the reduction of one of the Fe(III) ions to Fe(II). In addition, the 1 s $\rightarrow$ 3 d transition peak is evident 10 eV below the absorption edge. The peak is visible for both forms of the protein and exhibits intensities of 4.3 and 4.2% of the main absorption edge, thus suggesting that each iron atom is in a six-coordinate site of relatively low symmetry[72)].

Fourier transformed EXAFS spectra of the purple enzyme reveal three major peaks, assigned as follows: first shell, Fe–O(N); second shell, Fe—Fe and Fe–P(C); third shell, Fe—N(C) (imidazole). Due to interference by Fe–O (tyrosine) bonds at 1.8–1.9 Å, an Fe-μ-oxo bond could not be detected. The Fe—Fe distance of 3.0 Å lies within the range expected for a μ-oxo bridged structure. The spectrum of the purple, phosphate bound form also provides direct evidence for phosphate coordination to one of the iron atoms, with an Fe—P distance of 3.0 Å.

10. *Interaction With Phosphate and Other Perturbants*

The reduced (pink) form of uteroferrin is sensitive to a variety of agents which perturb its optical and/or magnetic properties. Investigations have focused on the interaction of the protein with the tetrahedral oxyanion phosphate, an inhibitor and a product of its acid phosphatase activity. Optical spectroscopy first showed that a species with an absorption maximum at 535–540 nm, still exhibiting enzymic activity, is rapidly generated following phosphate addition to the pink protein at pH 4.9[73]. Recent Mössbauer studies on ^{57}Fe reconstituted pink uteroferrin demonstrate a "purple enzyme-phosphate" complex, with a mixed valence center, to be a transient intermediate between fully reduced and fully oxidized forms[74]. This complex, with its Fe(II)–Fe(III) binuclear cluster, may account for the persistence of enzymic activity in uteroferrin turned purple by phosphate addition[74]. Yet to be explained, however, are the following observations concerning this "purple reduced uteroferrin-phosphate" intermediate: 1) it is EPR-silent, despite the mixed-valence state of its binuclear iron center[12,75] and 2) this species does not exhibit any contact-shifted resonances in the ^{1}H NMR spectrum such as those observed in the phosphate-free pink form[76]. Possible explanations for the absence of an EPR signal include unfavorable relaxation properties of the paramagnetic center such as might arise from an orbitally degenerate ground state and quantum mechanical mixing of states, inherently low transition probabilities arising from states not purely $S = 1/2$, or extreme g-tensor anisotropy making the signal too broad to be detected. Similarly, the absence of observable contact-shifted resonances may be due to a larger static magnetic susceptibility of the paramagnetic center at room temperature, resulting from a decrease in the spin-coupling constant and leading, in turn, to a contact frequency-shift beyond the range of the instruments used. Further study of this uteroferrin-phosphate intermediate is necessary.

In contrast to uteroferrin, the reaction of phosphate with pink (reduced) bovine spleen acid phosphatase has been reported to show a parallel loss of enzymic activity, shift in visible absorption maximum, and loss of EPR signal intensity[40]. These results suggest that phosphate binding is coupled to oxidation of the protein's binuclear iron center and do not provide evidence for a paramagnetic enzyme-phosphate intermediate[40]. Whether this is indicative of differences between the porcine and splenic enzymes or variabilities in experimental design is uncertain.

Upon standing in air, the absorption maximum of the uteroferrin-phosphate intermediate species changes further and stabilizes at 550 nm, the wavelength which is characteristic of the oxidized (purple) protein-phosphate complex. This final form of the protein is enzymically inactive, displaying Mössbauer parameters expected of an Fe(III)–Fe(III) center[74] and distinct from those observed for phosphate-free, oxidized uteroferrin[11]. Inorganic orthophosphate binds very tightly to purple uteroferrin or purple spleen phosphatase in a 1 : 1 complex, but may be removed by reduction of the protein followed by Sephadex chromatography[73]. Moreover, the *in vitro* preparation is similar to the initially isolated purple protein, which contains 1 mol of phosphate per mol of enzyme[40,73,74], in not exhibiting a ^{31}P NMR signal[76]. This suggests that the resonance line of phosphate is broadened beyond detectability, either because of proximity of phosphate to the binuclear iron center with its populated paramagnetic excited states or by irrotational binding to the protein. Recently, EXAFS spectroscopy has indicated an

Table 1. Effects of perturbants on pink uteroferrin's optical and EPR spectra

Perturbant	Shift in absorption (nm)	Change in EPR signal	Reference
Phosphate	510 → 535–540	Lose 90–95%	12, 73, 75
Pyrophosphate	510 → 536	Lose 90–95%	75
Arsenate	510 → 530	Lose 55%	75
Molybdate	510 → 515–520	Axial conversion	54, 75
Sulfate	510 → 505	Two new rhombic signals	75
Vanadate	510 → 520	Lose 100% and get new g_{ave} = 2.0 signal	9, 75
Fluoride	510 → 540 (Complex; depends on [Fl^-])	Complex (Axial signal superimposed on native g_{ave} = 1.74 signal)	75
Tartrate	None	None	75

Fe—P distance of 3 Å, thus providing direct evidence for phosphate interaction with the iron cluster in purple bovine spleen acid phosphatase[72)].

Pyrophosphate and arsenate produce changes in pink uteroferrin's optical spectrum which are similar to those induced by phosphate (Table 1)[75)]. Pyrophosphate obliterates over 90% of the EPR spectral intensity, but arsenate produces smaller decreases in the amplitude of the rhombic signal[75)]. Most likely, these anions share a common inhibitory mechanism involving oxidation or blocking of the iron center.

Molybdate, a structural analog of phosphate, is a potent inhibitor of pink uteroferrin's acid phosphatase activity, but has negligible effect on its optical spectrum and leaves the protein in its EPR-active form[75, 77)]. It converts the EPR spectrum of uteroferrin from a rhombic (g = 1.93, 1.75, 1.59) to an axial ($g_{\parallel}$ = 1.97, $g_{\perp}$ = 1.52) type which remains invariant to subsequent additions of phosphate, suggesting that both anions compete for the same binding site on the protein[75)]. With both ESEEM (Fig. 5A) and ENDOR (Fig. 5B) spectroscopy, which offer the advantage that interpretation does not depend on model compounds, a superhyperfine interaction of ^{95}Mo-molybdate with the S = 1/2

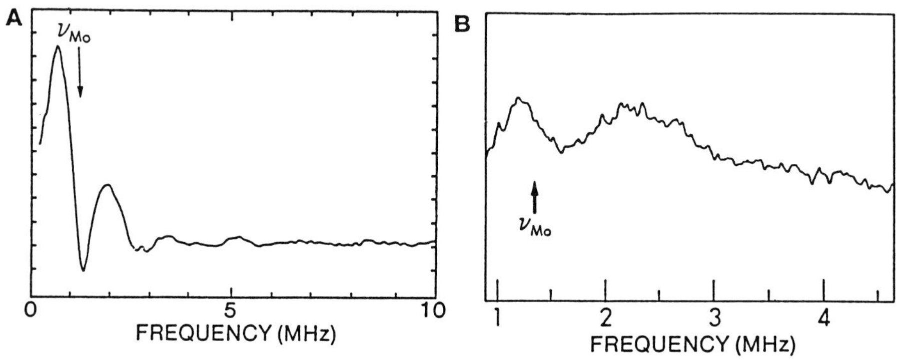

Fig. 5A, B. ESEEM (A) and ENDOR (B) spectra showing a superhyperfine interaction of ^{95}Mo-molybdate with the S = 1/2 iron center of pink uteroferrin. Both spectra, taken at the $g_{\perp}$ = 1.52 orientation, reveal a single pair of ^{95}Mo resonances centered at the ^{95}Mo Larmor frequency (ν_{Mo}) and separated by a hyperfine coupling of 1.2 MHz. (Adapted from Ref. 54)

iron center of the protein was recently detected[54]. A single pair of ^{95}Mo resonances centered at the ^{95}Mo Larmor frequency and separated by a hyperfine coupling constant of 1.2 MHz was observable. Therefore, a single monomeric species of molybdate (which exists in aqueous solution as a polyanion) is likely a ligand of the binuclear cluster. ESEEM and ENDOR studies of the deuterated uteroferrin-molybdate complex indicate that the binuclear site remains accessible to solvent despite the proximity of molybdate[54].

Other perturbants such as sulfate, vanadate, and fluoride produce more complex changes in the protein's optical and/or magnetic profile (Table 1). Sulfate, an anion which is structurally similar to phosphate but differs in net charge, does not produce a violet shift in the optical spectrum of reduced uteroferrin[75]. Its addition results in the conversion of the rhombic EPR signal to a complicated $g_{ave} = 1.75$ signal which may consist of two new overlapping rhombic spectra[75]. Vanadate forces the pink protein to an EPR-silent state, but itself serves as a one-electron acceptor to yield an EPR signal at $g = 2.0$ which is characteristic of the vanadyl (VO^{2+}) cation[75, 78]. Despite its oxidation, uteroferrin remains essentially pink, for reasons yet to be adduced. Fluoride causes a progressive loss of EPR signal amplitude which parallels a violet shift in the optical spectrum, but also induces formation of an axial component superimposed on the native EPR signal[75]. The mechanism of interaction between these agents and uteroferrin is largely obscure.

Tartrate-resistance (see below) was one of the first observations concerning the purple acid phosphatases with binuclear iron centers[33]. Consistent with this is the absence of any effect of tartrate on either the optical or EPR spectrum of pink uteroferrin, indicating that this anion interacts weakly, if at all, with the active binuclear iron center[75].

IV. Enzymatic Properties

Proteins possessing phosphatase activity are found in both the animal and plant worlds[79]. The division between acid and alkaline phosphatases is dependent on the pH optimum at which activity is displayed; pH 4.9–6.0 for acid phosphatases and pH ≈ 8 for alkaline phosphatases[80]. The acid phosphatases can be further classified into two subgroups based on their sensitivity to tartrate[79]. The tartrate-sensitive enzymes, present in all animal cells investigated except red blood cells, have a higher molecular weight ($>89{,}000$) than the tartrate-resistant proteins ($<65{,}000$) and are strictly orthophosphoric monoester hydrolases[79]. Of the tartrate-resistant proteins, the purple acid phosphatases are distinguished by their spin-coupled, redox active, binuclear iron center which is essential for enzymatic activity. Although the mechanism of enzyme-substrate interaction and the identity of their physiological substrate(s) remain obscure, substrate specificities and inhibitors of the purple acid phosphatases are well known.

A. Substrate Specificities

The first evidence for a unique class of enzymes, transcending conventional phosphatase classification, came from work on bovine spleen acid phosphatase[33, 79]. It was found that this "phosphoprotein phosphatase" was capable of removing the phosphorus from casein, possessing as well pyrophosphatase activity[33]. Although it was only slightly active towards β-glycerophosphate and aliphatic phosphates, the enzyme showed high activity towards phenylphosphate and related compounds. At this stage of investigation, protein-associated metals had not been detected.

With the discovery of uteroferrin and other purple acid phosphatases, it became apparent that all of these enzymes contain a bi-metallic active site which usually includes at least one iron atom. Substrate specificities of these proteins are nearly identical and agree closely with earlier descriptions[33]. Molecules containing the pyrophosphate linkage, such as nucleoside tri- and diphosphates, thiamine pyrophosphate, or inorganic pyrophosphate are hydrolyzed at rates similar to orthophosphate substrates. In particular, all of the enzymes show considerable activity towards *p*-nitrophenyl phosphate, the substrate generally used in enzyme assays[2]. Uteroferrin and bovine spleen phosphatase were also shown to be capable of hydrolyzing phosphate from proteins with phosphoserine residues[2,77], and phosphotyrosinase activity would not be unexpected. The purple acid phosphatases exhibit very little reactivity towards compounds with phosphoamide linkages or towards aliphatic substrates[13,41].

B. Activators and Inhibitors

Early descriptions of bovine spleen acid phosphatase have noted the ability of 2-mercaptoethanol to both activate the enzyme and cause a shift in its absorption maximum from 550 nm to lower wavelengths[33]. Other reducing agents, such as ascorbate, glutathione, dithiothreitol, and thioglycollate were subsequently shown to be effective as activators[32,39]. In general, reduction or activation of the purple acid phosphatases parallels a change in its color form violet to pink, while its enzymic activity is increased several-fold[32,39]. For experimental purposes, the reduced, EPR-active from of the protein is usually obtained by treatment of the phosphatase with ferrous iron and 2-mercaptoethanol or ascorbate. Reduction in the absence of ferrous iron results in substantial denaturation of the protein[38]. Through spectroscopic studies, it has been shown that the Fe(III)–Fe(III) center, present in the purple forms of uteroferrin and bovine spleen phosphatase, is converted to an Fe(III)–Fe(II) cluster by reducing agents in a one-electron process[9,11]. At present, it is not clear how the events in reduction, color change, and enzymic activation are related.

Although reduction and activation are synonymous for the vast majority of the purple acid phosphatases, several exceptions exist. The Fe–Zn forms of uteroferrin and bovine spleen phosphatase do not require prior reduction to exhibit enzymatic activity[9,36]. The Fe–Cu and Fe–Hg derivatives of uteroferrin also do not require activation and in fact, the Fe–Cu preparation is inactivated by reducing agents[26,36]. The recently described high molecular weight pink form of uteroferrin has enzymic properties identical to those of purple uteroferrin treated with 2-mercaptoethanol[29]. Finally, the sweet potato acid phosphatase, which may exist as an α_2 dimer with separate mononuclear iron centers, does not require the addition of reductant to promote its enzymatic activity[23].

Many reagents are capable of inhibiting the enzymic action of the purple acid phosphatases. They include: oxidants such as hydrogen peroxide and ferricyanide; tetrahedral oxyanions including phosphate, molybdate, vanadate, sulfate, and arsenate; sulfhydryl reagents such as heavy metal ions and *p*-chloromercuribenzoate; and fluoride ions, arsenite, nitrite, phosphonate, alloxan, and maleic acid[13, 32, 45]. As exceptions to the rule, the mixed-metal forms of uteroferrin are only slightly affected by phosphate and/or hydrogen peroxide[36]. Sodium dithionite is a strong inhibitor which acts by selectively removing the metal ions from the active site of the protein, thereby bleaching its color[38]. The apo- and one-iron species of uteroferrin, prepared by sodium dithionite treatment, display little or no activity towards *p*-nitrophenyl phosphate[38]. It is curious that neither azide nor cyanide, normally strong iron atom ligands, inhibits bovine spleen phosphatase or uteroferrin to any appreciable extent[32, 45].

In trying to understand the mechanism of enzyme-substrate interaction, much attention has been directed towards phosphate, both an inhibitor and a product of purple acid phosphatase activity. Keough et al.[73] first determined a K_d of 6 mM for phosphate binding to reduced uteroferrin. Additionally, they have shown that the rapid spectral shift upon phosphate addition is accompanied by a slower loss of enzymic activity, with $t_{1/2} = 51$ min. Pyrz et al.[74] have recently demonstrated that a reduced uteroferrin-phosphate complex is formed which has a pH-dependent visible absorption maximum ranging from 561–530 nm. This complex is converted to the oxidized phosphate complex with a first order rate constant of 4×10^{-3} min^{-1}, monitored by both spectral changes and loss of enzymatic activity[74]. These kinetic results suggest that in the initial stages of reaction, the enzyme-phosphate complex is in rapid equilibrium with free enzyme and phosphate. Air oxidation then proceeds to totally inactivate the enzyme-phosphate species, concomitantly shifting its absorption maximum to 550 nm, the wavelength which is characteristic of oxidized, purple uteroferrin.

Product inhibition studies on bovine cortical bone acid phosphatase indicate that its reaction with *p*-nitrophenyl phosphate involves a two-step hydrolytic transfer mechanism (pseudo Uni Bi), with *p*-nitrophenyl and phosphate as the first and second products released, respectively[19]. Moreover, bone phosphatase was inhibited by transition state analogs of phosphate, suggesting formation of a phosphoryl-enzyme intermediate in the reaction. The possibility exists that most, if not all, acid phosphatases catalyze their reaction by this two-step transfer mechanism[19]. ^{31}P and ^{17}O NMR results of Mn(III) Kintoki phosphatase, which show that the catalytic mechanism involves a transition state displacement and P–O cleavage, also support this idea[81].

The role of the Zn atom in *E. coli* alkaline phosphatase, which catalyzes the phosphorylation of serine 99 in the amino acid sequence to form the covalent enzyme-phosphate intermediate[82], is well understood. Unfortunately, knowledge of the molecular mechanisms underlying the enzymic activity of the purple acid phosphatases is far more rudimentary. Spectroscopic studies utilizing inhibitors and perturbants (Sect. III.B.10), such as phosphate and molybdate, indicate that substrate binds close to the binuclear iron cluster of uteroferrin and bovine spleen phosphatase[54, 72, 74]. Most likely, the substrate interacts with the redox active iron of the pair[75].

V. Biological Roles of the Purple Acid Phosphatases

A. *Intracellular Mammalian Phosphatases*

There is little reason to question the biological role as true phosphatases of intracellular purple acid phosphatases such as those from spleen[2], B-lymphocyte-derived cells[30], and bone cortex[35]. These enzymes exhibit high specific activities, are particularly vigorous in hydrolyzing protein phosphate esters, and are abundant within digestive organelles (lysosomes) of the cell[83]. All of these findings support a physiological phosphatase function. Such a function is most clearly displayed by the purple acid phosphatase of bone. Histochemical studies in the rat demonstrate high phosphatase activity resistant to tartrate, but inhibitable by dithionite (which can remove iron from the purple acid phosphatases)[38] in resorbing bone[84]. Such activity is probably representative of the purple acid phosphatase isolated from rat bone. Since the enzyme is particularly active toward acidic phosphoproteins believed to function in bone mineralization by trapping calcium ions, its abundance in resorbing bone likely acts to retard new calcification[84].

Two possible difficulties with the straightforward view that the purple enzymes function as true phosphatases might be considered. First, they display the interesting and poorly understood phenomenon of inhibition by inorganic phosphate, a reaction product of phosphatase activity. The apparent K_i for phosphate acting as a competitive inhibitor of the hydrolysis of *p*-nitrophenyl phosphate by uteroferrin is 3.2 mM at pH 4.9[73]. Very likely, a similar K_i is exhibited by the tissue phosphatases, in view of their similarities in spectroscopic and structural properties to uteroferrin. Intracellular inorganic phosphate, which may exceed 1 mM in concentration[85], may then act to restrain the phosphatase activity of spleen purple acid phosphatase and its intracellular relatives.

Second, most purple acid phosphatases require activation by reducing agents. Although intracellular metabolites such as NADH and NADPH have redox potentials sufficiently low to reduce the purple phosphatases, it is not yet clear whether they, or other reductants, are present in effective concentration in the lysosome-like organelles to which the enzymes are localized.

These are probably minor cavils, however, so that the prevailing evidence still supports an enzymic function of intracellular purple acid phosphatases.

B. *Uteroferrin*

More problematic and more controversial is the biological role of uteroferrin. Present in allantoic fluid in concentrations often exceeding 2 mg/ml (5×10^{-5} M), it is hard to imagine an enzymatic function for this protein in the physiological compartment in which it abounds. Conceivably, it is simply leaked into the allantoic fluid by hyperproducing lining cells of the allantoic membrane, for which uteroferrin functions as a lysosomal acid phosphatase[86]. An alternative possibility is that uteroferrin functions not as an enzyme, since it is almost inactive at the pH of the fluid in which it is found, but as a carrier of iron from sow to fetal pig[87].

In favor of this hypothesis is the rate at which uteroferrin is secreted into the allantoic sac, more than 2 g of protein carrying over 6 mg of iron per day at midterm pregnancy in the pig[87)]. This is sufficient to satisfy the developmental needs of the fetal pig until late pregnancy, when the rate of secretion of uteroferrin declines and other mechanisms, not yet delineated, must become operative. But how is iron bound to uteroferrin made available to the growing fetus?

One possibility, postulated by Roberts and coworkers[87)] on the basis of studies of ^{59}Fe exchange from uteroferrin to transferrin, is that uteroferrin relinquishes its iron to fetal transferrin, which then delivers to fetal cells synthesizing hemoglobin and other essential iron proteins. However, other studies of iron transfer from uteroferrin to transferrin, using an EPR method to detect Fe bound to transferrin and thus obviating the need for reductively labeling uteroferrin with radioactive iron, failed to demonstrate substantial transfer of iron between uteroferrin and transferrin[88)]. Nevertheless, an iron-transport function of uteroferrin is not excluded, since its mannose-rich oligosaccharide targets it to the mannose receptors of reticuloendothelial cells[3)]. Lysosomes of these cells may then degrade the protein to make its iron available for fetal needs. Whatever the mechanisms of transfer of iron from uteroferrin to transferrin, the observations that iron originally bound to uteroferrin is recovered in both fetal hemoglobin and fetal transferrin[87)] argue strongly for a role of uteroferrin in iron delivery to the fetal pig.

Both uteroferrin and bovine purple spleen acid phosphatase are competent to catalyze production of hydroxyl radical *in vitro* when superoxide anion is present[89)]. The generation of hydroxyl radical probably entails a Fenton-like reaction sequence dependent upon the easily reducible iron atom of the solvent-accessible binuclear cluster. No evidence is available to suggest that this reaction also occurs *in vivo*. However, the high mannose content of uteroferrin promotes uptake of the protein by reticuloendothelial cells of the fetal liver, ultimately directing the protein to lysosomes. Since these multifunctional organelles function in antimicrobial defense, it may be that the redox properties of uteroferrin are also exploited by the fetus in guarding against infection.

C. Purple Acid Phosphatases in Plants

Too little is known of the plant purple acid phosphatases to warrant speculation about their physiological role(s). Their relatively low abundance and high specific activity are most consistent with straightforward phosphatase function.

VI. Problems and Perspectives

Since the discovery of an oxo-bridged binuclear iron center in hemerythrin[90)], a similar structural feature has been implicated in proteins such as ribonucleotide reductase[50)], methane mono-oxygenase[91)], ferritin[92)], and the purple acid phosphatases. As noted earlier (Sect. III.B.2.), the presence of an oxo-bridged site in purple, oxidized uteroferrin and a hydroxo-bridged structure in the pink, reduced form of the enzyme is likely, based on spectroscopic comparisons to hemerythrin. However, an Fe–O–Fe vibration mode in the resonance Raman spectrum has yet to be detected. In the case of hemerythrin, the function of its iron center in the reversible binding of dioxygen and its role in oxygen transport are well established[90]. A comparable statement cannot be made for the purple acid phosphatases. In particular, the interaction of uteroferrin with phosphate, an inhibitor and product of its enzymic activity, still presents perplexities (Sect. III.B.10.). The "purple reduced uteroferrin-phosphate" intermediate, which is EPR-silent and does not exhibit any contact-shifted ^1H NMR resonances, requires further investigation. Possibly, experiments using high molecular-weight uteroferrin[29], which is inhibited but remains pink in the presence of phosphate, or use of the mixed metal forms of uteroferrin[9, 36] which remain EPR-active even with phosphate, may provide new pathways for inquiry into this reaction.

Among the persisting problems which circumvent understanding of the binuclear iron center of uteroferrin and the purple acid phosphatases are the identity of the ligands of each iron atom and how these change in nature or arrangement during redox chemistry. In addition, the mechanism of enzymic activity, as well as the physiological role(s) of the phosphatases, must be examined. Recent determination of the three-dimensional crystal structure of lactoferrin[93], an iron-binding protein, has provided new insights into the structure and function of the transferrins. Knowledge of the nature and location of the iron atoms has clarified and corrected previous inferences and speculations on the structure of the transferrins. A similar outcome may be hoped for in the case of uteroferrin and the purple acid phosphatases.

VII. References

1. Chen, T. T., Bazer, F. W., Cetorelli, J. J., Pollard, W. E., Roberts, R. M.: J. Biol. Chem. *248,* 8560 (1973)
2. Campbell, H. D., Zerner, B.: Biochem. Biophys. Res. Commun. *54,* 1498 (1973)
3. Saunders, P. T., Renegar, R. H., Raub, T. J., Baumbach, G. A., Atkinson, P. H., Bazer, F. W., Roberts, R. M.: J. Biol. Chem. *260,* 3658 (1985)
4. Herrmann, K. M., Shultz, J., Hermodson, M. A.: J. Biol. Chem. *255,* 7079 (1980)
5. Kuchta, R. D., Hanson, G. R., Holmquist, B., Abeles, R. H.: Biochemistry *25:* 7301 (1986)
6. Kennedy, M. C., Emptage, M. H., Dreyer, J.-L., Beinert, H.: J. Biol. Chem. *25,* 11098 (1983)
7. Dreyer, J. L.: Eur. J. Biochem. *150,* 145 (1985)
8. Sugiura, Y., Kuwahara, J., Nagasawa, T., Yamada, H.: J. Am. Chem. Soc. *109,* 5848 (1987)
9. Davis, J. C., Averill, B. A.: Proc. Natl. Acad. Sci. USA *79,* 4623 (1982)
10. Sinn, E., O'Connor, C. J., deJersey, J., Zerner, B.: Inorg. Chim. Acta *78,* L13 (1983)
11. Debrunner, P. G., Hendrich, M. P., deJersey, J., Keough, D. T., Sage, J. T., Zerner, B.: Biochim. Biophys. Acta *745,* 103 (1983)
12. Antanaitis, B. C., Aisen, P., Lilienthal, H. R.: J. Biol. Chem. *258,* 3166 (1983)
13. Antanaitis, B. C., Aisen, P.: Adv. Inorg. Biochem. *5,* 111 (1983)
14. Anderson, T. R., Toverud, S. U.: Calcif. Tissue. Res. *24,* 187 (1977)
15. Hara, A., Sawada, H., Kato, T., Nakayama, T., Yamamoto, H., Matsumoto, Y.: J. Biochem. (Tokyo) *95,* 67 (1984)
16. Robinson, D. B., Glew, R. H.: Clin. Chem. *26,* 371 (1980)
17. Lam, K. W., Yam, L. T.: Clin. Chem. *23,* 89 (1977)

18. McDowell, K. J., Sharp, D. C., Fazleabas, A., Roberts, R. M., Bazer, F. W.: J. Reprod. Fertil. [Suppl.] *32,* 329 (1982)
19. Lau, K.-H. W., Freeman, T. K., Baylink, D. J.: J. Biol. Chem. *262,* 1389 (1987)
20. Lam, K. W., Lee, P., Li, C. Y., Yam, L. T.: Clin. Chem. *26,* 420 (1980)
21. DiPietro, D. L., Zengerle, F. S.: J. Biol. Chem. *242,* 3391 (1967)
22. Beck, J. L., McConachie, L. A., Summors, A. C., Arnold, W. N., deJersey, J., Zerner, B.: Biochim. Biophys. Acta *869,* 61 (1986)
23. Hefler, S. K., Averill, B. A.: Biochem. Biophys. Res. Commun. *146,* 1173 (1987)
24. Sugiura, Yukio, Kawabe, Hideo, Tanaka, Hisashi, Fujimoto, Sadaki, Ohara, Akira: J. Biol. Chem. *256,* 10664 (1981)
25. Hunt, D. F., Yates, J. R. III, Shabanowitz, J., Zhu, N. Z., Zirino, T., Averill, B. A., Daurat-Larroque, S. T., Shewale, G., Roberts, R. M., Brew, K.: Biochem. Biophys. Res. Commun. *144(3),* 1154 (1987)
26. Buhi, W. C., Gray, W. J., Mansfield, E. A., Chun, P. W., Ducsay, C. A., Bazer, F. W., Roberts, R. M.: Biochim. Biophys. Acta *701,* 32 (1982)
27. Baumbach, G. A., Saunders, P. T., Bazer, F. W., Roberts, R. M.: Proc. Natl. Acad. Sci. USA *81,* 2985 (1984)
28. Couso, R., Lang, L., Roberts, R. M., Kornfeld, S.: J. Biol. Chem. *261,* 6326 (1986)
29. Baumbach, G. A., Ketcham, C. M., Richardson, D. E., Bazer, F. W., Roberts, R. M.: J. Biol. Chem. *261,* 12869 (1986)
30. Ketcham, C. M., Baumbach, G. A., Bazer, F. W., Roberts, R. M.: J. Biol. Chem. *260,* 5768 (1985)
31. Echetebu, Z. O., Cox, T. M., Moss, D. W.: Clin. Chem. *33,* 1832 (1987)
32. Davis, J. C., Lin, S. S., Averill, B. A.: Biochemistry *20,* 4062 (1981)
33. Glomset, J., Porath, J.: Biochim. Biophys. Acta *39,* 1 (1960)
34. Robinson, D. B., Glew, R. H.: J. Biol. Chem. *255,* 5864 (1980)
35. Anderson, T. R., Toverud, S. U.: Arch. Biochem. Biophys. *247,* 131 (1986)
36. Beck, J. L., Keough, D. T., deJersey, J., Zerner, B.: Biochim. Biophys. Acta *791,* 357 (1984)
37. Antanaitis, B. C., Aisen, P.: J. Biol. Chem. *259,* 2066 (1984)
38. Keough, D. T., Dionysius, D. A., deJersey, J., Zerner, B.: Biochem. Biophys. Res. Commun. *94,* 600 (1980)
39. Campbell, H. D., Dionysius, D. A., Keough, D. T., Wilson, B. E., deJersey, J., Zerner, B.: Biochem. Biophys. Res. Commun. *82,* 615 (1978)
40. Burman, S., Davis, J. C., Weber, M. J., Averill, B. A.: Biochem. Biophys. Res. Commun. *136,* 490 (1986)
41. Roberts, R. M., Bazer, F. W.: Steroid Induced Uterine Proteins. Beats, W. (ed.) Elsevier, Amsterdam (1980)
42. Averill, B. A., Davis, J. C., Burman, S., Zirino, T., Sanders-Loehr, J., Sage, J. T., Debrunner, P. G.: J. Am. Chem. Soc. *109,* 3760 (1987)
43. Antanaitis, B., Aisen, P.: in: Urushizaki, I., Aisen, P., Listowsky, I., Drysdale, J. (eds.) Structure and Function of Iron Storage and Transport Proteins, Elsevier, Amsterdam (1983)
44. Antanaitis, B. C., Aisen, P.: in: Xavier, A. V. (ed.), Frontiers in Bioinorganic Chemistry, VCH, Weinheim (1986)
45. Schlosnagle, D. C., Sander, E. G., Bazer, F. W., Roberts, R.: J. Biol. Chem. *251,* 4680 (1976)
46. Antanaitis, B. C., Strekas, T., Aisen, P.: J. Biol. Chem. *257,* 3766 (1982)
47. Lauffer, R. B., Antanaitis, B. C., Aisen, P., Que, L., Jr.: J. Biol. Chem. *258,* 14212 (1983)
48. Siamwiza, M. N., Lord, R. C., Chen, M. C., Tadahisa, T., Harada, I., Matsuura, H., Shimanouchi, T.: Biochemistry *14,* 4870 (1975)
49. Freier, S. M., Duff, L. L., Shriver, D. F., Klotz, I. M.: Arch. Biochem. Biophys. *205,* 449 (1980)
50. Sjoberg, B.-M., Graslund, A., Sanders-Loehr, J., Loehr, T. M.: Biochem. Biophys. Res. Commun. *94,* 793 (1980)
51. Sjoberg, B.-M., Loehr, T. M., Sanders-Loehr, J.: Biochemistry *21,* 96 (1982)
52. Salama, S., Spiro, T. G.: J. Am. Chem. Soc. *100,* 1105 (1978)
53. Antanaitis, B. C., Peisach, J., Mims, W. B., Aisen, P.: J. Biol. Chem. *260,* 4572 (1985)
54. Doi, K., McCracken, J., Peisach, J., Aisen, P.: J. Biol. Chem. *263,* 5757 (1988)
55. Gaber, B. P., Miskowski, V., Spiro, T. G.: J. Am. Chem. Soc. *96,* 6868 (1974)

56. Antanaitis, B. C., Aisen, P.: in: Xavier, A. V. (ed.), Frontiers in Bioinorganic Chemistry, VCH, Weinheim (1986)
57. Bertrand, P., Guigliarelli, B., Gayda, J. P.: Arch. Biochem. Biophys. *245,* 305 (1986)
58. Antanaitis, B. C., Aisen, P.: J. Biol. Chem. *257,* 5330 (1982)
59. Muhoberac, B. B., Wharton, D. C., Babcock, L. M., Harrington, P. C., Wilkins, R. G.: Biochim. Biophys. Acta *626,* 337 (1980)
60. Petersson, L., Graslund, A., Ehrenberg, A., Sjoberg, B.-M., Reichard, P.: J. Biol. Chem. *255,* 6706 (1980)
61. Sands, R. H., Dunham, W. R.: Quart. Rev. Biophys. *7,* 443 (1975)
62. Suzuki, M., Uehara, A., Endo, K.: Inorg. Chim. Acta *123,* L9 (1986)
63. Mockler, G. M., deJersey, J., Zerner, B., O'Connor, C. J., Sinn, E.: J. Am. Chem. Soc. *105,* 1891 (1983)
64. Antanaitis, B. C., Aisen, P.: J. Biol. Chem. *257,* 1855 (1982)
65. Antanaitis, B. C., Aisen, P., Lilienthal, H. R., Roberts, R. M., Bazer, F. W.: J. Biol. Chem. *255,* 11204 (1980)
66. Abragam, A., Bleaney, B.: Electron Paramagnetic Resonance of Transition Ions, Clarendon, Oxford (1970)
67. Que, L., Jr., Lipscomb, J. D., Zimmermann, R., Munck, E., Orme-Johnson, N. R., Orme-Johnson, W. H.: Biochim. Biophys. Acta *452,* 320 (1976)
68. Hendrickson, W. A., Co, M. S., Smith, J. L., Hodgson, K. O., Klippenstein, G. L.: Proc. Natl. Acad. Sci. USA *79,* 6255 (1982)
69. Que, L., Jr.: Coord. Chem. Rev. *50,* 73 (1983)
70. Elam, W. T., Stern, E. A., McCallum, J. D., Sanders-Loehr, J.: J. Am. Chem. Soc. *105,* 1919 (1983)
71. Kurtz, D. M., Jr., Sage, J. T., Hendrich, M., Debrunner, P. G., Lukat, G. S.: J. Biol. Chem. *258,* 2115 (1983)
72. Kauzlarich, S. M., Teo, B. K., Zirino, T., Burman, S., Davis, J. C., Averill, B. A.: Inorg. Chem. *25,* 2781 (1985)
73. Keough, D. T., Beck, J. L., deJersey, J., Zerner, B.: Biochem. Biophys. Res. Commun. *108,* 1643 (1982)
74. Pyrz, J. W., Sage, J. T., Debrunner, P. G., Que, L., Jr.: J. Biol. Chem. *261,* 11015 (1986)
75. Antanaitis, B. C., Aisen, P.: J. Biol. Chem. *260,* 751 (1985)
76. Doi, K., Gupta, R., Aisen, P.: J. Biol. Chem. *262,* 6982 (1987)
77. Roberts, R. M., Bazer, F. W.: Biochem. Biophys. Res. Commun. *68,* 450 (1976)
78. Chasteen, N. D.: Biol. Mag. Res. *3,* 53 (1981)
79. Hollander, V. P.: in: Boyer, P. D. (ed.), The Enzymes, 3rd edn., Academic Press, NY (1971)
80. Sundararajan, T. A., Sarma, P. S.: Biochem. J. *56,* 125 (1954)
81. Kawabe, H., Sugiura, Y., Tanaka, H.: Biochem. Biophys. Res. Commun. *103*, 327 (1981)
82. Coleman, J. E., Chlebowski, J. F.: Adv. Inorg. Biochem. *1,* 1 (1979)
83. Schindelmeister, J., Munstermann, D., Witzel, H.: Histochemistry *87,* 13 (1987)
84. Hammerstrom, L. E., Anderson, T. R., Marks, SC., Jr., Toverud, S. U.: J. Histochem. Cytochem. *31,* 1167 (1983)
85. Avison, M. J., Hetherington, H. P., Shulman, R. G.: Ann. Rev. Biophys. Biophys. Chem. *15,* 377 (1986)
86. Roberts, M. R., Bazer, F. W.: BioEssays *1,* 8 (1986)
87. Buhi, W. C., Ducsay, C. A., Bazer, F. W., Roberts, R. M.: J. Biol. Chem. *257,* 1712 (1982)
88. Doi, K., Antanaitis, B. C., Aisen, P.: J. Biol. Chem. *261,* 14936 (1986)
89. Sibille, J. C., Doi, K., Aisen, P.: J. Biol. Chem. *262,* 59 (1987)
90. Wilkins, P. C., Wilkins, R. G.: Coord. Chem. Rev. *79,* 195 (1987)
91. Woodland, M. P., Daulat, S. P., Cammack, R., Dalton, H.: Biochim. Biophys. Acta *873,* 237 (1986)
92. Chasteen, N. D., Antanaitis, B. C., Aisen, P.: J. Biol. Chem. *260,* 2926 (1985)
93. Anderson, B. F., Baker, H. M., Dodson, E. J., Norris, G. E., Rumball, S. V., Waters, J. M., Baker, E. N.: Proc. Natl. Acad. Sci. USA *84,* 1769 (1987)

Phosphines and Metal Phosphine Complexes: Relationship of Chemistry to Anticancer and Other Biological Activity

Susan J. Berners-Price and Peter J. Sadler

Department of Chemistry, Birkbeck College, University of London, Malet Street, London, WC1E 7HX, United Kingdom

Current interest in the biological chemistry of phosphines and their metal complexes ranges from the widespread application of PH_3 as a fumigant, to the clinical use of a Au(I)PEt_3 complex as an antiarthritic drug. Metal phosphine complexes also offer potential as heart-imaging agents and anticancer drugs. Data on cytotoxicity and anticancer activity are surveyed in detail with particular emphasis on diphosphines and their copper, silver and gold complexes. Chemistry of likely relevance to pharmacology is discussed: especially redox reactions (disulphide reduction, oxygen abstraction), radical reactions and metal binding. Much phosphine chemistry of relevance to biology has yet to be explored.

Abbreviations . . . 28
1 Introduction . . . 29
1.1 Scope of the Article . . . 29
1.2 Phosphorus in Biology . . . 30
1.2.1 Reduced Phosphorus in Bacteria? . . . 31
1.2.2 Comparison of P with As . . . 31
1.2.3 Organophosphorus Toxins . . . 31

2 The Chemistry of Phosphines . . . 32
2.1 General Properties . . . 32
2.2 Metal Phosphine Complexes . . . 33
2.2.1 Homogeneous Hydrogenation Catalysts . . . 35
2.2.2 Dinitrogen Reduction . . . 37

3 Phosphines in Biology and Pharmacology . . . 38
3.1 Gold(I) Complexes with Monodentate Phosphines . . . 38
3.1.1 Antiarthritic Activity of Au(I) Phosphine Complexes . . . 38
3.1.2 Cytotoxicity and Antitumour Activity of Au(I) Phosphine Complexes . . . 40
3.2 Anticancer Activity of Monodentate Phosphines and Their Complexes with Other Metals . . . 44
3.3 Tetrahedral, Chelated Au(I)Diphosphine Complexes . . . 46
3.3.1 Background . . . 46
3.3.2 Factors Influencing Chelate Ring Formation by Au(I) . . . 48
3.3.3 Ring-Closure Induced by Thiols and Blood Plasma . . . 52

© Springer-Verlag Berlin Heidelberg 1988

3.3.4 Reactions of $[Au(dppe)_2]Cl$ with Thiols and Cells . . . 53
3.3.5 Cytotoxicity and Antitumour Activity of Tetrahedral, Bis(Diphosphine) Au(I) Complexes . . . 57
3.4 Activity of Diphosphine Ligands, Bridged Au(I) and Other Metal Complexes . . . 64
3.4.1 Cytotoxicity and Antitumour Activity of Diphosphine Ligands . . . 64
3.4.2 Cytotoxicity and Antitumour Activity of Linear, Bridged Digold Diphosphine Complexes . . . 65
3.4.3 Other Metal dppe Complexes . . . 69
3.5 Tetrahedral, Chelated Ag(I) Diphosphine Complexes . . . 70
3.5.1 Preparation and Properties of Four-Coordinate Ag(I) Complexes . . . 70
3.5.2 Cytotoxicity and Antitumour Activity of Bis(Diphosphine)Ag(I) Complexes . . . 72
3.6 Copper(I) Phosphine Complexes . . . 73
3.6.1 Structural Chemistry of Cu(I) Phosphine Complexes and Chelation of Diphosphines . . . 73
3.6.2 Antitumour Activity of Cu(I) Diphosphine Complexes . . . 76
3.7 Mechanism of Cytotoxic Action . . . 77
3.8 Insecticidal Activity of PH_3 . . . 79

4 Reactions of Phosphines of Relevance to Biology . . . 80
4.1 Phosphines as Deoxygenation Agents . . . 80
4.2 Reactions of Phosphines and Au(I) Phosphine Complexes with Disulphides . . . 83
4.3 Radical Reactions of Phosphines . . . 85
4.4 Formation of Phosphonium Salts . . . 89
4.5 Binding of Phosphines to Haems . . . 90
4.6 Proton Affinities of Phosphines . . . 92

5 Technetium Phosphine Complexes: Myocardial Imaging . . . 92

6 Conclusions . . . 94

7 New Data Added at Proof Stage . . . 95
7.1 Metal Bis(Diphosphine) Complexes . . . 95

8 References . . . 97

Abbreviations

Diphosphine ligands:

dppm	bis(diphenylphosphino)methane
dppe	1,2-bis(diphenylphosphino)ethane
dppp	1,3-bis(diphenylphosphino) propane
dppb	1,4-bis(diphenylphosphino)butane
cis-dppey	*cis*-1,2-bis(diphenylphosphino)ethylene
depe	1,2-bis(diethylphosphino)ethane
eppe	1-(diethylphosphino)-2-(diphenylphosphino)ethane
dmfppe	1,2-bis(di(m-fluorophenyl)phosphino)ethane
dpfppe	1,2-bis(di(p-fluorophenyl)phosphino)ethane
dmpe	1,2-bis(dimethylphosphino)ethane
SGlu	β-D-thioglucose
$SGlu(Ac)_4$; (TATG)	tetraacetyl-β-D-thioglucose
GSH	glutathione (reduced form)
DNA	deoxyribonucleic acid
RNA	ribonucleic acid
ip	intraperitoneal
sc	subcutaneous
iv	intravenous
ILS	increase in lifespan
MTD	maximum tolerated dose
T/C	Ratio of response in treated animals (T) to tumour progression in untreated animals (C) (expressed as a percentage). Note that ILS = (T/C − 100)%

1 Introduction

1.1 Scope of the Article

There is increasing interest in the biological chemistry of phosphines and their metal complexes, Table 1. PH_3 is a fumigant widely used against stored product commodities in many third world countries. Auranofin is a gold(I)phosphine complex which has recently been introduced into the clinic for treating rheumatoid arthritis. Auranofin also has limited antitumour activity in animals and this has lead to the discovery of other phosphines with broad spectrum anticancer activity. 1,2-bis(diphenylphosphino)ethane (diphos) and its Au(I), Ag(I) and Cu(I) complexes are particularly effective.

Table 1. Phosphines and metal phosphine complexes of current interest in biology and medicine

Phosphine	Application/Interest	Comment
PH_3	Fumigant for stored commodities (pesticide)	Generated *in situ:* moisture + metal phosphide
$Et_3PAu(TATG)$	Orally-active antiarthritic agent (auranofin) (TATG = tetraacetyl-β-D-thioglucose)	Approved for clinical use by FDA 25 May 1985 (SKF, "Ridaura")
R_3P-Au-X	High cytotoxicity	Limited antitumour activity
Ph_2P-Y-PPh_2	Broad spectrum antitumour activity	Activity related to structure (highest for Y = CH_2CH_2, diphos)
$XAu(Ph_2P(CH_2)_nPPh_2)AuX$	Cytotoxicity, broad spectrum antitumour activity	Activity related to structure highest for n = 2, Au − x = $Au(III)Cl_3$ also active Ag(I)complex inactive
$[M(Ph_2P(CH_2)_nPPh_2)_2]X$	Cytotoxicity, broad spectrum antitumour activity	M = Au(I), Ag(I) or Cu(I) n = 2,3 or CH=CH
Trans-$[Tc(dmpe)_2Cl_2]^+$	Diagnostic nuclear medicine, heart imaging agent	Positively-charged lipophilic complex acts as transport agent for ^{99m}Tc into myocardial cells

Phosphines are well known to chemists. The ligands themselves are widely used in organic synthesis (e.g. the Wittig reaction) and transition metal phosphine complexes have been studied extensively primarily for their applications as hydrogenation catalysts (e.g. Wilkinson's catalyst). There are several recent reviews of various aspects of phosphine and metal phosphine chemistry[1)] together with others mentioned elsewhere. In this article we have attempted to relate this knowledge of the diverse chemistry of phosphines to the types of reactions that are likely to take place in biological systems and may account for their biological activity. We have focussed our attention on cytotoxicity and antitumour activity.

1.2 Phosphorus in Biology

In mammalian systems the chemistry of phosphorus is entirely that of the P(V) oxidation state and the whole of biology revolves around phosphate and its derivatives: phosphate esters (polyphosphates such as ATP, phosphoamino acids, phosphosugars), phosphate precipitation (mineralised tissues, apatite) and phosphate solubilisation (utilisation of stores)[2]. Why is P(V) the only oxidation state observed? Firstly phosphorus in the form of soluble phosphates are readily available to all forms of life whereas reduced forms are not available. Secondly, from the oxidation state diagram (Fig. 1) it is evident that P(V) is a very stable oxidation state and very strong reducing agents are required to reduce

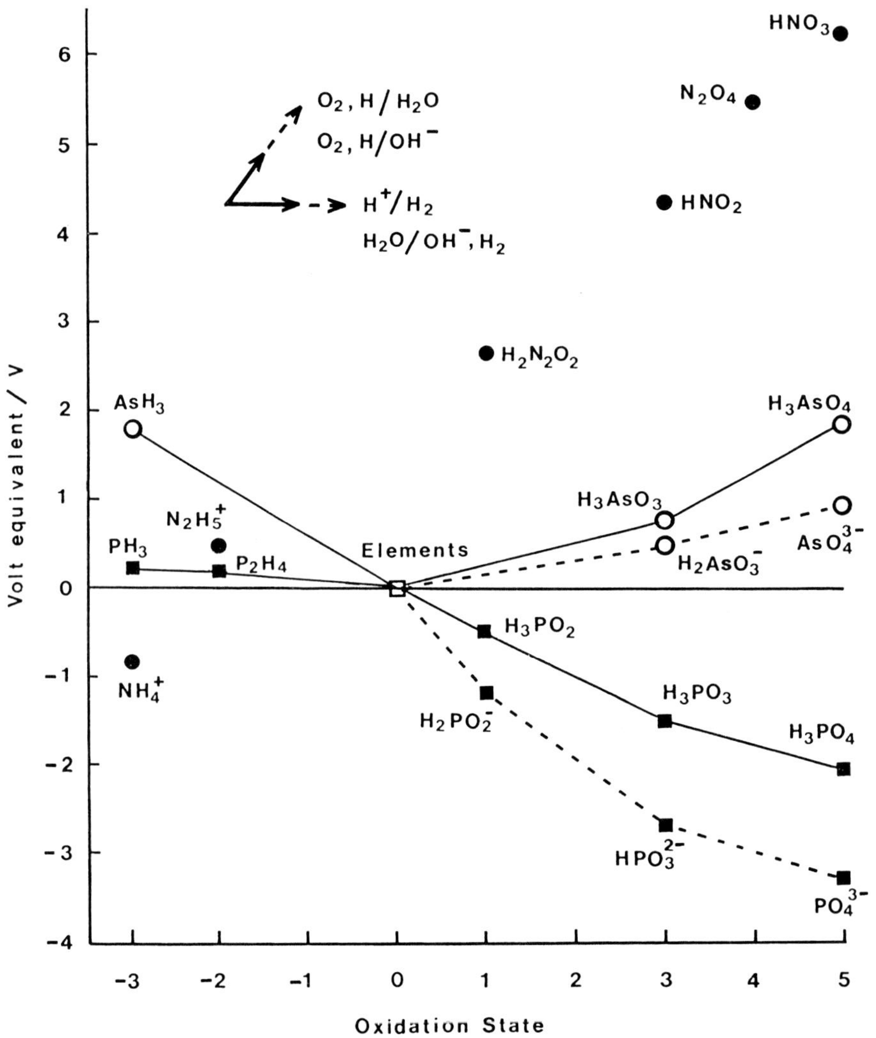

Fig. 1. An oxidation state diagram for P and As in acid (——) and alkaline (-----) solutions and selected data for N (acidic solution). Data from Ref. 3

PO_4^{3-} (E^0 PO_4^{3-}/PH_3 ca -0.4 V)[3]. Conversely, even weak oxidising agents will oxidise PH_3 all the way to H_3PO_4.

1.2.1 Reduced Phosphorus in Bacteria?

Several anaerobic bacteria are now known to contain ferredoxins[4] with potentials that are low enough (e.g. -0.4 V) to make them candidates for phosphate reduction. Therefore it seems thermodynamically possible that anaerobic microorganisms could synthesise reduced P compounds, for example, PH_3. The literature on this appears to be sparse, however there are hints and debates that this may be the case.

The natural phenomenon known as "Will-o'-the wisp": flickering lights observed in darkness over peat-bogs, swamps and marshes, has been ascribed to the spontaneous ignition of PH_3 evolved with other gases during the decomposition of organic materials under waterlogged conditions[5,6]. Tsubota[7] reported detection of phosphite and hypophosphite through microbial reduction of phosphate in flooded paddy soils. More recent studies[8] have failed to detect PH_3 evolution from water-logged soils. The identification of bacteria capable of reducing phosphate may provide a unique opportunity for studying the natural biochemistry of reduced phosphorus and requires investigation. Even though the thermodynamic considerations suggest that phosphate reduction is feasible, little appears to be known about the kinetics of the system. The redox potential data for P[3] are based on thermochemical calculations rather than on measured equilibria.

Consideration of the oxidation state diagram (Fig. 1) shows that PH_3 is a very strong reducing agent. Oxidation to phosphate is an 8-electron process. It may be predicted on thermodynamic grounds that, in mammalian systems, reduced forms of phosphorus will be very reactive, and that oxidation pathways are likely to dominate their biological chemistry.

1.2.2 Comparison of P with As

Elsewhere in this article organophosphines will be compared briefly with organoarsines. It is instructive to compare the oxidation state diagram of As with P (Fig. 1). The contrast is apparent: As(V) is more easily reduced than P(V). It would not be surprising if the metabolism of AsH_3 differed from PH_3. Methylated arsenic species such as $CH_3AsO_3H_2$ and $(CH_3)_2AsO_2H$ appear as metabolites in mammals exposed to both As(V) as arsenate, and As(III) as arsenite. This implies that reduction of As(V) to As(III) can occur *in vivo,* since As(V) cannot be methylated[9]. Such reductions have been demonstrated for some bacteria[10,11]. Methylation of As(III) in model systems is thought to occur via arsenic coordination to cobalt of cobaloxime[9] and may be similar for Vitamin B_{12} pathways. No similar methylation pathways appear to have been identified in phosphorus biochemistry.

1.2.3 Organophosphorus Toxins

There is a class of synthetic phosphorus compounds that are often mentioned in a biological context. These are the organophosphorus poisons (neurotoxins and pesticides). They all have the general structure **1**

1

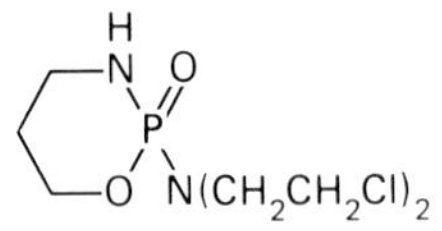

2

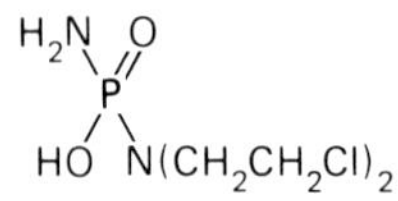

3

where G and G′ are groups which are difficult to displace from P (e.g. alkoxy, dialkylamino or alkyl), X is O or S, and Y is a fairly good anionic leaving group (e.g. F, p-NO_2-C_6H_4O). The mechanism of action involves inhibition of enzymes involved in nerve function (usually acetylcholinesterase) by phosphorylation as a result of displacement of Y by a functional group at the active site of the enzyme[2]. The biochemistry of these compounds appears to be very different to the trivalent organophosphine compounds and will not be discussed further here. However, it is relevant to mention the related compound cyclophosphamide, **2**, which has anticancer activity. The active form of the drug is thought to be **3**, which is formed by hydroxylation in the liver, and the mechanism probably involves alkylation of DNA via the alkyl halide[12].

2 The Chemistry of Phosphines

2.1 General Properties

In this article the term "phosphine" is used to describe both PH_3 and its organo-substituted derivatives PH_2R, PHR_2 and PR_3. We will be concerned mostly with tertiary substituted phosphines PR_3 where R is alkyl or aryl. These include the class of bridged diphosphines $R_2P(CH_2)_nPR_2$.

The fundamental characteristic of all PR_3 compounds is the presence of a lone pair of electrons on P. These compounds therefore can behave both as bases and nucleophiles. The greater size and lower electronegativity of P with respect to N leads to higher polarizability and hence higher nucleophilic reactivity than analogous nitrogen compounds. In addition P has the ability to expand its valence shell to ten electrons and thus trivalent phosphorus compounds can also behave as electrophiles. However, electrophilic reactivity is generally found only for PR_3 compounds containing electron-withdrawing substituents (e.g. Cl, F) and will not be discussed further.

The high nucleophilic reactivity of tertiary phosphines, and the strong bonds which P forms with C, N, O or S, have meant that they have become one of the most widely used class of reagents in organic synthesis. The various applications of phosphines in organic chemistry have been reviewed in great detail[2,13,14]. Our aim is to highlight the type of reactions that phosphines undergo that may have a counterpart in a biological system, for example reactions with disulphides, activated oxygen species and the formation of phosphonium salts (see Sects. 4.1, 4.2 and 4.4).

One of the most dominant features of the chemistry of tertiary phosphines is that the driving force of many of their reactions is the formation of the very strong phosphoryl P

= O bond. The energies of the P = O bonds in tertiary phosphine-oxides lie in the range 535 to 582 $kJmol^{-1}$ [15]. The high bond strength arises from overlap between a filled 2p orbital on O and a vacant 3d orbital on P. In contrast, PH_3 shows little tendency to behave as a nucleophile and does not undergo autoxidation or oxygen abstraction reactions[16]. The oxide OPH_3, if it exists, is certainly less stable than the tertiary phosphine oxides. It may be predicted that O-abstraction reactions will play an important role in the biological chemistry of tertiary phosphines, whereas PH_3 has an alternative 8-electron oxidative pathway to phosphate. In Sect. 4.1 some possible biological substrates that could undergo deoxygenation reactions with phosphines are considered. One similarity in the biological chemistry of PH_3 and PR_3 is their ability to bind strongly to transition metals.

2.2 *Metal Phosphine Complexes*

Tertiary phosphines are amongst the most commonly encountered ligands in transition metal complexes. There are examples of complexes of PR_3 ligands with virtually every transition metal[17, 18]. This is probably a reflection of the electronic versatility of phosphines: they form kinetically-stable complexes with metals in both high (e.g. + IV) and low e.g. − I) oxidation states.

As far as the biological chemistry of metal phosphine complexes is concerned, our interest in the properties of phosphines coordinated to metals arises for two reasons. Firstly, because the biological (in particular cytotoxic) activity of phosphines could involve coordination to metals (e.g. Fe, Cu or Zn) in critical sites. This point is discussed further in Sects. 3.7 and 4.5. Secondly, alteration of the substituents on PR_3 can cause substantial changes in the reactivity of transition-metal complexes. Therefore, by understanding the factors that influence the reactivity it may be possible to design rationally metal-phosphine complexes as chemotherapeutic agents.

The nature of P–M bonds in metal complexes of tertiary phosphines has been the subject of much controversy. Phosphines are frequently classified as π-acceptor ligands along with others such as CO. It was orginally thought that the good bonding properties of PR_3 ligands were attributable to d–π backbonding involving filled metal d-orbitals interacting with empty 3d orbitals on phosphorus, which reinforced the P–M σ-bond. The ligand PF_3 does have much in common with CO and the photoelectron spectrum of $Ni(PF_3)_4$ is consistent with the occurrence of π-bonding[19]. However, it has been argued that for all other simple tertiary phosphines dπ–dπ backbonding is not required to explain the observed physical properties[20, 21]. The issue of π-bonding is still not resolved but it has been suggested that bonding is clear at two extremes: metals in oxidation states II or higher form essentially pure σ bonds with PR_3 ligands whereas metals in oxidation state 0 or below form combined σ with π bonds to PF_3, PCl_3 and $P(OPh)_3$[22]. For other complexes the tendency for π-bonding to occur will be greatest for metals in low oxidation states. The ability of phosphines to stabilise low oxidation states is an important feature of their chemistry.

The relative σ-donor powers of tertiary phosphines is also an area of controversy. It has usually been assumed that σ-donor ability decreases in the order:

$Me_3P > Me_2Ph > MePH_2P > PPh_3$.

Table 2. Some literature data on the properties of PH_3 and some tertiary phosphines

Phosphine	IP[a] (ev)	pK_a	PA[b] (kcal/mol)	Cone angle[c] (degrees)
PH_3	10.58[d], 9.96[e]	−14[f]	187.9[g]	87
PMe_3	8.58[h], 8.65[i], 8.62[j]	8.65[k]	223.5[l], 228.0[g]	118
PMe_2Et	–	8.62[k]	–	–
$PMeEt_2$	–	8.62[k]	–	–
PEt_3	8.31[c]	8.69[k]	–	132
P^iPr_3	8.05[c]	–	–	160
P^nBu_3	8.11[c]	8.43[k]	–	132
P^tBu_3	7.71[c]	11.40[m]	–	182
PEt_2Ph	–	6.25[k]	–	136
PMe_2Ph	8.37[h], 8.32[j], 8.45[n]	6.49[k]	226.0[l]	122
$PMePh_2$	8.28[j], 8.07[h]	4.65[l, o]	226.7[l]	136
PPh_3	7.80[j], 7.88[p], 7.92[q], 8.10[h]	2.73[k, m]	226.7[l]	145
$P(4MeOC_6H_4)_3$	–	4.57[m]	–	145
$P(4FC_6H_4)_3$	–	1.97[m]	–	–
$P(cC_6H_{11})_3$	–	9.70[k], 9.65[m]	–	170

[a] lone pair ionization energy; [b] gas-phase proton affinity; [c] Ref. 23 (see Fig. 2); [d] Ref. 24; [e] Ref. 25; [f] Ref. 26; [g] Ref. 27; [h] Ref. 28; [i] Ref. 29; [j] Ref. 30; [k] Ref. 31; [l] Ref. 32; [m] Ref. 33; [n] Ref. 34; [o] based on $pK_a = 7.85 - 2.67\ \Sigma\sigma^*$ (Ref. 31); [p] Ref. 35; [q] Ref. 36

Indeed, pK_a measurements indicate that the donor ability of PR_3 towards protons in solution decreases with increasing substitution of phenyl groups (Table 2). However, Puddephatt and coworkers[28, 30)] have suggested that this order should actually be reversed on the basis of the relative lone pair ionisation potentials and proton-affinities of these phosphines in the gas phase (see Table 2). The role of solvent in determining pK_a's is discussed in Sect. 4.6.

Altering the substituents on tertiary phosphines can cause substantial changes in the chemical and physical properties of their transition metal complexes, including the strength of the M–P bond. Before 1970, almost everything was rationalised in terms of electronic effects of the phosphine substituents, but since then it has been recognised that steric effects are often as important, and dominate in many cases. Tolman[23)] introduced the ligand cone angle, Θ, (Fig. 2) as a measure of its steric bulk, and a large number of parameters including molecular structures, rate and equilibrium constants and NMR chemical shifts have been correlated with ligand cone angles. The cone angles of some PR_3 ligands are tabulated in Table 2. Steric and electronic effects are often strongly interrelated. For example, increasing the steric bulk of the ligand opens the C–P–C angles,

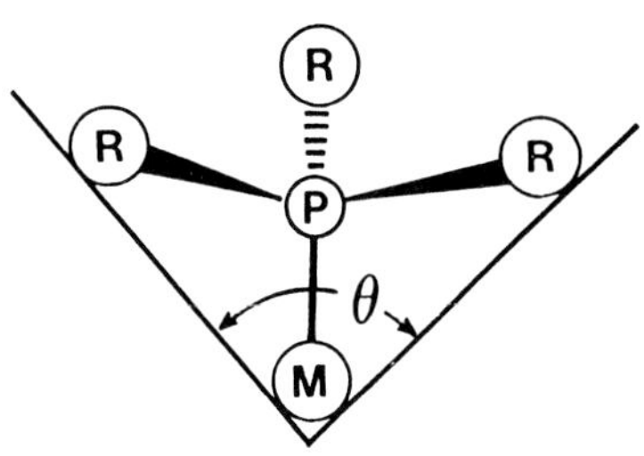

Fig. 2. Schematic representation of a phosphine cone angle Θ (based on Ref. 23)

thereby increasing the percentage of s-character in the P–C bonds and decreasing it in the lone pair. This is reflected by the decrease in the lone pair ionisation energies:

$PH_3 > PMe_3 > PEt_3 > P^iPr_3 > P^tBu_3$ (see Table 2).

However, the percent s-character in the phosphorus lone pair will also increase with increasing electronegativity of the substituents on P.

Steric effects dominate the reactivity of many metal-phosphine complexes. The presence of several bulky phosphines in the same coordination sphere often facilitates ligand dissociation, producing a vacant coordination site. For example, steric effects dominate the dissociation equilibria of $Ni(PR_3)_4$[23] and $Pd(PR_3)_4$[37]. The extent of ligand dissociation for the equilibria:

$$Pd(PR_3)_4 \rightleftarrows Pd(PR_3)_3 + PR_3 \rightleftarrows Pd(PR_3)_2 + PR_3$$

increases in the order:

$$PMe_3 \sim PMe_2Ph \sim PMePh_2 < PEt_3 \sim PPh_3 < P^iPr_3 < PPh(^tBu)_2.$$

Complexes of weakly-held ligands can be favoured by increasing the size of PR_3. The reaction:

$$[Ni(PR_3)_4] + N_2 \rightarrow [(N_2)Ni(PR_3)_3] + PR_3$$

proceeds for PEt_3, but not PMe_3, because the bulkier phosphine is more readily displaced[38].

We will discuss later how the ease of ligand dissociation appears to play a crucial role in the cytotoxicity of metal phosphine complexes. The majority of metal homogeneous hydrogenation catalysts contain phosphine ligands and ligand dissociation appears to play a role in their catalytic function. Changes in the phosphine substituents can also lead to a selective recognition of substrates for hydrogenation: a step that may be paralleled by target recognition of a phosphine or a metal phosphine complex in a cell. It is therefore relevant to discuss the chemistry of metal phosphine catalysts.

2.2.1 Homogeneous Hydrogenation Catalysts

A very large number of organometallic complexes have been found to act as hydrogenation catalysts[20]. They are used in the large scale syntheses of organic compounds because hydrogenation can be achieved at ambient temperature, and low H_2 pressures. The substrates are generally olefins but other functional groups including acetylenes, aldehydes and ketones are also reduced.

Most of the catalysts contain rhodium, ruthenium or iridium with PPh_3 ligands. The phosphine ligands can play a number of roles. For example, they can confer solubility in the organic solvents used for the hydrogenation reactions. Secondly, they can stabilise both low and high oxidation states of the metal allowing it to undergo facile oxidative addition of hydrogen, followed by facile reductive elimination of the reduced substrate. The third, and probably most important role of the phosphine, involves steric interactions. Steric effects can provide substrate selectivity, for example, by directing the least hindered double-bond of a polyolefin to the coordination site.

The majority of the homogeneous olefin hydrogenation catalysts can be divided into two classes: monohydride catalysts which have a single M–H group, and dihydride catalysts which have two adjacent hydrides (*cis* MH_2), present at some stage of the catalytic cycle. Monohydride catalysts promote the isomerisation of olefins, and usually exhibit selectivity for hydrogenation of conjugated olefins and terminal rather than internal olefins. An example is $HRuCl(PPh_3)_3$. The PPh_3 ligands inhibit coordination of other large ligands so that the complex is coordinately unsaturated and readily reacts with the substrate. The ligands also direct the least hindered (terminal) olefin to the metal.

Dihydride complexes can catalyse the *cis* addition of hydrogen, without olefin isomerisation. An example is $RhCl(PPh_3)_3$ ("Wilkinson's Catalyst"), which is the most commonly utilised, and consequently most widely studied, homogeneous hydrogenation catalyst. Considerable effort has been devoted to understanding the mechanism of the catalytic cycle and these mechanistic studies have been reviewed[20, 39]. Dissociation of a PPh_3 ligand gives the 3-coordinate Rh(I) intermediate, $RhCl(PPh_3)_2$. This undergoes rapid hydrogenation to form the coordinately-unsaturated Rh(III) complex $RhCl(H)_2(PPh_3)_2$ which undergoes olefin addition. Reductive-elimination of the alkene product regenerates $RhCl(PPh_3)_2$. The rate of hydrogenation of cyclohexene by $RhCl(PR_3)_3$ catalysts decreases in the order $PPh_3 > PEtPh_2 > PEt_2Ph > PEt_3$[40], reflecting the more facile dissociation of the bulkier ligands.

Steric interactions also play a role in asymmetric homogeneous catalytic hydrogenation. By the use of a catalyst containing a chiral phosphine it is possible to hydrogenate a prochiral unsaturated substrate to give high yields of the required enantiomer. Most of the asymmetric catalysts studied to date are Rh(I) complexes of the type $[Rh(PR_3^*)_2(diene)_2]^+$ (where PR_3^* is a chiral phosphine)[20]. The highest enantioselectivity has been achieved with chiral disphosphine ligands (P–P*) rather than monophosphines. This is attributed to reaction of the chelated complex via a different mechanism. Hydrogenation of $[Rh(PR_3)_2(diene)]^+$ and $[Rh(P–P)_2(diene)]^+$ gives rise to the intermediates (**4**) and (**5**) respectively:

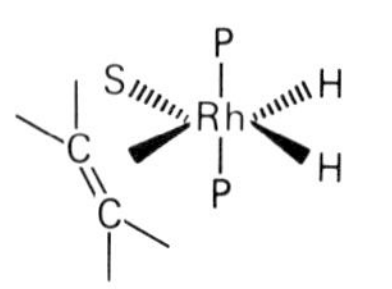

4

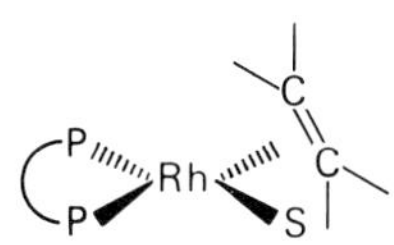

5

Phosphines avoid becoming *trans* to hydride groups whenever possible. Oxidative-addition of hydrogen is *cis,* and the monophosphine complex can form an intermediate in which the two phosphines are *trans* to each other. This then adds olefin to form intermediate **4**. The diphosphine complex cannot form an intermediate where phosphine is not *trans* to hydride and this changes the equilibrium for oxidation-addition of hydrogen. The mechanism proceeds via an "olefin" rather than a "hydride" path, so that complex **5** is an intermediate. Hydrogenation of **5** produces the intermediate complex **6**:

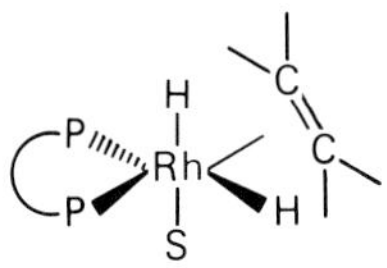

6

The hydride *trans* to the phosphine migrates to the olefin and this *trans* destabilisation probably contributes to the rapid rate of reaction. High enantioselectivity ($<90\%$) has been restricted to hydrogenation of prochiral α-acetylaminocinnamic derivatives and a variety of chiral phosphines have been used. The highest enantioselectivity was obtained with the dppe-analogues (–)DIPAMP (**7**) and (S,S)-Chiraphos (**8**)

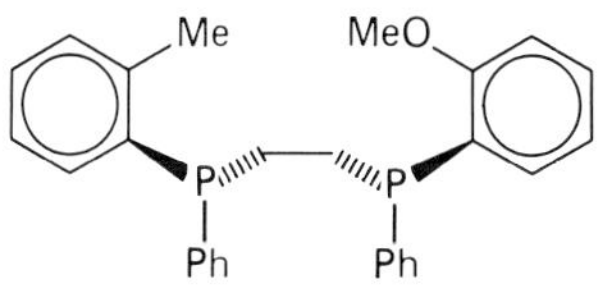

7

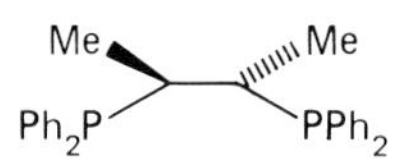

8

It is evident that by substituting bidentate tertiary phosphine ligands for monophosphines it is possible to cause major changes in the reactivity of a metal-phosphine complex. We discuss in other sections the contrasting profile of biological activity of chelated diphosphine complexes compared to related monophosphine complexes.

2.2.2 Dinitrogen Reduction

It is relevant to briefly discuss the molybdenum phosphine complexes that have been proposed as models for the enzyme nitrogenase[41)]. These complexes have a number of features in common with some of the homogeneous hydrogenation catalysts. Dinitrogen complexes of Mo are formed with phosphine ligands which stabilise the Mo(O) oxidation state. Chatt and coworkers have compared the reactivity of the complexes *trans*-$[M(N_2)_2(dppe)_2]$ and *cis*- and *trans*-$[M(N_2)_2(PMe_2Ph)_4]$ (where M is Mo or W) in the presence of strong acids[42)]. Treatment of both monophosphine complexes with sulphuric acid in methanol at room temperature reduced the coordinated N_2 to NH_3. The yields were 20–36% reduction of one N_2 for the Mo complex and 90% for the W complex. In contrast, when either the Mo or W dppe complexes were treated with H_2SO_4 the product was *trans*-$[M(HSO_4)(NNH_2)(dppe)_2]$. It was suggested that these differences may be due to the fact that the chelated dppe ligands do not readily dissociate. For reduction and protonation to proceed beyond the hydrazide level the metal must release electron density into the N_2H_2 ligand. For the tetrakis complexes the monodentate phosphine can readily dissociate and be replaced by a harder oxo ligand (e.g. SO_4^{2-}). This stabilises the higher oxidation state so that Mo (or W) can deliver all six of its electrons to dinitrogen. However, this oxidative degradation destroys any catalytic potential of the system.

In recent work, Pickett and Talarmin[43)] have shown that protolysis of *trans*-$[W(N_2)_2(dppe)_2]$ can be made to proceed beyond the hydrazido stage by coupling protonation with electronation using controlled-potential electrolysis at a Hg-pool cathode:

$$trans\text{-}[W^{0}(N_2)_2(dppe)_2] \xrightarrow[\text{thf, }-N_2]{TsOH \cdot H_2O} trans\text{-}[W^{IV}(NNH_2)TsO(dppe)_2]^{+}$$

$$2\,e^{-}(\text{electrolysis})/N_2$$

$$NH_3$$

The electrolysis produces ca. 0.23 mol NH_3 per mol *trans*-[W(NNH_2)TsO(dppe)$_2$] together with a small amount of N_2H_4 (ca. 0.015 mol) and consumes 2 mol electrons. The system is cyclic since the starting complex (ca. 0.9 mol) is regenerated. This sequence of protonation-electronation steps provides a chemical precedent for the proposal[42a)] that biological N_2 fixation involves successive protonation/electronation reactions of N_2 bound to a transition metal centre in an enzyme. By varying the substituents of the diphosphine (e.g. introducing H-bonding or H^+ donating groups) it should be possible to stabilize various intermediate reduced and protonated states of N_2 and learn more about the course of such reactions in model systems.

3 Phosphines in Biology and Pharmacology

3.1 Gold(I) Complexes with Monodentate Phosphines

3.1.1 Antiarthritic Activity of Au(I) Phosphine Complexes

Gold(I) thiolate compounds (e.g. aurothiomalate "Myocrisin") have been in clinical use for the treatment of rheumatoid arthritis since the 1930's[44,45)]. There have been conflicting reports over the years on the efficacy of gold therapy, but some believe that gold compounds are amongst the few drugs in clinical use that actually induce remission of the disease. The lack of enthusiasm for gold therapy has arisen primarily as a result of the severe toxic side effects that occur in ca. 40% of patients. Gold(I) thiolate complexes are water-soluble and are usually administered by large monthly intramuscular injections. Nephrotoxicity is a common problem caused by accumulation of gold in the kidneys following a long course of treatment.

In the mid 1960's Sutton and colleagues at Smith Kline & French began a search for orally-active gold compounds. They hoped that these would allow blood gold levels to be maintained by a low daily dose, thus preventing the accumulation of gold in tissues which follows large injected doses. A series of lipophilic gold(I) phosphine complexes R_3PAuX (X = Cl, or thiosugar), with linear 2-coordinate structures, were found to exhibit oral antiarthritic activity in animal models of arthritis[46)].

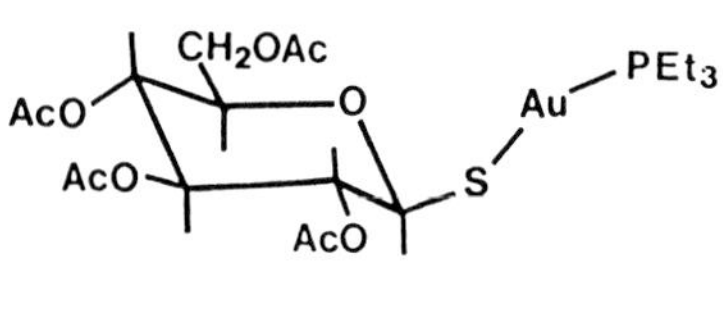

9

The nature of the phosphine ligand (R_3P) influenced the extent of oral absorption of the complex and antiarthritic activity was found to be closely related to the gold serum concentrations (Fig. 3). The maximum absorption and activity occurred when the phosphine was Et_3P. Et_3PAuCl was initially selected for clinical trial but diarrhoea proved to be a troublesome side-effect and it was abandoned in favour of the tetraacetylated thioglucose derivative, auranofin (**9**). Under the tradename "Ridaura" this was approved by the FDA for clinical use in the USA in May 1985.

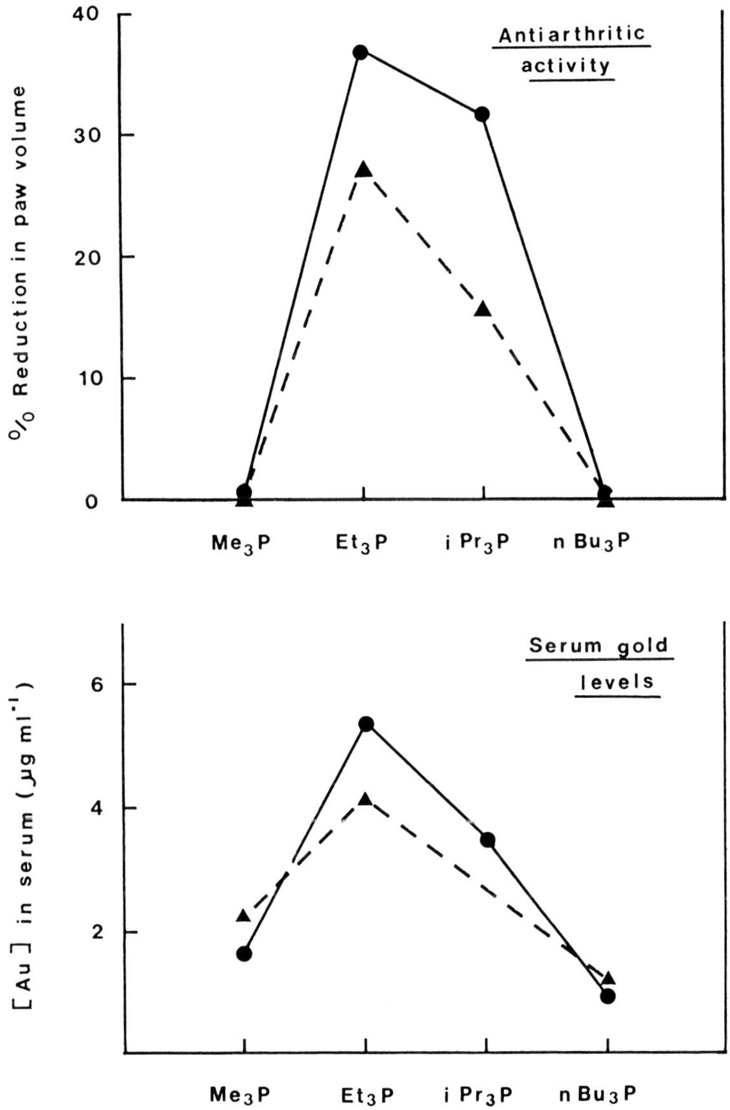

Fig. 3. The correlation between the antiarthritic activity (in rat model, assessed by the reduction in paw volume of uninjected leg on day 16, oral doses of compound equivalent to 10 mg Au kg^{-1}) and serum Au concentration for the series of complexes R_3PAuCl (●) and $R_3PAuSGlu$(▲) (where SGlu is β-D-thioglucose). Data from Ref. 46

Table 3. Comparative pharmacokinetic data for auranofin and aurothiomalate (based on Ref. 47)

	Auranofin	Aurothiomalate
Oral absorption	ca. 25%	<1%
Blood: cell-bound	ca. 40%	<1%
protein-bound	ca. 60%	ca. 99%
Steady-state blood [Au]	3–3.5 μM (6 mg oral/day)	15–36 μM (50 mg im./week)[a]
Excretion: faecal	85%	30%
renal	15%	70%
Body retention of one dose after 6 months	<1%	25–42%

[a] im. = intramuscular injection

The results of clinical trials on auranofin to date[47, 48] suggest that its efficacy approaches that of gold(I)thiomalate and it is relatively well tolerated in most patients. Auranofin appears to have a contrasting pharmokinetic profile to aurothiomalate (Table 3). The higher percentages of oral absorption, cell-associated gold and faecal compared to renal excretion, can all be attributed to the high lipophilicity of the complex.

The mechanism(s) of antiarthritic activity of gold is still a matter of debate but it probably involves Au(I) binding to protein sulphydryl groups. The Au-P bond in auranofin is stable in most model reactions but studies with ^{195}Au, ^{35}S and ^{32}P-radiolabelled auranofin in dogs[49] have shown that ^{35}S and ^{32}P are excreted more rapidly than ^{195}Au, and Et_3PO has been identified in the urine of auranofin-treated patients[47]. Once the phosphine has been released the products of auranofin metabolism could be similar to the products from the metabolism of gold(I) thiolate drugs. Thus the phosphine ligand influences the distribution of gold *in vivo* but it is not known if it also contributes to the antiarthritic activity of the complex. However, auranofin has additional pharmacological properties that are not observed for gold(I) thiolate complexes, and these are almost certainly attributable to the presence of the Et_3P ligand.

3.1.2 Cytotoxicity and Antitumour Activity of Au(I) Phosphine Complexes

Lorber and coworkers demonstrated that auranofin exhibited antiproliferative effects against cultured human cancer cells[50]. They also reported that auranofin increased the survival times of mice with P388 leukaemia[51]. A number of Au(I) complexes of Ph_3P have also been reported to have anticancer activity[52, 53].

Mirabelli and coworkers have evaluated the antitumour activity of auranofin in 15 tumour models in mice[54]. They reported that it was active only against ip P388 leukaemia and required ip administration for activity. However, they found that auranofin exhibited potent cytotoxicity against a variety of tumour cell lines *in vitro*. It did not alter cell cycle distribution or preferentially kill cells in logarithmic growth. DNA, RNA and protein syntheses were inhibited non-selectively at concentrations which were acutely lethal to cells. The cellular response was rapid and cells treated with auranofin at concentrations as low as 1 μM for 2 h showed extensive morphological changes including surface membrane blebbing, cell rounding and membrane lysis. There appeared to be a direct relationship between cellular association of auranofin-derived gold and the cytotoxic effect.

The antitumour activity and cytotoxicity of a large number of gold(I) complexes have also been evaluated[55]. Au(I) thiolate complexes exhibit very low cytotoxicity to cancer

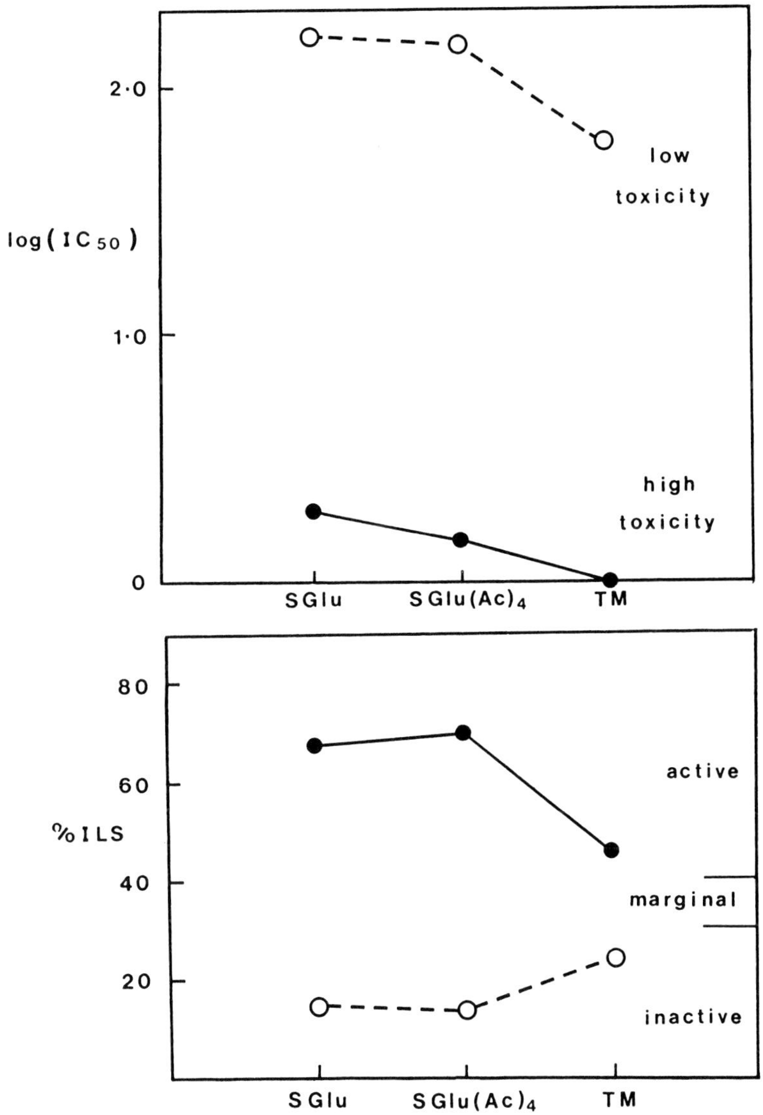

Fig. 4. The comparative cytotoxicity (*top:* clonogenic assay, 2 h treatment of B16 mouse melanoma cells) and antitumour activity (P388 leukaemia, ILS at the maximum tolerated dose) of gold(I) phosphine complexes (●) Et_3PAuSR where SR = SGlu (β-D-thioglucose), $SGlu(Ac)_4$ (tetraacetyl-β-D-thioglucose) or TM (thiomalate), and the non-phosphine analogues $[AuSR]_n$ (○). Data from Ref. 55

Table 4. The *in vitro* cytotoxicity and *in vivo* antitumour activity of auranofin analogues. (data from Ref. 55)

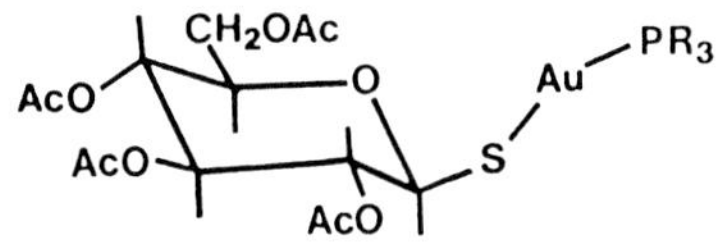

PR_3	IC_{50} (μM)[a] B 16 cells	MTD μmol/kg[b]	ILS max, %[c] P 388 leukaemia	
PEt_3[d]	1.5	18	70	
$PEt_2(O^iPr)$	2	17	90	ACTIVE
$PEt_2(OEt)$	1	14	70	
$P(NMe_2)_3$	2	8	60	
PEt_2BuOH	8	17	58	
PEt_2Ph	2	13	55	
P^iPr_3	4	14	46	
PMe_3	2	9	45	MARGINAL
PPh_3	4	7	36	
$PEtPh_2$	4	6	32	

[a] In vitro cytotoxic potency to B 16 melanoma cells assessed by a clonogenic assay; IC_{50} is the concentration required to reduce the number of colonies formed by 50% following a 2 h exposure to the compound; [b] maximally tolerated dose in mice bearing ip. P 388 leukaemia on a daily × 5 ip. regimen; [c] % increase in lifespan relative to non-treated tumour-bearing mice;
[d] Auranofin

cells *in vitro* and are totally inactive against ip P 388 leukaemia (Fig. 4). Introduction of a phosphine ligand gives rise to potent cytotoxicity and *in vivo* antitumour activity (Fig. 4). A large number of gold(I) phosphine complexes R_3PAuX have been evaluated and all exhibit potent cytotoxicity to tumour cells *in vitro*. Variations of the phosphine substituents (R_3P) modulate the antitumour activity *in vivo* (Table 4 and Fig. 5) as do changes in the *trans* ligand X (Fig. 5). It is noticeable that antitumour activity is lowest for the complexes that contain a good leaving group (e.g., X = Cl or NO_3), although cytotoxic potency is unaffected. These complexes are very reactive towards thiols and it may be predicted that the *trans* ligand will be rapidly displaced by –SH groups *in vivo* so that the R_3PAu– binds strongly to proteins (e.g. in serum) and is prevented from reaching intracellular targets. Auranofin (X = TATG, tetraacetyl-β-D-thioglucose) was one of the most active complexes tested and TATG would be a poor leaving group under aqueous conditions. However, the cytotoxic potency of auranofin to cultured tumour cells *in vitro* is reduced ten-fold when the culture-media contain serum proteins[54)], suggesting that activity is still lost through gold binding to proteins. Serum is not merely an aqueous solution but contains strongly hydrophobic compartments, e.g. lipoproteins; in these TATG may be readily displaced.

Potent *in vitro* cytotoxic activity and *in vivo* antitumour activity has been observed only for Au(I) complexes with coordinated phosphine ligands.[1] Complexes of type YAuCl, where Y = R_2S, pyridine, or cyclooctene have a low cytotoxicity and, at best, a

1 An exception is $K[Au(CN)_2]$ which exhibits antitumour activity in several animal models[52)]. Gold(I) may act as a delivery agent for toxic cyanide.

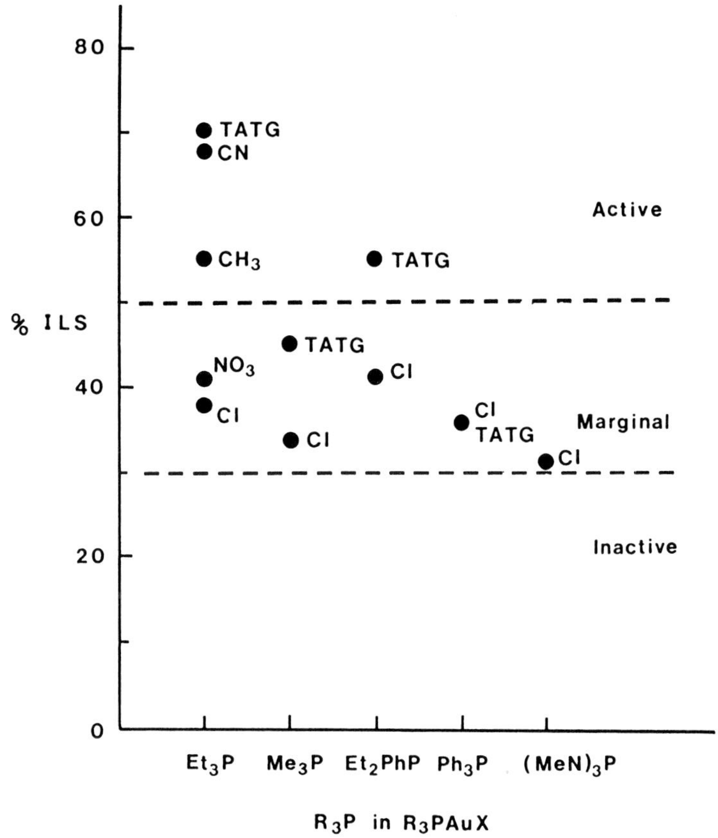

Fig. 5. Dependence of the antitumour activity (P388 leukaemia in mice) of R_3PAuX complexes on the phosphine R_3P and the *trans* ligand X. Complexes with the more strongly bound *trans* ligands TATG (tetraacetyl-β-D-thioglucose), CN or CH_3 tend to be most active. Data from Ref. 55

marginal antitumour activity *in vivo,* although much lower in potency than the phosphine complexes. The arsine complex $Et_3AsAuCl$ is potently cytotoxic *in vitro* but inactive as an antitumour agent[55)].

What is the role of the phosphine in the cytotoxicity of these Au(I) complexes? As described above, gold(I) phosphine complexes are lipophilic whereas polymeric Au(I) thiolates do not readily enter cells. The cytotoxicity may be related to the lipophilicity of the phosphine ligands which allows the gold to penetrate the cell membrane and bind to reactive –SH groups. The observed differences in activity on changing the phosphine substituents may reflect the extent of cellular association of the different complexes. It is known that the proliferation of cancer cells can be inhibited by –SH inhibitors[50)]. An alternative possibility is that the phosphines themselves produce the cytotoxic effect. Clinical studies with auranofin have demonstrated that PEt_3 is released from Au(I) and is oxidised[47)]. Could Au(I) be delivering the cytotoxic phosphines to cellular targets? If this is the case then differences in the reactivity of the various phosphines would be expected to have a profound effect on the cytotoxicity of their Au(I) complexes.

3.2 Anticancer Activity of Monodentate Phosphines and Their Complexes with Other Metals

The National Cancer Institute (NCI, Bethesda) have tested over 150 phosphines (PR_3) for anticancer activity. With the exception of a number of diphosphine ligands (which we discuss in detail in Sect. 3.4.1), none of the tested tertiary phosphine ligands exhibited more than at best marginal activity (PPh_2Me). However, it is probable that many of the phosphines would have been rapidly oxidised *in vivo* (or even in the testing medium), and so the data may actually refer to phosphine-oxides. Coordination to Au(I) may protect the ligands from unfavourable oxidation.

Many metal phosphine complexes have been tested for anticancer activity by the NCI and in general, it appears that they do not exhibit significant antitumour activity as exemplified by the data on the complexes shown in Table 5. However, we suggest that there is a need to re-test many complexes of this type under carefully controlled conditions. In some cases the complexes may be air-sensitive and in others the solvent (vehicle) used for administration appears to be inappropriate (e.g. H_2O for Wilkinson's catalyst). There can be no certainty that the species actually tested correspond to the formulae. For example it is known that in solution $RhCl(PPh_3)_3$ reacts with O_2 to give more soluble dioxygen complexes[39]. In addition, inactivity in one tumour system (L 1210 lymphoid leukaemia in the case of Table 5) does not necessarily imply lack of activity in other models. It is notable from the high doses used in tests that many have a low toxicity. This may be attributable to a combination of low solubility and instability. Speciation studies of these complexes in the test media, in blood plasma and cell culture media are needed. It is notable that the PPh_3 analogue of *cis*-$[PtCl_2(NH_3)_2]$ (cisplatin), i.e. *cis*-$[PtCl_2(PPh_3)_2]$, is inactive. PPh_3 does not possess a P–H bond and an N–H linkage appears to be an almost obligatory requirement in active amine analogues of cisplatin[142, 143]. The inactivity may also be a consequence of the high *trans* labilising power of P compared to N and the high thermodynamic stability of Pt–P bonds, so that the phosphine itself is not sufficiently labile to be the cytotoxic agent when the complex is delivered to cells. It may be possible to change kinetic lability by changing oxidation state. For instance, the Pt(0) complex, **10**, is active in ip P 388 leukaemia (T/C 138% at 25 mg/kg).

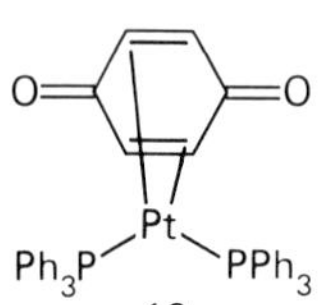

10

However, it is inactive in many other tumour systems including L 1210 leukaemia (NCI data). Pt(0) like Au(I) is a $5d^{10}$ metal ion. If the antitumour activity of the Au(I) phosphine complexes is attributable to the presence of the phosphine ligands, then the kinetic lability of the complexes is likely to be crucial, so that the ligand can be released at the target site.

As discussed above, the cytotoxicity of auranofin to cultured tumour cells *in vitro* was significantly reduced when the culture media contained serum proteins[54]. This suggested to us that in order to obtain a broader spectrum of antitumour activity it would be

Table 5. The following complexes have been found to be *inactive* (T/C < 125%) against ip. L1210 lymphoid leukaemia in mice (data supplied by the National Cancer Institute). The vehicle of administration and dose range tested (see footnote)[a] are given in brackets.

$Cr(CO)_5(PR_3)(PR_3 = PPh_3, P^nBu_3, P(OPh)_3)$(cm, s/t, 50–200)[a, b] $Cr(CO)_4(PPh_3)_2$ (s/t, 100)
$Mo(CO)_5PPh_3$ (s/t, c, 100) $Mo(1,10\text{-phen})(PPh_3)_2$ (s/t, c, 50, 25[c])

$Mn(CO)_4(Me)(PPh_3)$ (c, 100)
$ReCl_3(PMe_2Ph)_3$ (s/t, 200), $ReCl_3(O)(PPh_3)_2$ (s/t, 12–50)

$Fe(CO)_2(PF_3)_3$ (s, 25–50), $Fe(CO)_3(PF_3)_2$ (s, 25–50), $Fe(CO)_4(PPh_3)$ (c, 400),
$Fe(cp)I(CO)(PPh_3)$ (s/t, 100)
$Ru(CO)_2Cl_2(PPh_2(CH = CHPh))_2$ (c, 50–100), $Ru(CO)(CF_3CO_2)_2(PPh_3)_2$ (s/t, 100–200)
$OsH(Br)(CO)(PPh_3)_3$ (s/t, 100)

$Co(dmg)_2(py)(P^nBu_3)$ (s/t, 100–200), $CoCl_2(PPh_3)_2$ (s/t, 50–100)[c], $Co(cp)I_2(PPh_3)$ (u, 150)
$RhCl(PPh_3)_3$ (s, 100–400), $Rh(NO)(PPh_3)_3$ (s, 100–400), $RhH(PPh_3)_4$ (c, 100),
$RhCl(CO)(PPh_3)_2$ (s/t, 200), $RhCl_2(NO)(PPh_3)_2$ (c, 100–200), $[Rh(cod)(PPh_3)_2]PF_6$ (s/t, 100),
$[Rh(PMePh_2)_4]PF_6$ (s/t, 100)
$IrCl(CO)(PPh_3)_2$ (c, 200)

$NiCl_2(PPh_3)_2$ (s, s/t, c, 6.25–100)[d], $NiCl_2(P^nBu_3)_2$ (c, 27)[e],
$Ni(NO_3)_2(P^nBu_3)_2$ (s/t, 100), $NiI_2(PPh_3)_2$ (s/t, 200), $Ni(SCN)_2(PPh_3)_2$, (g, 400)
$NEt_4[NiBr_3(PPh_3)]$ (c, 50–100), $Ni(CO)_2(PPh_3)_2$ (c, 100)
trans-$[PdCl_2(PPh_3)_2]$ (u, 400), *trans*-$[PdCl_2(P^nBu_3)_2]$ (o, 100)[f],
trans-$[PdBr_2(PMePh_2)_2]$ (s, 100), *trans*-$[PdCl_2(PMePh_2)_2]$ (s, 75), $[PdCl(P^nBu_3)(\mu\text{-}Cl)]_2$ (g, 100)

cis-$[PtCl_2(PPh_3)_2]$ (w, 10), *trans*-$[PtCl_2(PPh_3)_2]$ (s/t, 100),
cis-$[PtI_2(PPh_3)_2]$ (s/t, 6.3), *cis*-$[Pt(N_3)_2(PPh_3)_2]$ (s/t, 6.3–100)
trans-$[PtCl_2(P^nBu_3)_2]$ (w, 12.5), *trans*-$[Pt(SCN)_2(P^nBu_3)_2]$ (w, 6.3)
trans-$[Pt(NO_3)_2(P^nBu_3)_2]$ (w, 50), *trans*-$[PtBr_2(P^nBu_3)_2]$ (w, 10),
trans-$[PtI_2(P^nBu_3)_2]$ (w, 1.9) *trans*-$[Pt(Ph)_2(P^nBu_3)_2]$ (w, 12.5),
trans-$[Pt(SR)_2(PPh_3)_2]$ (R = ^iPr, Ph) (w/t, 6.3[f]–400)
trans-$[PtCl(H)(PPh_3)_2]$ (w, 6.3), *trans*-$[PtI(Me)(PPh_3)_2]$ (w, 6.5),
trans-$[PtX(H)(PPh_3)_2]$ (X = CN, Cl, N_3, I) (s/t, 400),
trans-$[PtCl_2(PPh_2cyc)_2]$ (w, 50), *trans*-$[PtCl_2(PMe_2Ph)_2]$ (w, 11.3)
cis-$[PtCl_2(CO)(PR_3)]$ ($PR_3 = PPh_3, PMePh_2$) (w, 10–12),
cis-$[PtCl_2(C_2H_4)(PPh_3)]$ (w, 3.1), *trans*-$[PtI_2(CO)PPh(cyc)_2$ (w, 5),
$[PR_3ClPt(\mu\text{-}Cl)]_2$ ($PR_3 = PPh_3$ (w, 0.75), $PPh(cyc)_2$ (w, 25), P^nBu_3 (w, 8) $P(ptol)_3$ (s, 12.5) $Pt(PF_3)_4$ (u, 50), $Pt(PPh_2(ptol))_2(PhC = CPh)$ w/t, 12.5)

$Cd(PPh_3)_2Br_2$ (s/t, 50–100), $Cd(PPh_3)_2I_2$ (c, 25–100)
$Hg(PPh_3)_2Cl_2$ (c, 12.5–25)

$SnCl_4(PPh_3)$ (c, 100)

[a] w H_2O, s saline, t Tween 80, c hydroxypropylcellulose, g gum acacia, o olive oil, sesame oil or peanut oil, cm carboxymethylcellulose, u unknown; doses are in mg/kg given by intraperitoneal (ip.) injection;
[b] Also inactive in sc adenocarcinoma 755 and sc sarcoma 180 (in mice);
[c] P388 leukaemia;
[d] Also inactive in ip. B16 melanoma and ip. P388 leukaemia (in mice);
[e] Also inactive in sc adenocarcinoma 755 (in mice); [f]Administered sub-cutaneously.

necessary to reduce the reactivity of Au(I) towards thiols, whilst at the same time retaining sufficient kinetic lability in the Au–P bond. This can be achieved with chelated Au(I) diphosphine complexes.

3.3 Tetrahedral, Chelated Au(I) Diphosphine Complexes

3.3.1 Background

In the vast majority of its complexes Au(I) exhibits linear 2-coordination[56]. As a consequence, ligand-exchange reactions occur very readily via an associative mechanism and a 3-coordinate transition-state. Facile ligand exchange reactions with thiols probably play an important role in the antiarthritic activity of many Au(I) compounds, but they may also bring about a reduction in the cytotoxic potency of Au(I) phosphine complexes. It seemed likely that tetrahedral Au(I) complexes containing chelated diphosphine ligands would be more stable with respect to ligand-exchange because they would have to react via a ring-opening mechanism (Fig. 6) (5-coordinate Au(I) species are unknown).

However, both four-coordination and chelation have been little explored in Au(I) chemistry. There are examples from the early literature of bis-chelated Au(I) complexes with rigid bidentate As and P containing ligands (Table 6), but from more recent studies involving *flexible* bidentate phosphines no evidence for four-coordinate chelated complexes was reported[62–64]. In general the characterised products were reported to have bridged digold or annular structures with linearly coordinated Au(I) (Fig. 7). Complexes of the type $XAu(Ph_2P(CH_2)_nPPh_2)AuX$ have been characterised where n = 1 – 12[62, 65]. The bridged-digold structure has been confirmed by X-ray crystallography where X is Cl and n = 1[66], n = 2[67, 68], n = 3[69] and *cis*-CH=CH[70].

The crystal structure of $[Au_2dppm_2Cl_2]$ showed an 8-membered Au_2P_4 ring[71].

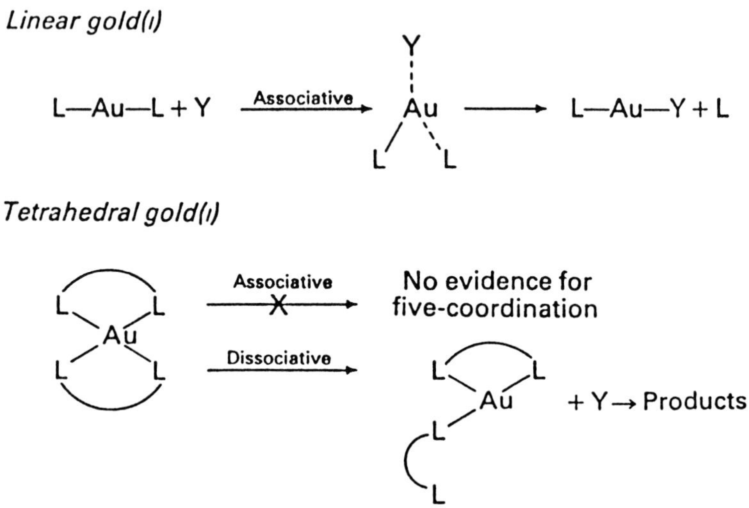

Fig. 6. The contrasting possible exchange mechanism for linear and tetrahedral bis-chelated gold(I) complexes

Table 6. Results of some early studies on four-coordinate Au(I) complexes with rigid aromatic phosphine and arsine ligands

Structure		Comments	Ref.
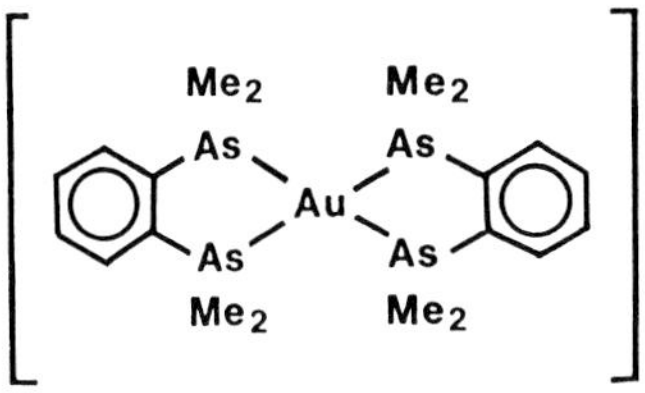	X	Preparation, X = Hal, ClO_4 X-ray powder photographs Preparation, conductivity X = Cl, ClO_4, Br, I, CuI_2, AuI_2, picrate	57 58 59
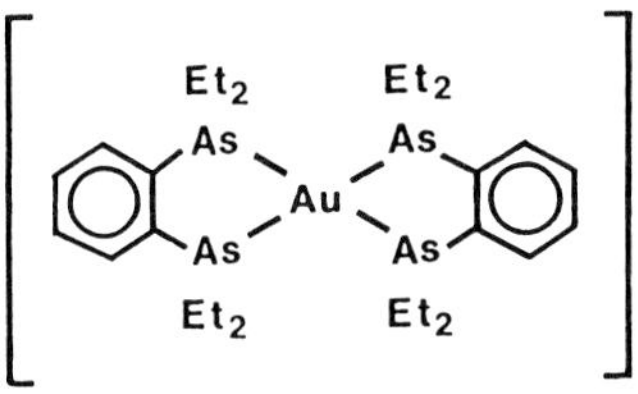	I	X-ray diffraction Au–As 2.48 Å As–Au–As 90°, 120°	60
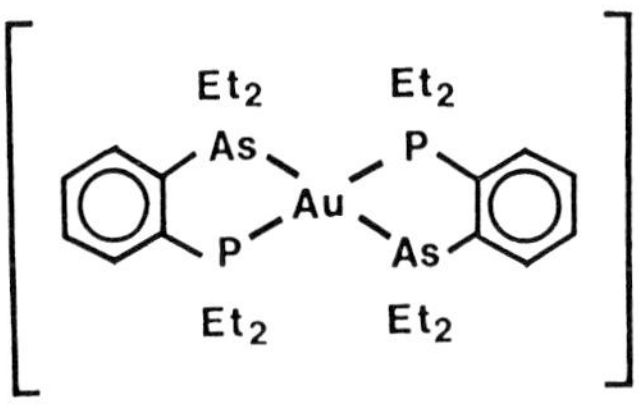	X	X = optically-active anion optical isomers not resolved	60
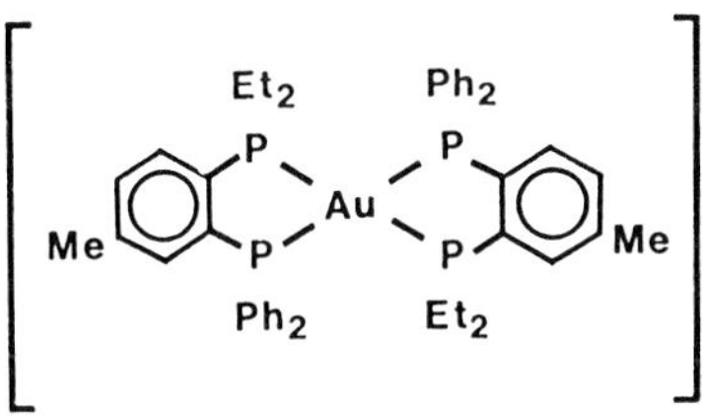	X	X = I, optically-active anion optical isomers not resolved	61

In the early 1980's the existence of tetrakis (monophosphine)Au(I) complexes began to be established. Complexes of the type $[Au(PR_3)_4]X$ were isolated and characterised for the ligands PMe_3[72], PPh_3[73, 74], and $PMePh_2$[75]. ^{31}P NMR solution studies of gold(I) monophosphine complexes in the presence of free PR_3 demonstrated that complexes with different Au: P ratios undergo rapid ligand exchange above 213 K[76–78]:

$$R_3PAuX + PR_3 \rightleftarrows (R_3P)_2AuCl + PR_3 \rightleftarrows (R_3P)_3AuX + PR_3 \rightleftarrows [(R_3P)_4Au]X.$$

Four-coordinate complexes were not formed for very bulky phosphines such as $P(C_6H_{11})_3$, P^nBu_3, P^nBu_2Ph and P^iPr_2Ph.

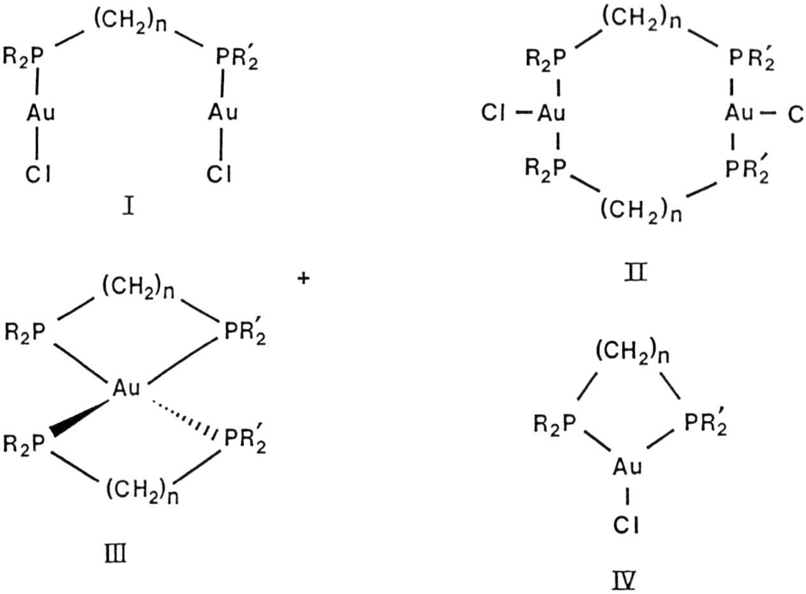

Fig. 7. Structures of some possible Au(*I*) diphosphine complexes: (*I*) bridged digold, (*II*) annular, (*III*) tetrahedral, (*IV*) chelated 3-coordinate complex

3.3.2 Factors Influencing Chelate Ring Formation by Au(I)

^{31}P NMR studies[79)] of the diphosphine-bridged digold complex $ClAu(Ph_2P(CH_2)_2PPh_2)$ AuCl in the presence of free dppe demonstrated that the bis-chelated complex $[Au(dppe)_2]^+$ was present in solution at Au : P ratios as low as 1 : 1.5 (Fig. 8). Furthermore, the four-coordinate complex exhibited remarkably high kinetic stability: separate sharp peaks for free dppe and $[Au(dppe)_2]^+$ were observed in the ^{31}P NMR spectrum at *room temperature* (Fig. 8). The crystal structure of the SbF_6^- salt (Fig. 8) showed a slightly flattened tetrahedral geometry for Au(I), the P-Au-P angles range from 85 to 130° (regular tetrahedral geometry is prevented by the small bite of the ligand). The four Au–P bond lengths are approximately equal and slightly shorter than observed for the tetrakis(monophosphine)Au(I) complexes (Table 7). Bates and Waters[80)] have determined the crystal structure of the Cl^- complex (Fig. 8) and reported similar bond-lengths and angles for the $[Au(dppe)_2]^+$ cation (Table 7).

The ^{31}P NMR spectrum of the mixed ligand complex $[Au(dppe)(depe)]^+$ (Fig. 9) established the existence of a tetrahedral bis(diphosphine) species in solution. The A_2B_2 multiplet pattern was resolved in the spectrum at room temperature. The two bond $^{31}P(Et_2)$-Au-$^{31}P(Ph_2)$ coupling was 52 Hz, showing that any ring opening and closure occurs at a rate $\ll 52\ s^{-1}$.

The stability of the bis-chelated complexes $[Au(R_2P(CH_2)_2PR'_2)_2]^+$ appears to be dependent on the nature of the phosphine-substituents (R and R'). The ^{31}P NMR titration of ClAu(depe)AuCl with depe (R=R' =Et) in $CDCl_3$ is shown in Fig. 8. In contrast to the dppe titration, the bis-chelated complex $[Au(depe)_2]^+$ (15.8 ppm) was not present in solution until the Au : depe ratio exceeded 1 : 1. It was the only species present at a 1 : 2

Table 7. Bond lengths and angles in some four-coordinate copper(I) and gold(I) phosphine complexes

Complex	M–P/Å	P–M–P	Ref.
Copper(I)			
$[Cu(dppe)_2]CF_3CO_2 \cdot$ toluene	2.31–2.37	89–127°	113
$[Cu(dppe)_2][Cu(Ar)_2]$	2.30–2.32	88–121°	123
$(N_3)(dppe)Cu(\mu\text{-}dppe)Cu(dppe)(N_3)$	2.32 (bridging) 2.29, 2.32 (chelated)	90–120°	125
$Cl(dppe)Cu(\mu\text{-}dppe)Cu(dppe)Cl \cdot$ 2 acetone	2.28 (bridging) 2.29, 2.31 (chelated)	89–122°	114
Gold(I)			
$[Au(PPh_3)_4]BPh_4$	3.95 (long bond) 2.39–2.41 (trigonal unit) 2.50, 2.56 (tetrahedral − 150 °C)	 116–121°	74
$[Au(PPh_2Me)_4]PF_6$	2.45	105, 118°	75
$[Au(dppe)_2]SbF_6 \cdot$ acetone	2.39–2.42	85–130°	79
$[Au(dppe)_2]Cl \cdot 2\,H_2O$	2.38–2.41	86–131°	80

Au : depe ratio, and it showed comparable kinetic stability to $[Au(dppe)_2]^+$, as shown by the appearance of a separate peak for free depe with addition of excess ligand. This difference can be attributed to a higher stability of the complex with a 1 : 1 Au : P–P ratio[79)].

It can be seen from Fig. 8 that a single sharp peak was observed at a 1 : 1 Au : depe ratio, and this remained sharp on cooling the solution to 230 K indicating that it corresponds to a single stable species. This could have a monomeric (chelated) structure or an annular structure (Fig. 7). Strong evidence for an annular structure was obtained by using the ligand eppe (R = Ph, R′ = Et)[81)]. The ^{31}P NMR spectrum of a D_2O solution containing a 1 : 1 Au : eppe ratio consisted of two overlapping AA′BB′ multiplets which could be assigned unequivocally to the two isomers of the annular complex $[Au_2(Ph_2P(CH_2)_2PEt_2)_2]^{2+}$ (Fig. 10). The isomer containing two $Ph_2P\text{-}Au\text{-}PEt_2$ linkages predominated by 2 : 1 over the isomer containing both $Ph_2P\text{-}Au\text{-}PPh_2$ and $Et_2P\text{-}Au\text{-}PEt_2$ linkages. The stability of the annular complexes showed a solvent dependence so that in $CDCl_3$ solutions two broad ^{31}P NMR resonances were observed[81)]. Similarly, in aqueous solutions containing a 1 : 1 Au : dppe ratio a single peak attributable to $[Au_2(dppe)_2]^{2+}$ was observed. Steric interactions between bulky phenyl substituents may destabilise the annular complexes with respect to chelated species. It seems likely that in aqueous solvents the equilibrium

$$[Au_2Cl_2(P\text{–}P)_2] \rightleftarrows [Au_2(P\text{–}P)_2]^{2+} + 2\,Cl^-$$

will lie to the right so that the gold atoms are linearly coordinated, whereas in $CDCl_3$ it will lie to the left so that each gold is essentially 3-coordinate. From molecular models it is apparent that this distortion towards a trigonal-planar geometry gives rise to severe steric crowding for Ph, but not Et substituents. It was presumed that the high thermodynamic stability of the annular depe complex accounts for the observation that $[Au(depe)_2]^+$ decomposed unless stabilised by a large counter-anion (e.g. SbF_6^-, or PF_6^-) as $[Au_2Cl_2(depe)_2]$ was identified as one of the products. Both $[Au(dppe)_2]^+$ and $[Au(eppe)_2]^+$ were readily isolated with Cl^- as the counteranion[81)].

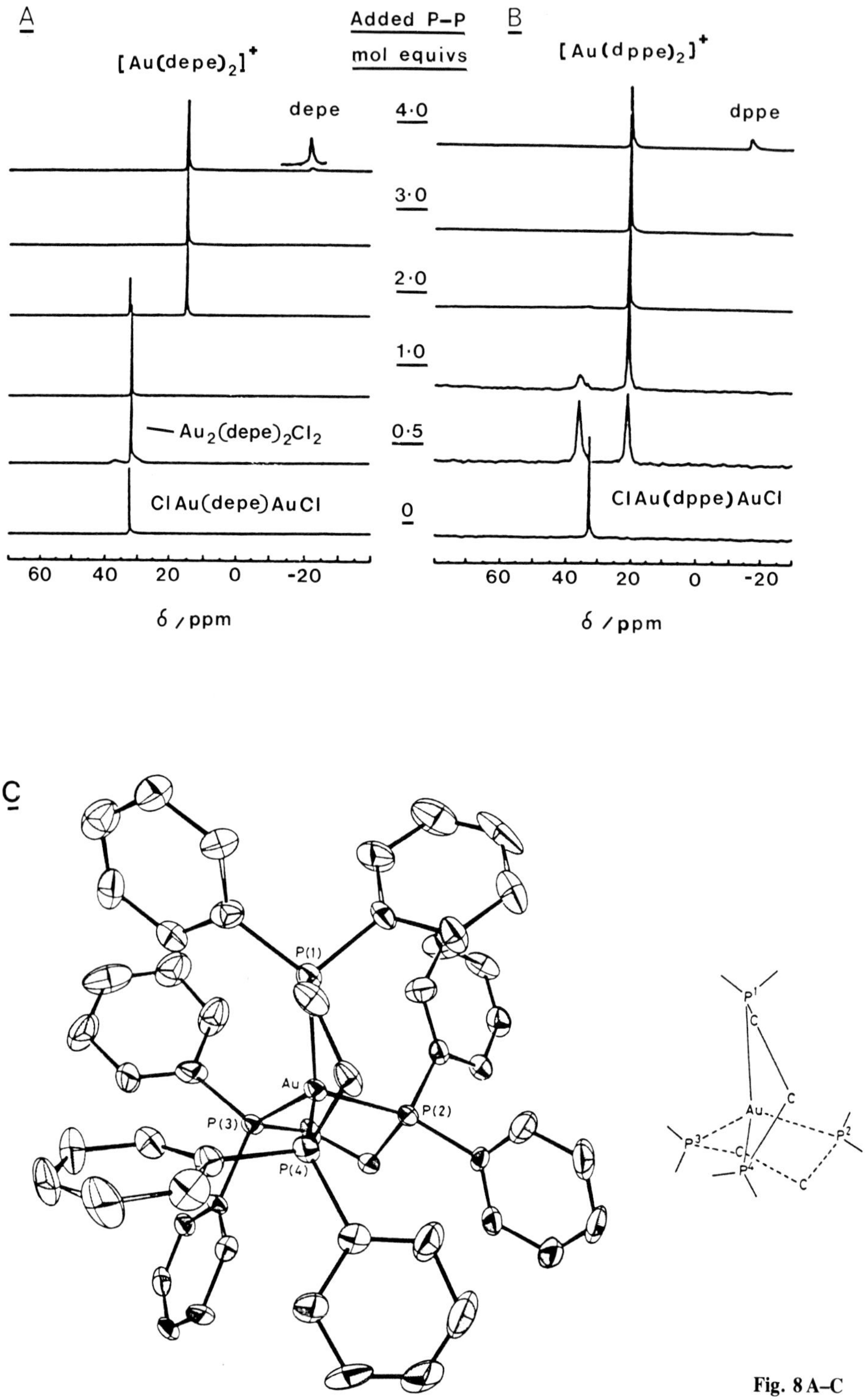
A
[Au(depe)$_2$]$^+$
depe
Added P–P
mol equivs
4·0
3·0
2·0
1·0
0·5
0
— Au$_2$(depe)$_2$Cl$_2$
ClAu(depe)AuCl
60 40 20 0 -20
δ / ppm
B
[Au(dppe)$_2$]$^+$
dppe
ClAu(dppe)AuCl
60 40 20 0 -20
δ / ppm
C
P(1)
P(2)
P(3)
P(4)
Au

Fig. 8 A–C

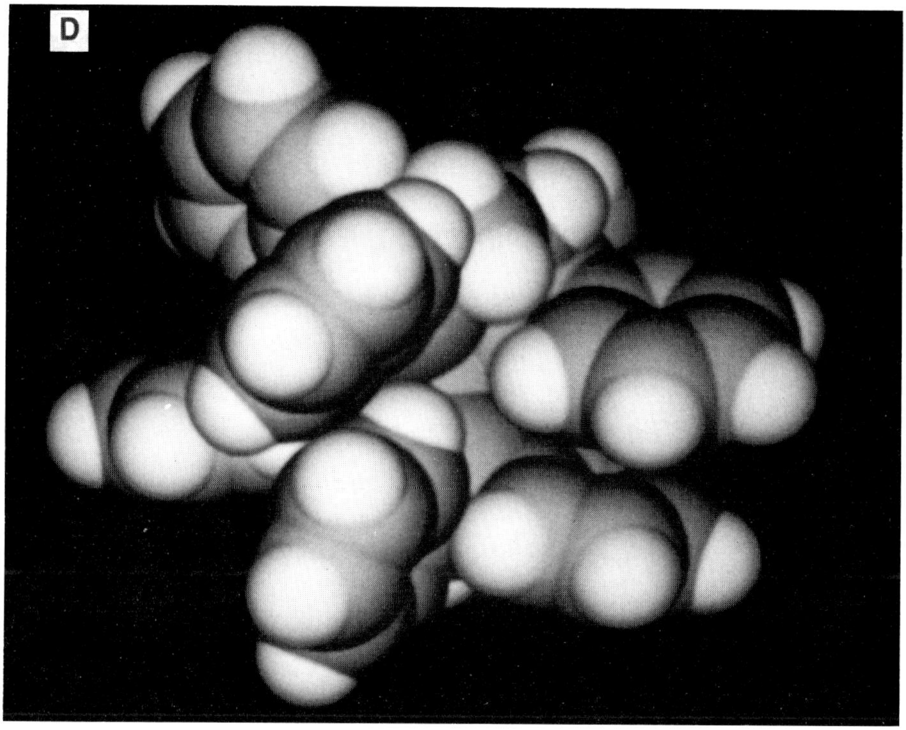

Fig. 8 A–D. $^{31}P\{^1H\}$ NMR spectra of the complexes ClAu(P–P)AuCl in $CDCl_3$, in the presence of 0–4 mol equivalents of free ligand (P–P), where P–P is depe ($Et_2P(CH_2)_2PEt_2$) (**A**), and dppe ($Ph_2P(CH_2)_2PPh_2$) (**B**). Note the appearance of peaks corresponding to the bis-chelated complexes $[Au(P–P)_2]^+$ at Au : P–P ratios >2 : 1 (dppe) and >1 : 1 (depe). The annular complex $Au_2(P–P)_2Cl_2$ is observed as a stable species at a 1 : 1 Au : P–P ratio where P–P is depe but not dppe (based on Ref. 79). (**C**) The molecular structure of the $[Au(dppe)_2]^+$ cation in $[Au(dppe)_2]SbF_6 \cdot$ acetone (from Ref. 79). (**D**) A space filling model of $[Au(dppe)_2]^+$ in $[Au(dppe)_2]Cl \cdot 2\,H_2O$ generated by computer graphics (with the help of Drs. I. Tickle and R. E. Norman and data from Ref. 80) showing the shielding of Au(I) by the two chelated dppe ligands

The effect of the size of the chelate ring on the formation of stable bis-chelated complexes was investigated by ^{31}P NMR titrations of complexes $ClAu(Ph_2P(CH_2)_nPPh_2)$ AuCl with free diphosphine for the ligands dppm (n = 1), dppp (n = 3), dppb (n = 4) and *cis*-$Ph_2PCH{=}CHPPh_2$ (dppey)[81]. The ligands able to form 5- or 6-membered chelate rings (i.e. dppey and dppp), showed analogous behaviour to dppe. The bis-chelated complexes were observed in solution at very low Au : P ratios, and underwent slow exchange with free ligand on the NMR time scale. $[Au(dppey)_2]^+$ had the highest stability (both kinetic and thermodynamic) of all the bis-chelated complexes in our series, which was consistent with the more rigid nature of the ligand. The annular complexes were not observed as stable species in $CDCl_3$ solutions at a 1 : 1 Au : L–L ratio. The tetrahedral complexes were isolated with Cl^- as the counter-anion. There was no evidence for the formation of 4- or 7-membered chelate rings from similar titrations of the dppm and dppb complexes, and in both cases the annular complexes were observed as

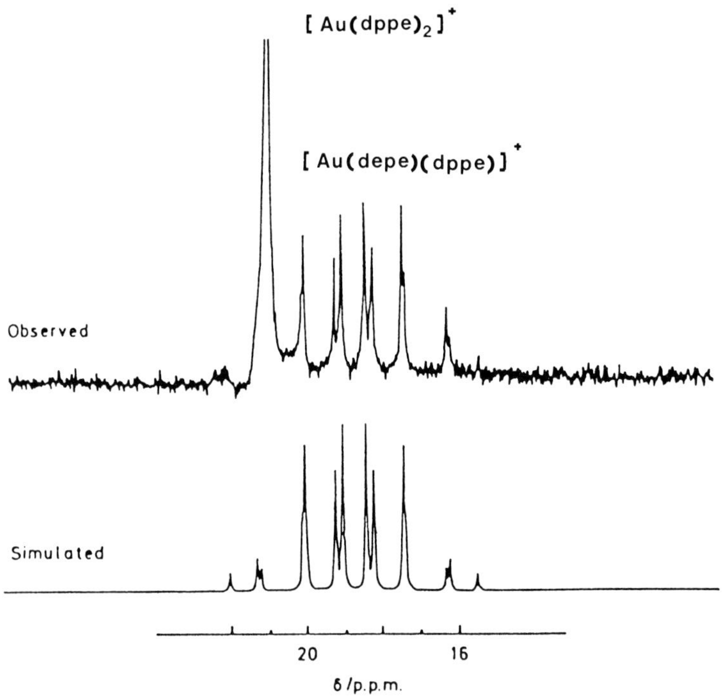

Fig. 9. $^{31}P\{^1H\}$ NMR spectrum of a solution of ClAu(depe)AuCl in $CDCl_3$ with 3 mol equiv. of added dppe at 302 K. The A_2B_2 pattern corresponds to the bis-chelated complex $[Au(dppe)(depe)]^+$ and confirms that the complex is tetrahedrally coordinated in solution. Note that the trans-gold $^2J(^{31}P(Ph) - {}^{31}P(Et))$ coupling constant (52 Hz) is resolved at room temperature (based on Ref. 79)

stable species at a 1 : 1 Au : L–L ratio. The titration of ClAu(dppb)AuCl with dppb is shown in Fig. 11.

The formation of bis-chelated complexes is most favourable for phenyl-substituted diphosphines which can form either 5- or 6-membered chelate rings.

3.3.3 Ring-Closure Induced by Thiols and Blood Plasma

It is now becoming increasingly evident that not only are these four-coordinate Au(I) complexes exceptionally stable but also the formation of chelate rings can contribute to the driving force for some unusual reactions. For instance, Bates and Waters[80] isolated $[Au(dppe)_2]Cl$ from the reaction of ClAu(dppe)AuCl with Na_2S. Thiols can induce the conversion of bridged digold complexes XAu(dppe)AuX into $[Au(dppe)_2]^+$. ^{31}P NMR studies[82] showed that the addition of 2 mol equivalents of sodium thioglucose (SGlu) to ClAu(dppe)AuCl converted it to (GluS)Au(dppe)Au(SGlu). With further addition of NaSGlu the bis-chelated complex $[Au(dppe)_2]^+$ was formed. The reaction appeared to be complete at a Au : SGlu ratio of ca 1 : 2 (Fig. 12). Similarly, the addition of glutathione (GSH) at pH 7 to XAu(dppe)AuX, where X is Cl or SGlu induced the formation of $[Au(dppe)_2]^+$ [82]. These reactions represent novel chemistry for Au(I); firstly, because they must involve displacement of P from Au(I) by S, and secondly, because transfer of

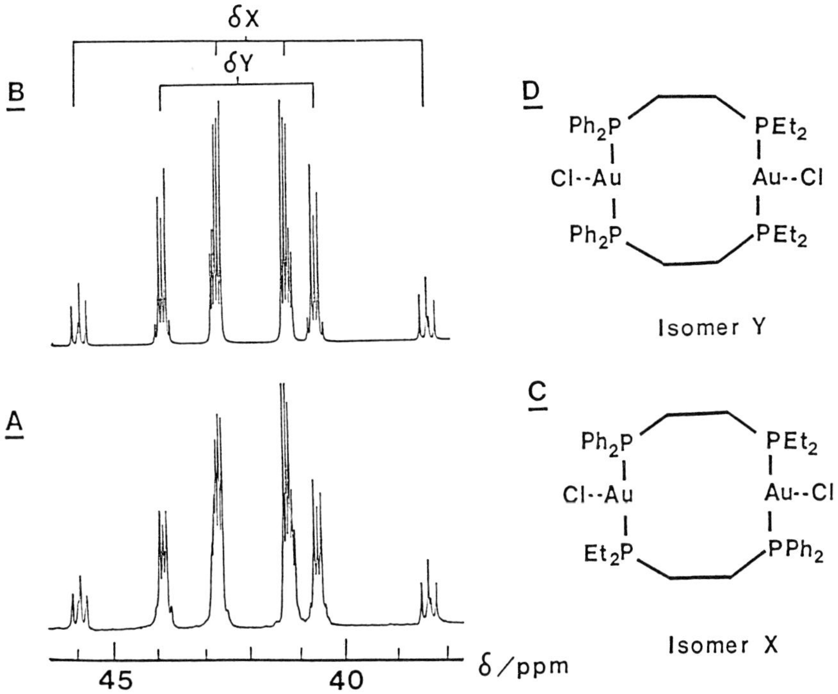

Fig. 10 A–D. Observed (**A**) and simulated (**B**) $^{31}P\{^1H\}$ NMR spectra of the two isomers of $[Au_2(Ph_2P(CH_2)_2PEt_2)_2]Cl_2$ in D_2O at 300 K. The observation of two overlapping AA′BB′ multiplets corresponding to the two possible isomers (X and Y) shows that the complex with a 1 : 1 Au : eppe ratio has an annular structure in solution (based on Ref. 81)

dppe from one bridged digold molecule to another must occur. It is notable that in the crystalline state there are pairs of ClAu(dppe)AuCl molecules held together by short intermolecular Au-Au interactions[67, 68] (Fig. 13). The same feature appears in the crystal structure of ClAu(dppp)AuCl[69]. If similar interactions exist in solution they could facilitate dppe transfer. Alternatively, digold annular complexes could be formed in solution. A third possibility is that thiolate bridged species could be intermediates. Au-S-Au linkages can be very stable and this could explain why the ring-closure reactions of ClAu(dppe)AuCl induced by SGlu required prior formation of the bis(thiolato) complex. These ring-closure reactions may have a special significance in the molecular pharmacology of Au(I) diphosphine complexes because complexes of the type XAu(dppe)AuX have anticancer activity and we have observed that they undergo conversion into $[Au(dppe)_2]^+$ in blood plasma. This is discussed further in Sect. 3.4.2.

3.3.4 Reactions of $[Au(dppe)_2]Cl$ with Thiols and Cells

In contrast to ClAu(dppe)AuCl, the bis-chelated complex $[Au(dppe)_2]Cl$ does not react significantly with thiols in aqueous solution[83]. In addition, ^{31}P NMR studies have shown that the complex remains essentially intact in human plasma[82–84] (Fig. 14). It was readily transferred from plasma to red blood cells but no ^{31}P NMR signal was observed[84]

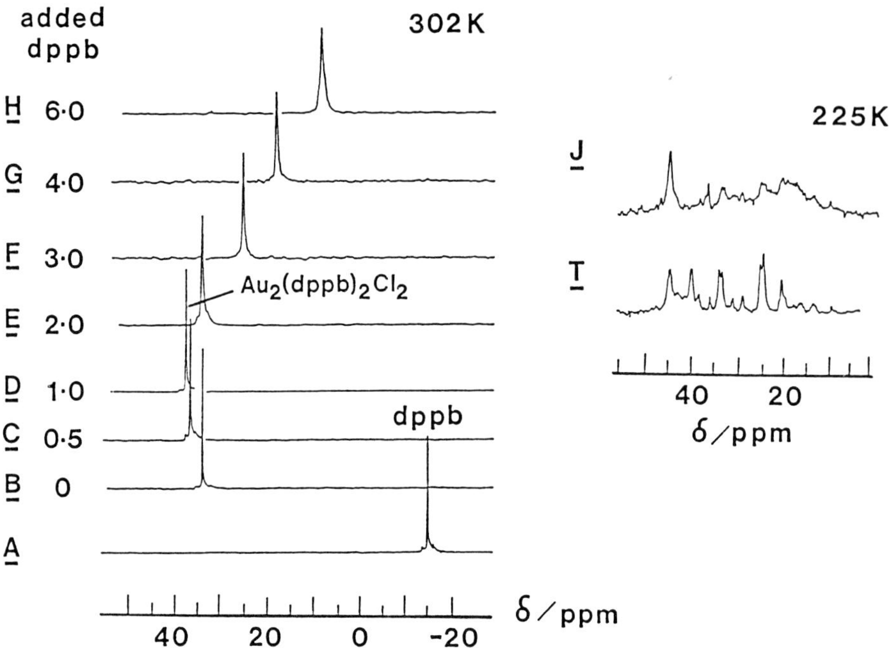

Fig. 11. $^{31}P\{^1H\}$ NMR spectra at 302 K of ClAu(dppb)AuCl in DMF *(B)* in the presence of 0–6 mol equiv. of free dppb ($Ph_2P(CH_2)_4PPh_2$) (*C* to *H*). *I* and *J* are the spectra with 2.0 and 3.0 mol equiv. of added dppb at 225 K. The bis-chelated complex $[Au(dppb)_2]^+$ is not observed as a stable species. The complexity of the low (225 K) temperature spectra indicate that the products at a 1 : 2 Au : dppb ratio are polymeric (based on Ref. 81)

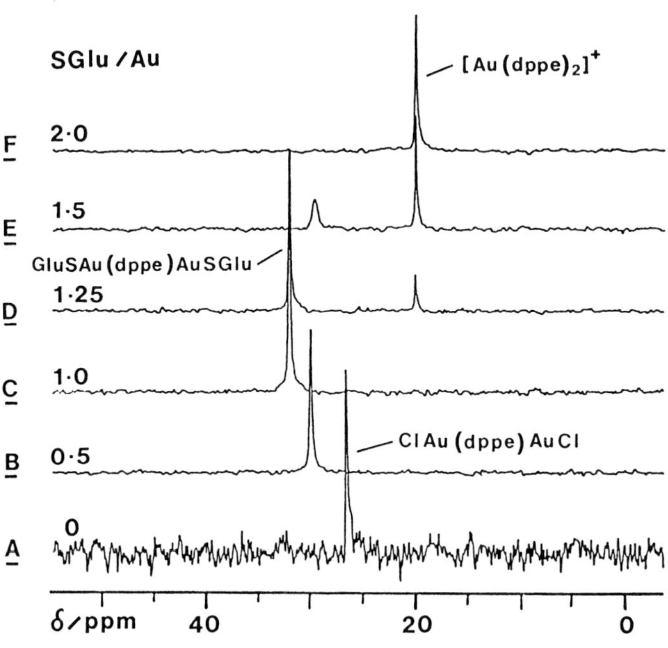

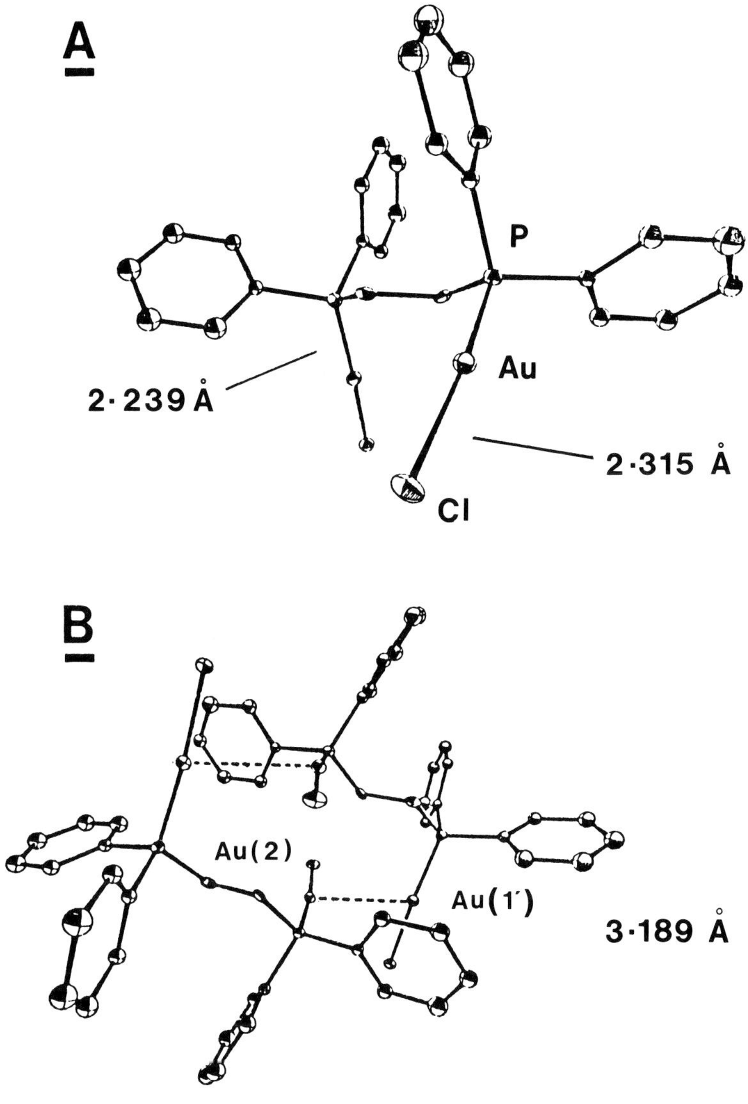

Fig. 13 A, B. The crystal structure of ClAu(μ-dppe)AuCl **(A)**. In the crystal these molecules form dimers **(B)** with short intermolecular Au–Au contacts (3.189 Å) (based on Ref. 67)

◀ **Fig. 12.** $^{31}P\{^{1}H\}$ NMR spectra of ClAu(dppe)AuCl in MeOH : H_2O (1 : 1 v/v) in the presence of 0–4 mol equivalents of NaSGlu (where SGlu is β-D-thioglucose). Note the formation of the bis-chelated complex $[Au(dppe)_2]^+$ at Au : SGlu ratios > 1 : 1 (based on Ref. 82)

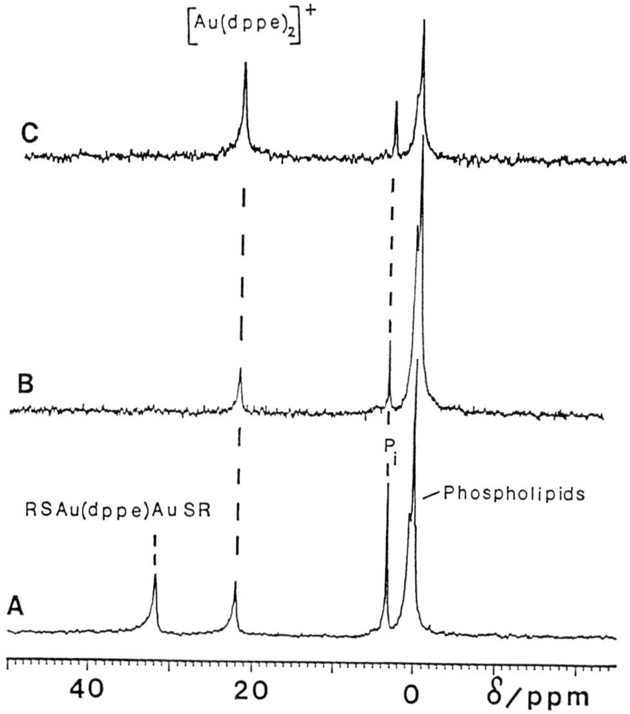

Fig. 14 A–C. 81 MHz $^{31}P\{^{1}H\}$ NMR spectra of **(A)** human plasma treated with 1 mM GluSAu (μ-dppe)AuSGlu (total accumulation period 14 h); **(B)** bovine serum treated with 0.2 mM GluSAu (μ-dppe)AuSGlu (total accumulation period 16.5 h); **(C)** human plasma treated with 2.2 mM $[Au(dppe)_2]Cl$. Note that the bridged digold complex has been totally converted into the bis-chelated complex $[Au(dppe)_2]^+$ at the lower concentration (based on Ref. 82)

(Fig. 15). On addition of sodium dodecyl sulphate (SDS) to either red cells or erythrocyte ghosts (red cell membranes) treated with $[Au(dppe)_2]Cl$, ^{31}P signals attributable to the membrane phospholipids and $[Au(dppe)_2]^+$ appeared (Fig. 15). These results strongly suggest that the tetrahedral complex remains intact within the cell and the majority binds within the membrane and is mobilised by disrupting the membrane structrue with SDS. ^{1}H spin-echo NMR studies of $[Au(dppe)_2]Cl$-treated red cells indicate that the complex does not bind significantly to intracellular glutathione[84)]. Auranofin, has been shown to bind to albumin (at cysteine-34) in serum[85)], and also to glutathione inside red cells[86)]. Thus by imposing four-coordination on Au(I) the high thiol reactivity is considerably reduced. The antitumour properties of the series of bis-chelated Au(I)diphosphine complexes have been investigated, and a summary of the results is given in the following section.

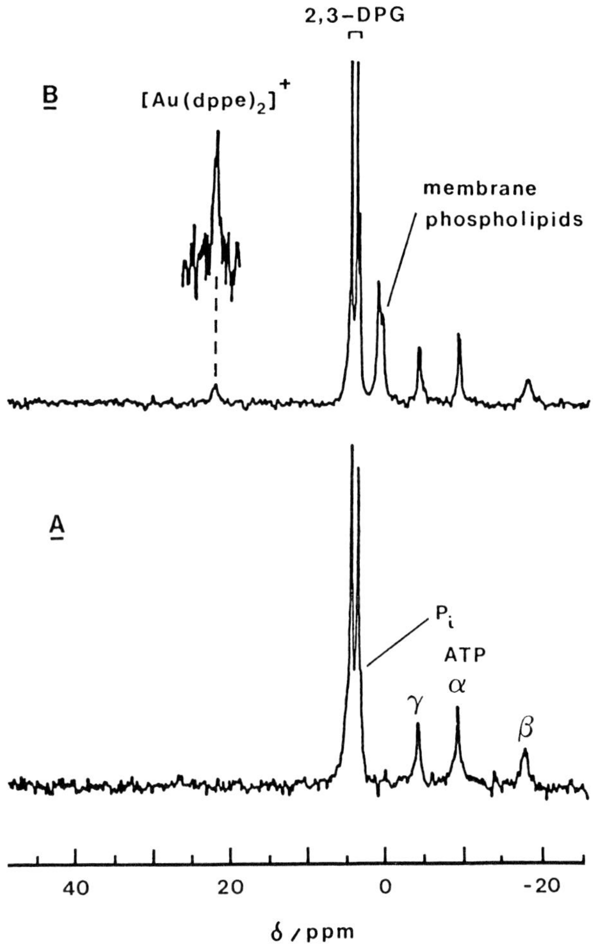

Fig. 15 A, B. 81 MHz $^{31}P\{^{1}H\}$ NMR spectra of human red blood cells following treatment with $[Au(dppe)_2]Cl$ before (**A**) and after (**B**) the addition of sodium dodecyl sulphate (SDS). Note that the resonances for $[Au(dppe)_2]^+$ and the membrane phospholipids are resolved only after treatment with SDS indicating that the complex is immobilised (and the ^{31}P NMR resonance broadened beyond detection) by binding within the membrane (based on Ref. 84)

3.3.5 *Cytotoxicity and Antitumour Activity of Tetrahedral, Bis(Diphosphine)Au(I) Complexes*

$[Au(dppe)_2]Cl$ is cytotoxic to tumour cells *in vitro* at micromolar concentrations. The concentrations required to inhibit the clonogenic capacity[2] of Bl6 melanoma and P388 leukaemia cells by 50% (IC_{50}) following a 2 h exposure are 2 μM and 6 μM respectively. *Cis*-$[PtCl_2(NH_3)_2]$ is cytotoxic to a variety of cell types within a similar concentration range[142 b)]. In contrast to the behaviour of auranofin, the cytotoxic effects of $[Au(dppe)_2]Cl$ are not immediate. The survival curves (Fig. 16) show a shoulder at low doses, with subsequent exponential decreases in survival at higher concentrations[83)]. The

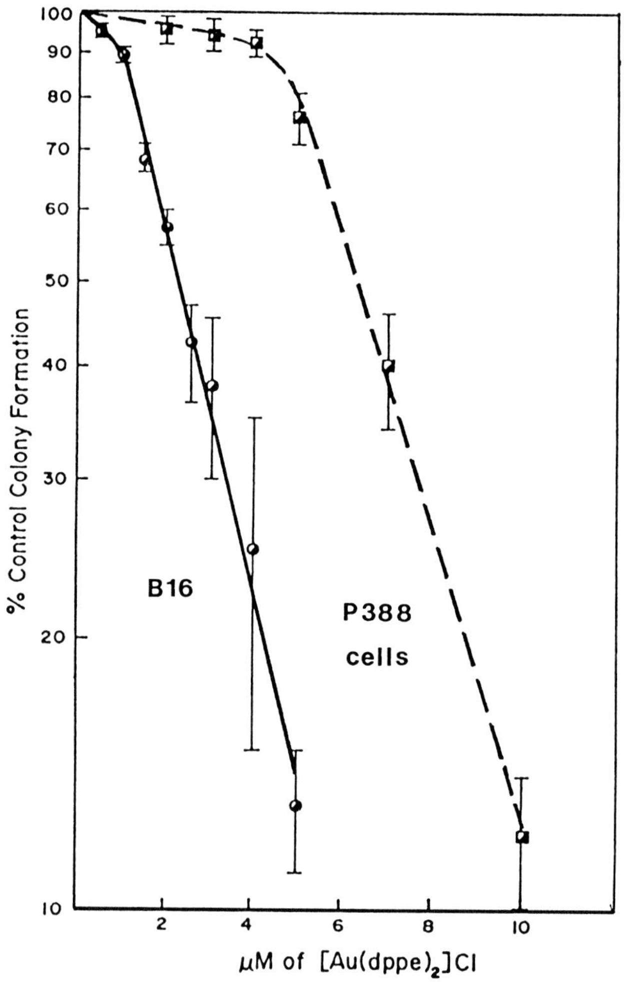

Fig. 16. Cytotoxic activity of [Au(dppe)$_2$]Cl against B16 melanoma (◑) and P388 leukaemia (◪) cells (assesed in a clonogenic assay following a 2 h exposure) (based on Ref. 83)

dose of [Au(dppe)$_2$]Cl required to produce acute cell death is greater than 20-fold of that required to inhibit the clonogenic capacity[2] of the cells. Macromolecular synthesis in B 16 cells is inhibited by [Au(dppe)$_2$]Cl in both a concentration and time dependent manner. Protein synthesis appears to be preferentially inhibited relative to DNA and RNA synthesis. It has been suggested that the effects observed on macromolecular synthesis may be the result of chromatin damage because alkaline-elution studies show that the complex produces DNA-protein crosslinks and DNA-single strand breaks when incubated with L 1210 cells. Single-strand breaks are significant only at superlethal concentrations

2 In the clonogenic assay cells are grown on plate surfaces, treated with drugs, washed, then incubated for 5 days in fresh medium. The viability is measured by the ability of a cell to form a colony of greater than 50 cells (see Ref. 54). The IC_{50} value is the concentration of drug required to reduce the number of colonies formed by 50% relative to controls.

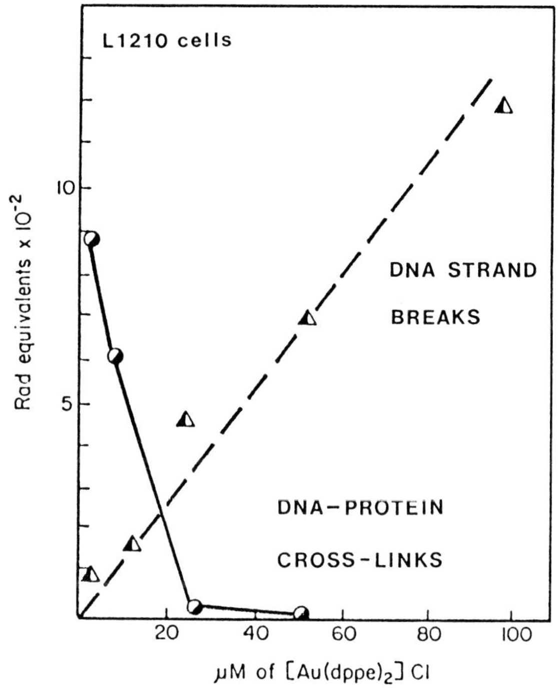

Fig. 17. DNA strand breaks (△) and DNA-protein crosslinks (◐) produced by $[Au(dppe)_2]Cl$ in L1210 cells after a 1 h treatment at 37 °C (based on Ref. 83)

of $[Au(dppe)_2]Cl$, whereas DNA-protein crosslinks are observed within the cytotoxic concentration range of the complex (5 μM) (Fig. 17)[83)].

The differences in the cytotoxic effects of $[Au(dppe)_2]Cl$ and auranofin indicate that different mechanisms are involved. The cytotoxic effects of $[Au(dppe)_2]Cl$ may be mediated by interference with the reproductive capacity of the cell whereas those produced by auranofin appear to be acutely mediated. An important difference between the two complexes is that the cytotoxic potency of $[Au(dppe)_2]Cl$ is not significantly reduced by the presence of serum proteins in the culture media consistent with the high stability of the complex in the presence of thiols, and in serum[83)].

The antitumour activity of $[Au(dppe)_2]Cl$ has been evaluated in a variety of tumour models in mice. It produces a dose-dependent prolongation of lifespan in mice bearing ip P388 leukaemia (Fig. 18)[83)]. When administered by intraperitoneal (ip) injection at its maximally tolerated dose (2.9 μmol/kg/day × 5) $[Au(dppe)_2]Cl$ gave an average of 87% increase in lifespan over 21 separate dose-response studies. It is less toxic to mice when given intravenously (iv.), subcutaneously (sc.) or orally, and is inactive by these routes of administration. It is also inactive against systemic (iv. inoculated) P388 leukaemia by either ip. or iv. administration. Other tumours sensitive to $[Au(dppe)_2]Cl$ are ip. M5076 reticulum sarcoma, ip. L1210 leukaemia, ip. B16 melanoma and sc. adenocarcinoma 16/c. The spectrum of antitumour activity of $[Au(dppe)_2]Cl$ is shown in Table 8.

$[Au(dppe)_2]Cl$ exhibits no significant loss in activity against a subline of P388 leukaemia which is resistant to cisplatin. Moreover, $[Au(dppe)_2]Cl$ and cisplatin can be administered concurrently at their respective maximum tolerated doses to tumour-bearing mice, with no lethality[83)]. This combination is more effective against moderately advanced P388 leukaemia than cisplatin alone. These results provide indications that the

Table 8. Spectrum of activity of dppe and Au(I) complexes in transplantable murine tumours[a]. Data from Refs. 83 and 89

Tumour model	Route of administration	dppe	$[Au(dppe)_2]Cl$	XAu(dppe)AuX X = Cl	XAu(dppe)AuX X = S (glucose)
ip. Tumours					
P 388 leukaemia	ip.	++	++	++	++
	iv.	b	–	b	b
	sc.	b	–	b	b
	po.	b	–	b	b
L 1210 leukaemia	ip.	++	+	+	b
M 5076 reticulum cell sarcoma	ip.	++	++	++	++
B 16 melanoma	ip.	+	+	+	+
B 16 melanoma-F 10	ip.	–	–	–	–
Lewis lung carcinoma	ip.	++	b	–	b
Madison lung carcinoma	ip.	++	b	–	b
Colon carcinoma 26	ip.	–	–	–	–
iv. Tumours					
P 388 leukaemia	ip.	–	–	–	–
	iv.	b	–	b	b
sc. Tumours					
Mammary adenocarcinoma 16/c	ip.	+	+	++	+
	iv.	b	+	b	b
M 5076 reticulum cell sarcoma	ip.	–	?	–	–
ADJ-PC 6 plasmacytoma	ip.	++	?	++	b
Mammary adenocarcinoma 13/c	ip.	–	–	–	–
Colon carcinoma 26	ip.	–	–	–	–
Colon carcinoma 07/A	ip.	–	–	–	–
B 16 melanoma	ip.	–	b	–	b
Lewis lung carcinoma	ip.	–	b	–	b
Madison lung carcinoma	ip.	–	b	–	b

[a] ++ indicates >50% ILS in ip. tumours and >90% tumour growth inhibition in sc. tumour models. + indicates >30% ILS in ip. tumours and >75% tumour growth inhibition in sc. tumour models. – indicates <30% ILS in ip. tumours and <75% tumour growth inhibition in sc. tumour models. ? indicates irreproducible activity

[b] Not reported

mechanism, or site, of action of $[Au(dppe)_2]Cl$ is different to that of *cis*-$[PtCl_2(NH_3)_2]$ (cisplatin) (see Sect. 3.7).

Other Au(I) bis-chelated diphosphine complexes are also highly cytotoxic to B 16 melanoma cells *in vitro* (Table 9)[87]. The IC_{50} concentrations (measured by a clonogenic assay) for complexes containing phenyl-substituted phosphines and 5- or 6-membered chelate rings are between 1 and 4 μM. The ethyl-substituted complex $[Au(depe)_2]PF_6$ inhibits the cloning efficiency of B 16 cells at a concentration almost 4-fold greater than $[Au(dppe)_2]Cl$. Furthermore, $[Au(depe)_2]PF_6$ is inactive against ip. P 388 leukaemia in mice, whereas the complexes containing phenyl-substituted phosphines exhibit comparable activity to $[Au(dppe)_2]Cl$ at similar maximally-tolerated doses (Table 9)[87]. These complexes are active also against ip M 5076 reticulum cell sarcoma and B 16 melanoma (Table 9). The mixed-ligand complex $[Au(eppe)_2]Cl$ (which contains both Ph and Et phosphine substituents) is slightly less active against ip. P 388 leukaemia than the com-

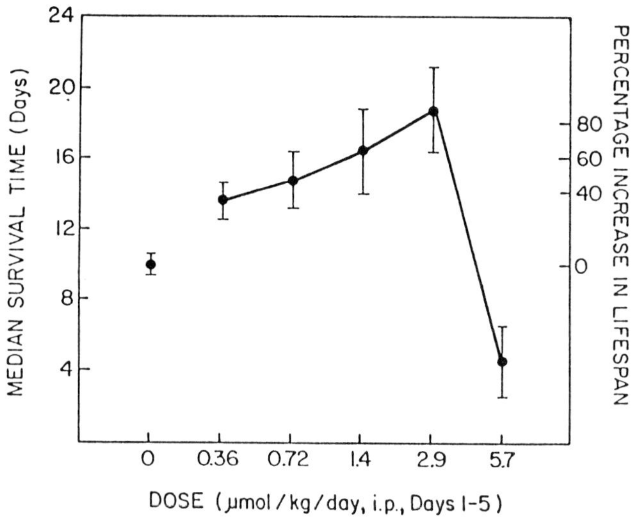

Fig. 18. Dose-response curve for $[Au(dppe)_2]Cl$ in mice bearing ip. P388 leukaemia (treatment on days 1 to 5 with a single daily ip. dose). Each point is the average of median survival time at a particular dose in 21 experiments (from Ref. 83)

Table 9. Activity of Au(I), Ag(I) and Cu(I) diphosphine complexes in mice bearing ip. P388 leukaemia, ip. M 5076 reticulum cell sarcoma and ip. B 16 melanoma and *in vitro* cytotoxic activity to B 16 melanoma cells. Data from Refs. 83, 87, 105 and 129

Compound	MTD[a] μmol/kg/day[a]	P 388	M 5076 ILS[b] %	B 16	IC_{50} [c] μM
$[Au(dppe)_2]Cl$	3	83 ± 24[d]	57 ± 15[e]	38 ± 9[e]	4.5
$[Au(dppe)_2]NO_3$	3	90 ± 17[e]	i	i	4
$[Au(dppey)_2]Cl$	1.5	92 ± 26[f]	69	35	2
$[Au(dppp)_2]Cl$	3	89 ± 28[h]	116	34 ± 6[h]	0.6
$[Au(eppe)_2]Cl$	4	54 ± 16[g]	46	i	5
$[Au(depe)_2]PF_6$	4	40/30	i	i	17
$[Au(dpfppe)_2]Cl$	3	55/50	i	i	i
$[Au(dmfppe)_2]Cl$	10	45/55	i	i	i
$[Ag(dppe)_2]NO_3$	3	89/90	48	i	1
$[Ag(depe)_2]NO_3$	1	50/40	i	i	4
$[Ag(dppey)_2]NO_3$	5	60/40	i	i	i
$[(CuCl)_2(dppe)_3]$	2	100/115	60	54	3
$[Cu(eppe)_2]Cl$	2	48/45	i	i	i
$[Cu(dppey)_2]Cl$	3	66 ± 20[g]	40	42	i
$[Cu(dppp)_2]Cl$	2	89 ± 14[g]	i	32/47	i
cis-$[PtCl_2(NH_3)_2]$	6	125[j]	i	i	0.9

[a] Maximally-tolerated dose for B6D2F mice on an ip. qD × 5 regimen
[b] Maximum increase in lifespan produced in mice bearing ip. tumour; figures separated by / represent data generated in separate experiments. A drug is considered to be active if it produces >50% ILS in P388 leukaemia and >25% ILS in the other two tumour systems
[c] Concentration that inhibits cloning efficiency of B 16 melanoma cells by 50% on a 2 h exposure
[d] Based on 33 different experiments; [e] 5 experiments; [f] 7 experiments; [g] 4 experiments;
[h] 3 experiments; [i] Not reported; [j] Ref. 55

Table 10. Antitumour activity of diphosphine ligands (and some As and S analogues) against ip. P 388 leukaemia in mice (from Ref. 89)[a]. Those producing an increase in lifespan ⩾ 70% are indicated by *

General structure	Active	Marginal	Inactive
$Ph_2P(CH_2)_nPPh_2$	n=2*,[b], n=3*, *cis*–CH=CH* (CHMe)$_2$*, CH$_2$CHMe* 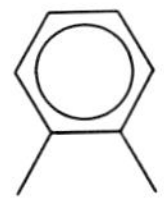	(CH$_2$)$_n$ n=1[c], n=4, n=5	n=0[c], *trans*–CH=CH[c] –C≡C–[c], n=6[c] 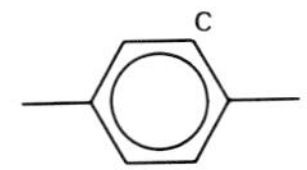
$R_2P(CH_2)_2PR'_2$	Ph*,[b], C$_6$D$_5$*, cyclohexyl* 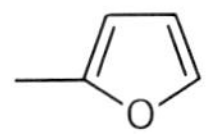 , 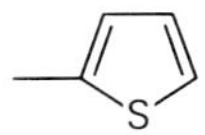 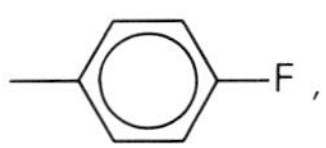 , 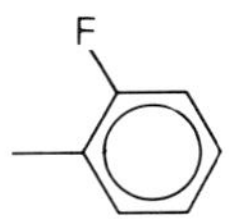	R,R′= 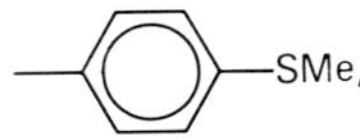 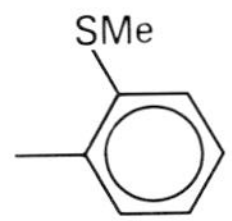 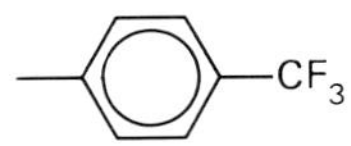 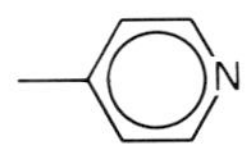	CH$_2$Ph, Et[c], (R=Et, R′=Ph)[c], 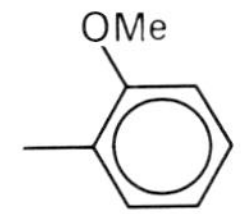 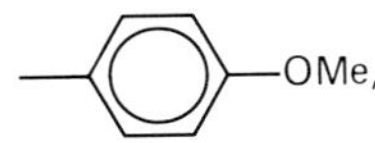 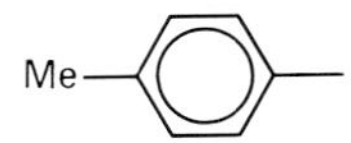 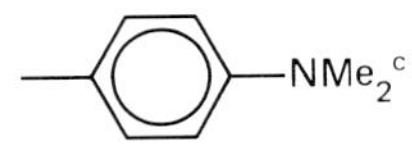 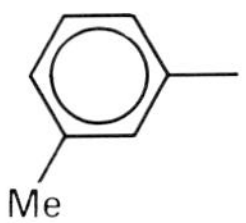 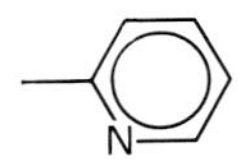

$Ph_2Y(CH_2)_2Y'Ph_2$	$Y=Y'=P$*[b]	Y,Y'=	$Y=P$, $Y'=As$; $Y=Y'=As$ $Y=Y'=S$[c,d]
Other	$PhP\{CH_2)_2PPh_2\}_2$		$\{Ph_2P(CH_2)_2PPh(CH\text{–}_2\}_2$ $P\{(CH_2)_2PPh_2\}_3$ $CH_3C\{CH_2PPh_2\}_3$[c]

[a] The compounds were administered by ip. injection for 5 days at their maximum tolerated dose (MTD). A compound is considered to be active if it produces an increase in lifespan > 50% and inactive if the ILS is < 30%. The MTD were generally in the range 50–80 μmol/kg

[b] dppe

[c] MTD > 200 μmol/kg

[d] one phenyl group coordinated per S atom

plexes containing only phenyl substituents at a comparable dose. The m- and p-fluorophenyl substituted analogues of $[Au(dppe)_2]Cl$ exhibit marginal activity in this tumour model.

The similarity in the spectrum of antitumour activities of the phenyl-substituted bis-chelated Au(I) diphosphine complexes can be attributed to their similar chemical reactivities[87]. As discussed in Sect. 3.3.2, the four-coordinate chelated complexes exhibit greatly enhanced kinetic and thermodynamic stabilities with respect to linear 2-coordinate Au(I) complexes such as auranofin. Their lower reactivity towards ligand exchange, particularly with thiols, may allow the complexes to reach cellular target without significant binding to –SH sites in proteins. The reduction in cytotoxic potency (and antitumour activity) on replacing phenyl substituents by ethyls can be rationalised in terms of the lower thermodynamic stability of $[Au(depe)_2]^+$ which appears to be destabilised with respect to the annular complex $[Au_2(depe)_2]^{2+}$ (see Sect. 3.3.2). In addition alkyl-substituted phosphines react more readily with protein disulphide bonds than those with aryl-substituents (see Sect. 4.2). We noted[87] that whereas $[Au(dppe)_2]Cl$ did not cleave disulphide bonds in model systems and remained essentially intact in serum, $[Au(eppe)_2]Cl$ slowly cleaved disulphide bonds in serum albumin with release of the phosphine oxide. This may account for the reduced antitumour activity of $[Au(eppe)_2]Cl$ *in vivo* compared to $[Au(dppe)_2]Cl$.

3.4 Activity of Diphosphine Ligands, Bridged Au(I) and Other Metal Complexes

3.4.1 Cytotoxicity and Antitumour Activity of Diphosphine Ligands

The antitumour activity of 1,2-bis(diphenylphosphino)ethane (dppe) was first reported by Struck and Shealy[88]. They were investigating the biological activity of 2-haloethylphosphine derivatives which may be considered as phosphorus analogues of nitrogen mustards. Dppe was isolated as a biproduct from the synthesis of $Ph_2P(CH_2)_2Cl$. It was reported to be cytotoxic to Eagles KB cells *in vitro* (at concentrations >0.1 μg/ml, 0.25 μM). Its antitumour activity *in vivo* was evaluated in 3 tumour models. It was active in Sarcoma 180 at a dose of 250 mg (0.63 mmol)/kg/day but the activity was not reproducible. It exhibited marginal activity in adenocarcinoma 755 at 400 mg (1.0 mmol)kg/day and was inactive in L 1210 leukaemia at the same dose.

Mirabelli, Johnson and coworkers[89, 90] have re-investigated the *in vitro* and *in vivo* cytotoxicity of dppe. They reported that in a clonogenic assay the concentration required to reduce the survival of B 16 melanoma cells by 50% following a 2 h exposure to dppe was 60 μM[90]. They reported that dppe has a spectrum of activity in tumour models in mice[89] (Table 8). When administered ip for 5 days to mice bearing ip P 388 leukaemia it reproducibly gave 100% increase in lifespan at its maximally tolerated dose of 50 μmol/kg. It also exhibited reproducible activity against ip. M 5076 reticulum cell sarcoma, ip. L 1210 leukaemia, B 16 melanoma, Lewis lung carcinoma, Madison lung carcinoma, sc. mammary carcinoma 16/c, and ADJ-PC 6 plasmacytoma.

Many analogues of dppe have also been evaluated for antitumour activity against ip. P 388 leukaemia[89]. These results are summarised in Table 10. In general replacement of phenyl- substituents for other groups resulted in a reduction in antitumour activity. Only the perdeuterophenyl or cyclohexyl analogues had activity approaching that of dppe. Substitution of one or both phenyl groups for ethyls (i.e. the ligands eppe and depe), results in complete loss of activity. The mono- and di-arsine analogues of dppe were relatively toxic to mice but devoid of antitumour activity. The disulphido analogue is also inactive.

Although potency *in vivo* is not markedly affected by increasing the length of the diphosphine P–P bridge, antitumour activity is generally reduced. The data listed in Table 10 are illustrated in Fig. 19. It is notable that activity is greatest for the ligands able to form 5- or 6-membered chelate rings, i.e., when the carbon bridge contains 2- or 3-carbon atoms or cis-ethylene. The rigid trans ethylene-, acetylene- and 1,4-phenyl-bridged ligands are inactive and markedly less toxic to mice[89]. These data suggest that the mechanism of action may involve chelation of metal ions *in vivo*. This is discussed further in Sect. 3.7.

It must be remembered that many of these diphosphines will be very readily oxidised under biologically relevant conditions. Although it may be possible to interpret these results in terms of differences in structure of the diphosphines, the biological activities may also reflect differences in the ease of oxidation and consequent detoxification. The apparent lower toxicity and reduced spectrum of antitumour activity of dppe in the earlier study[88] seems likely to have been due to the partial oxidation of the tested compound. Available data on phosphine oxides show them to be *inactive* as antitumour agents (Ref 89 and NCI data). The bisoxide of dppe, $Ph_2(O)PCH_2CH_2P(O)Ph_2$, is inactive against P 388 leukaemia, and has a lower toxicity in mice: maximum tolerated dose (MTD) 300 μmol/kg, compared to 50 μmol/kg for dppe[89].

$[Au(dppe)_2]Cl$ exhibits a comparable level of antitumour activity to dppe, but its cytotoxic potency in mice is 25-fold greater than the free ligand and it is highly cytotoxic to cultured cells *in vitro*. This suggests that coordination to gold(I) protects the ligand from unfavourable oxidation prior to delivery to cellular targets. Linear bridged digold diphosphine complexes are also considerably more potent cytotoxic agents than the free ligands.

3.4.2 *Cytotoxicity and Antitumour Activity of Linear, Bridged Digold Diphosphine Complexes*

The bridged digold diphosphine complex ClAu(dppe)AuCl has been tested for antitumour activity in the same tumour models as dppe[89]. In parallel with $[Au(dppe)_2]Cl$ it exhibited comparable activity to dppe against ip. P 388 leukaemia, but it was considerably more potent. The maximally-tolerated dose was 7 μmol/kg/day. Similarly, the complex was considerably more cytotoxic to tumour cells *in vitro* than the free ligand: in a clonogenic assay the concentration of ClAu(dppe)AuCl required to reduce the suvival of B 16 melanoma cells by 50% was 8 μM, (compared to a value of 60 μM for dppe[89, 90]).

More than 50 digold-diphosphine complexes have been synthesised and screened for antitumour activity at SK & F Laboratories. In general, the gold complexes are potent cytotoxic agents to B 16 cells *in vitro*, whereas the respective free ligands have, at best, marginal cytotoxic activity. The complexes were all evaluated against ip. P 388 leukaemia in mice. These results are summarised in Table 11. The majority of the complexes tested were 5–10 fold more potent than the appropriate diphosphine but the level of antitumour activies (i.e. ILS values) were similar. In only a few cases were the activities of ligands

Table 11. Antitumour activity of digold-diphosphine complexes $XAu(R_2P(CH_2)_nPR_2)AuX$ (and some As and S analogues) against ip. P 388 leukaemia in mice (from Ref. 89)[a]. Those producing an increase in lifespan ⩾ 70% are indicated by *

General structure	Active	Marginal	Inactive
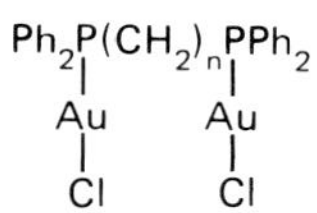	n = 1[b], n = 2*, *cis*-CH = CH* n = 3, n = 5	$(CH_2)_n$ n = 4, n = 6, *trans*-CH = CH[b] $-(CHMe)_2$, $-CH_2CHMe-$	—C≡C— [b]  [b]
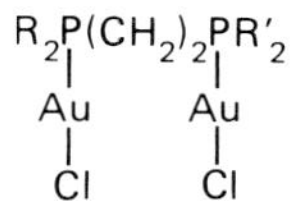	Ph*, C_6D_5, 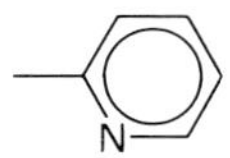 cyclohexyl* 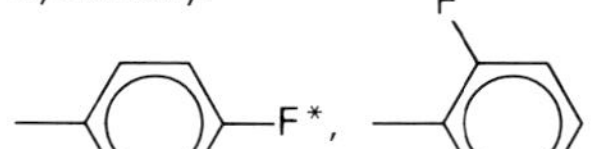 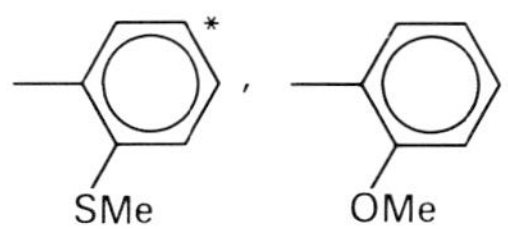	*R,R'* = 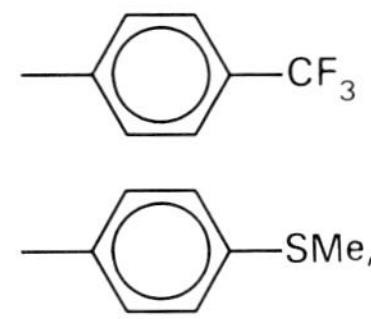	$-CH_2Ph$, Et[b], 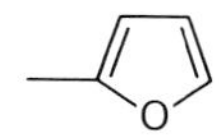 (R = Et, R' = Ph), 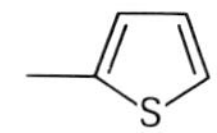 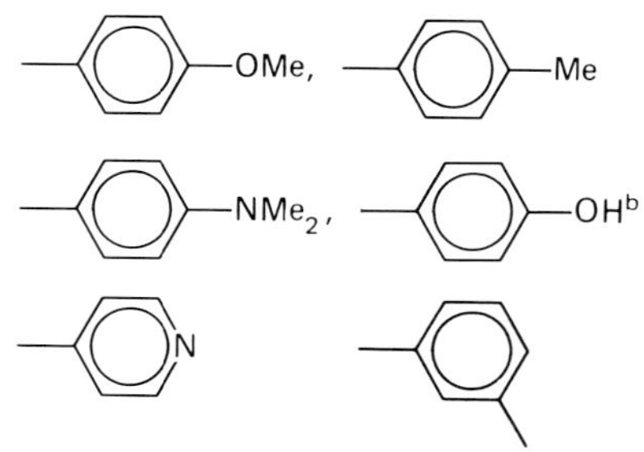
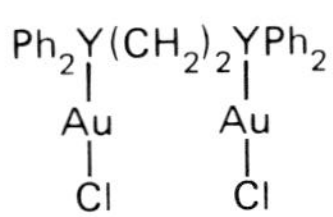	Y = Y' = P*	*Y,Y'* =	Y = P, Y' = As; Y = Y' = As Y = Y' = S[b,c]

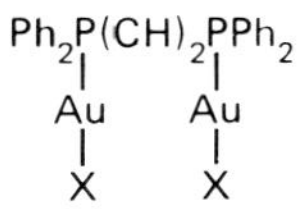	Cl*, Br*, OAc*, SCF_3, SCSOEt, SGlu*, SGal* SMan*, dppe($2Cl^-$)*, $SCH_2CH(NH_2)COOEt$, $PEt_3(2NO_3^-)$*	*X* SCN, $SCSNEt_2$	$SGlu(Ac)_4$, $SMan(Ac)_4$ SeGlu, $SeGlu(Ac)_4$
Other	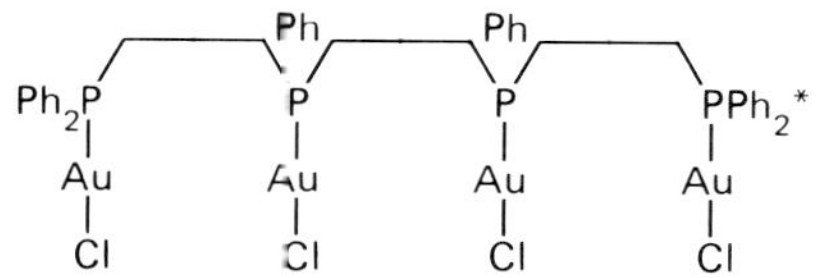	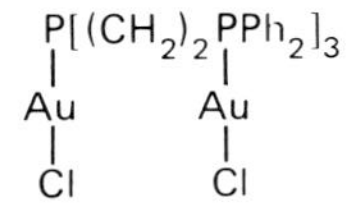	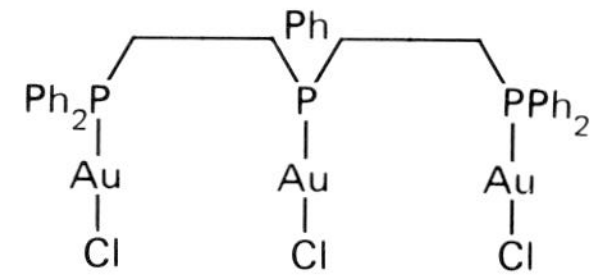 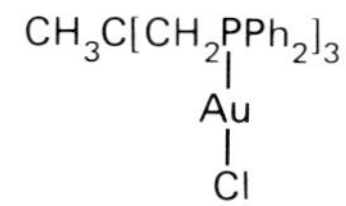

[a] The compounds were administered by ip. injection for 5 days at their maximum tolerated dose (MTD). A compound is considered to be active if it produces an increase in lifespan $>50\%$ and inactive if the ILS is $<30\%$. The MTD were generally in the range 4–10 μmol/kg

[b] MTD >25 μmol/kg

[c] One phenyl group coordinated to each S atom

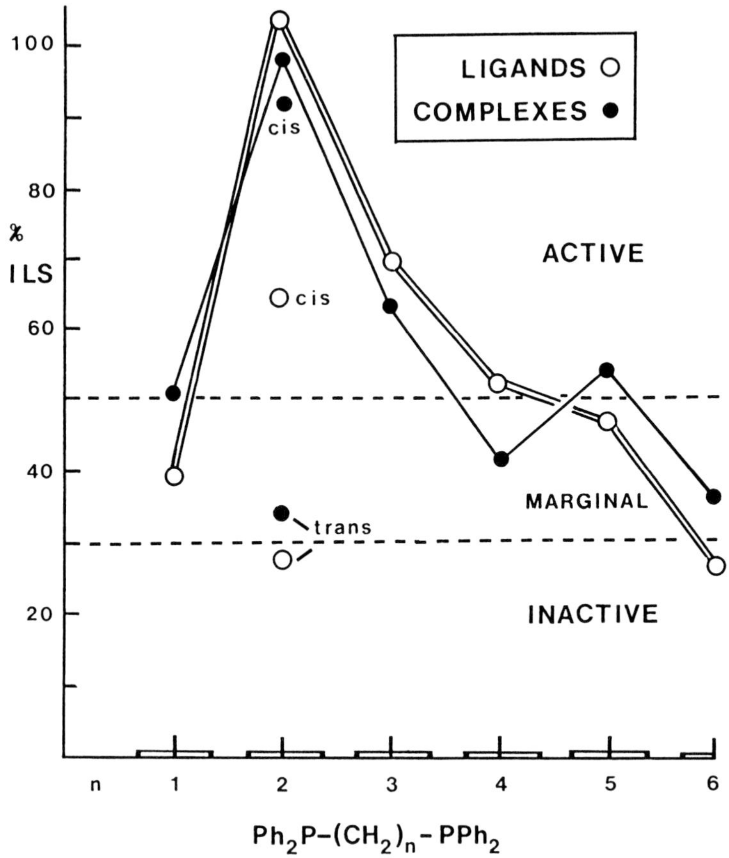

Fig. 19. Effect of number of carbons in the P–P bridge on activity of diphosphine ligands and bridged digold complexes $ClAu(\mu\text{-}Ph_2P(CH_2)_nPPh_2)AuCl$ against P388 leukaemia in mice. Data from Ref. 89. The points labelled *cis* and *trans* correspond to the two isomers of $Ph_2PCH = CHPPh_2$

and complexes significantly different. For example, the 2-pyridyl and o-methoxyphenyl analogues of dppe are inactive whereas the digold complexes exhibit good activity.

It is notable that there is a similar trend between the degree of antitumour activity of the complexes and the length of the diphosphine (P–P) bridge to that observed for the ligands alone (Fig. 19). In addition, in the series XAu(dppe)AuX, the level of antitumour activity is influenced by the nature of the ligand X trans to P (Table 11).

This structure-activity relationship may be related to whether the bridged digold complexes can be converted into ring-closed species *in vivo*.[82] As discussed in Sect. 3.3.3, the complexes XAu(dppe)AuX, where X is Cl or thioglucose (SGlu), are converted into $[Au(dppe)_2]^+$ by thiols[82]. Moreover, a similar conversion occurred when GluSAu(dppe)AuSGlu was added directly to blood plasma. At a concentration of 0.2 mM the bridged digold complex appeared to be totally converted into $[Au(dppe)_2]^+$ (Fig. 14)[82]. The mechanism of the reaction has not yet been investigated but presumably involves plasma thiols such as the Cys-34 residue of serum albumin. It is notable that in the series XAu(dppe)AuX, complexes generally have high antitumour activity if the ligand X is a good leaving group (e.g. X = Cl, Br, OAc, SMan and SGlu). The acetylated thioglucose ($SGlu(Ac)_4$) and thiomannose ($SMan(Ac)_4$) derivatives are inactive. This trend is in contrast to that found for Au(I)phosphine complexes of the type R_3PAuX: antitumour activity is lowest when X is a good leaving group, and highest for the tetraacetylthioglucose complexes (e.g. auranofin, **9**) (Fig. 5). As discussed earlier this may be related to rapid displacement of the ligand X by thiols resulting in the R_3PAu– moiety binding strongly to serum proteins. On the other hand, for the series XAu(dppe)AuX, metabolism into $[Au(dppe)_2]^+$ would be expected to occur more readily if the ligand (X) can be readily displaced. It is not yet known if the tetraacetylthioglucose derivative undergoes conversion into the bis-chelated species in blood plasma. The 1 : 1 Au : dppe complex $[Au_2(dppe)_2Cl_2]$ has high antitumour activity and, as discussed in Sect. 3.3.2, the annular complex is destabilised with respect to the bis-chelated complex and so it may be converted into $[Au(dppe)_2]^+$ *in vivo*.

The cellular pharmacology of the bridged digold complex GluSAu(dppe)AuSGlu (where SGlu is thioglucose)[91] is very similar to that of $[Au(dppe)_2]^+$ which is consistent with the ring-closed species being the major reaction product of the bridged digold complex in serum. The *in vitro* cytotoxic potency of GluSAu(dppe)AuSGlu was assessed against the murine cell lines B 16 melanoma, P 388 leukaemia and L 1210 leukaemia, and against the human cell lines HT-29 colon carcinoma and HL 60 promyelocytic leukaemia. The concentrations required to inhibit the clonogenic capacities of these cells by 50% were 4, 11, 11, 5 and 7 μM respectively. The dose-survival curves were similar to those produced by $[Au(dppe)_2]^+$ (Fig. 16) with a shoulder at low concentrations where no significant decrease in survival occurred. Moreover, the cytotoxic effects were not acute. The concentration required to reduce cell viability was 10-fold higher than that required to produce an equivalent reduction in their clonogenic capacity. In an alkaline-elution assay GluSAu(dppe)AuSGlu produced DNA single-strand breaks in L 1210 cells in a dose-dependent manner. The breaks were significant within the cytotoxic range of the complex. Macromolecular synthesis was strongly inhibited, but in contrast to the behaviour of $[Au(dppe)_2]^+$, DNA synthesis was inhibited preferentially to RNA or protein syntheses. The spectra of antitumour activity of both ClAu(dppe)AuCl and GluSAu(dppe)AuSGlu against transplantable animal tumours are similar to that of $[Au(dppe)_2]Cl$ (Table 8).

3.4.3 Other Metal dppe Complexes

A few other metal complexes of dppe have been tested for antitumour activity against ip. P 388 leukaemia in mice[89]. The dimeric Au(III) complex $Cl_3Au(\mu\text{-dppe})AuCl_3$ produced comparable activity to ClAu(dppe)AuCl, but was slightly less toxic (MTD = 16 μmol/kg). Au(III) complexes are often strong oxidising agents[44] and it is possible that both ClAu(dppe)AuCl and $[Au(dppe)_2]^+$ may be products of the metabolism of the complex *in vivo*. The Rh(I) complex $RhCl(PPh_3)(dppe)$ and the complexes $[MCl_2(dppe)]$ where M is Pt(II), Pd(II), and Ni(II) were all inactive against P 388 leukaemia.

The Ag(I) complex $(AgNO_3)_2$ (dppe), although toxic to mice (MTD 8 μmol/kg) was inactive as an antitumour agent. However, some bis-chelated Ag(I) diphosphine complexes do have *in vivo* anitumour activity.

3.5 Tetrahedral, Chelated Ag(I), Diphosphine Complexes

Ag(I) ions have long been known to possess potent antimicrobial properties, but the use of silver compounds as pharmaceuticals has been hampered by their unfavourable chemical properties. For example, many Ag(I) complexes are light sensitive, and the insolubility of AgCl often presents problems when compounds come into contact with physiological fluids. Over 100 silver compounds have been evaluated for anticancer activity by the National Cancer Institute, and only 5 have shown at best marginal activity[52]. The burn-treatment compound silver sulphadiazene is insoluble and applied topically to burn wounds. Its antibacterial action may be attributable to the slow release of Ag^+ at the wound site[92].

By slowing down the rapid-ligand exchange reactions common in Ag(I) chemistry it seemed likely that a silver complex could be designed that would no be precipitated by Cl^- and thus might have *in vivo* antibacterial and anticancer activity. Bis-chelated Ag(I) diphosphine complexes appeared to be good candidates for investigation.

3.5.1 Preparation and Properties of Four-Coordinate Ag(I) Complexes

In contrast to the chemistry of Au(I), four coordination is commonly found for Ag(I)phosphine complexes. This is often achieved by Ag-(hal)-Ag bridges such as in Et_3PAgX (where X is Cl, Br or I) which have tetrameric cubane structures[93, 94]. Linear coordination is generally found only with very bulky ligands and non-coordinating anions such as in the complexes $[Ag(P(mesityl)_3)_2]PF_6$[95] and $[Ag(P(NMe_2)_3)_2]BPh_4$[96]. Tetrakis monophosphine complexes $[Ag(PR_3)_4]X$ are usually isolable only with large non-coordinating anions such as $[Ag(PPh_3)_4]X$, where X is ClO_4, BrO_3 and NO_3[97]. Halide generally binds with the exclusion of the fourth phosphine ligand[98]. $[Ag(PMe_3)_4]Cl$ is a notable exception[99].

^{31}P NMR studies have demonstrated the high kinetic lability of phosphines in $[Ag(PR_3)_4]^+$ complexes: Ag-P spin-spin coupling is seen only at temperatures below 203 K[98]. The rate of Ag–P bond rupture in $[Ag(P(p\text{-}tol)_3)_4]^+$ was reported to be $>2000\ s^{-1}$ at 313 K.

Investigations of Ag(I) complexes with bidentate phosphine ligands have been far less extensive. The rigid bidentate ligand $o\text{-}C_6H_4(PEt_2)_2$ was reported to form a bis-chelated complex $[Ag(P\text{–}P)_2]I$[61] and $[Ag(dppe)_2]SO_3CH_3$ was prepared from $AgSO_3CH_3$[100]. There was no indication of the stability and ligand-exchange properties of these complexes. The ligand $Ph_2PCH_2PPh_2$ (dppm) does not appear to readily form chelate rings to Ag(I). The isolated complexes contain bridging dppm, as in the bridged dinuclear complex $(AgOAc)_2(\mu\text{-dppm})$[101] or the annular complexes $[Ag_2(dppm)_2X_2]$ (X = OAc[102] or NO_3[103]). The Ag-P bonds are labile in these dppm complexes so that Ag-P spin-spin coupling is not resolved above 203 K[101, 102]. ^{31}P NMR studies[104] have shown that the bridged dinuclear complex AcOAg(μ-dppe)AgOAc exists in $CDCl_3$ solution at a AgOAc : dppe ratio of 2 : 1, and in common with the dppm complex Ag-P spin-spin coupling is resolved only at low temperatures (< 221 K). However, in contrast, at a 1 : 2 AgOAc : dppe ratio, two overlapping doublets (attributable to unequal $^1J(^{31}P\text{-}^{109}Ag)$ and $^1J(^{31}P\text{-}^{107}Ag)$ coupling constants) are resolved in the ^{31}P NMR spectrum at 300 K (Fig. 20). Furthermore, the addition of excess dppe caused only a slight broadening of

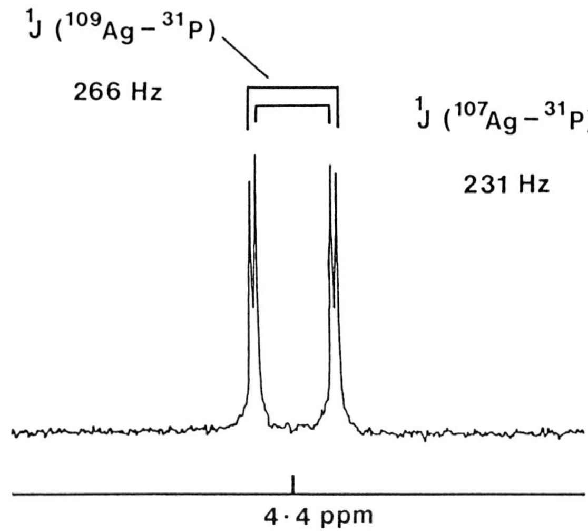

Fig. 20. The 24.2 MHz $^{31}P\{^1H\}$ NMR spectrum of $[Ag(dppe)_2]$ OAc in $CDCl_3$ at 302 K. The resolved $^{107/109}Ag-^{31}P$ couplings show that the complex is tetrahedral (see also data in Ref. 104)

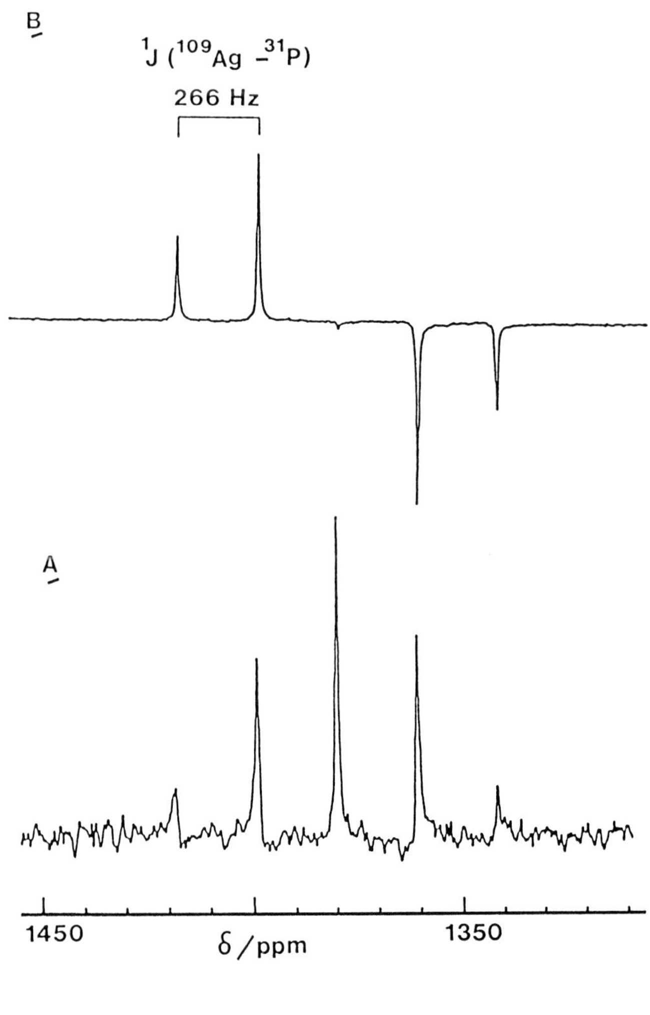

Fig. 21 A, B. 13.97 MHz ^{109}Ag NMR spectrum of $[Ag(dppe)_2]NO_3$ **(A)**, and the sensitivity-enhanced $\{^{31}P\}-^{109}Ag$ INEPT spectrum **(B)** (2048 scans compared to 12, 111 scans in **(A)**) (based on Ref. 104)

the multiplet resonance and the appearance of a slightly broadened peak for free dppe, indicating relatively slow exchange between free and bound ligand on the NMR time scale. The ^{109}Ag NMR spectrum confirms that the species is the bis-chelated complex $[Ag(dppe)_2]^+$ (Fig. 21). Both ^{109}Ag and ^{107}Ag are very insensitive NMR nuclei and there have been relatively few reports of their use. However, since $[Ag(dppe)_2]^+$ exhibited well resolved ^{31}P-^{109}Ag coupling, ^{109}Ag spectra were obtained relatively quickly via phosphorus-irradiation INEPT experiments (Fig. 21)[104)].

Bis-chelated Ag(I)diphosphine complexes $[Ag(P–P)_2]NO_3$, with 5- or 6-membered chelate rings, were readily prepared by reacting $AgNO_3$ with 2 mol equivalents of the appropriate diphosphine for the ligands dppe, depe, eppe, cis-dppey and dppp. They were characterised by a combination of ^{31}P and ^{109}Ag NMR spectroscopy. In parallel with the analogous bis-chelated Au(I) complexes they exhibited greatly enhanced kinetic and thermodynamic stabilities with respect to complexes with monodentate phosphines[104)]. In addition they are not decomposed by light and $[Ag(dppe)_2]NO_3$ was found to be stable in the presence of Cl^- ions. Ligand exchange reactions with thiols appear to be more thermodynamically favourable for $[Ag(dppe)_2]NO_3$ than the analogous Au(I) complex. In the presence of 20 mol equivalents of glutathione (GSH), dppe was displaced from the Ag(I) coordination sphere. In a similar reaction, $[Ag(dppey)_2]NO_3$ did not react appreciable with glutathione[105)].

3.5.2 *Cytotoxicity and Antitumour Activity of Bis(Diphosphine)Ag(I) Complexes*

$[Ag(dppe)_2]NO_3$ exhibits comparable antitumour activity against ip. P388 leukaemia to $[Au(dppe)_2]Cl$. It is also potently cytotoxic to B16 melanoma cells *in vitro*[105)]. The ethyl-substituted complex $[Ag(depe)_2]NO_3$, exhibits potent cytotoxic activity *in vitro* and marginal antitumour activity in vivo[105)]. This is in contrast to the behaviour of the Au(I) depe complex and suggests that the cytotoxic properties of the Ag(I) complexes may be due partly to the presence of the Ag(I) ion.

Neither $[Ag(dppe)_2]NO_3$ nor $[Ag(depe)_2]NO_3$ exhibited significant antibacterial activity, when evaluated against 12 bacterial strains using a microtitre assay[105)]. However both complexes exhibited antifungal activity against 3 strains of *C. albicans* when evaluated in a defined medium (yeast culture broth/lysine). $[Ag(depe)_2]NO_3$ had comparable antifungal potency to fungizone (Amphotericin B), the positive control in the assay. In a complex medium (Sabouraud's: dextrose/digests of casein and fresh meat) the Ag(I) complexes were totally inactive[105)]. Furthermore, the cytotoxic potency of $[Ag(depe)_2]NO_3$ against B16 melanoma cells *in vitro* was reduced ca. 5-fold when 10% foetal calf serum was added to the cell culture medium. These observations can be rationalised in terms of Ag(I) binding to –SH groups in serum proteins. In view of the lower reactivity of $[Ag(dppey)_2]NO_3$ towards thiols in model reactions this complex may offer greater potential as an antibacterial or antifungal agent. The complex is active against ip. P388 leukaemia in mice[105)].

3.6 Copper(I) Phosphine Complexes

3.6.1 Structural Chemistry of Cu(I) Phosphine Complexes and Chelation of Diphosphines

The cytotoxic potencies of dppe and related diphosphines are potentiated by the presence of Cu(II) salts[90)]. In view of the possibility that chelation of copper *in vivo* may be involved in the mechanism of action, we need to understand the nature of the copper phosphine species that may be formed.

Cu(II) phosphine complexes are unlikely to be products of the reaction of phosphines with Cu(II) ions *in vivo*. There are very few known examples of Cu(II) phosphine complexes in the literature: Cu(II) salts are generally reduced readily by phosphines to Cu(I). An exception is a series of complexes $Cu(hfac)_2PR_3$, where hfac is hexafluoroacetylacetonate, and PR_3 is PPh_3, PPh_2Me, $PPhMe_2$, Et_3P and P^nBu_3[106, 107)]. These were prepared strictly in the absence of H_2O and O_2 and were reported to be very unstable. Moreover, the addition of more than one equivalent of PPh_3 to $Cu(hfac)_2$ in methanol produced the Cu(I) complex $Cu(hfac)(PPh_3)_2$.

There have been some investigations of the oxidation and reduction products of Cu(II) salts with tertiary phosphines but few have been carried out under conditions relevant to biology. The nature of the products is highly dependent on the experimental conditions, in particular the Cu(II) : PR_3 ratio. For example the products of the reaction of neat Et_3P with anhydrous $CuCl_2$ have been characterised[108)]. At low $Et_3P : CuCl_2$ ratios (0.5) the products were "fused salts": $[Et_3PCl][CuCl_2]$ and $[Et_3PCl][Cu_2Cl_3]$. These react with excess Et_3P to produce Et_3PCl_2 and the Cu(I) complexes $CuCl(PEt_3)_n$ (n = 1 and 3). When H_2O is present the phosphorane presumably will hydrolyse to the phosphine oxide. Indeed, Cu(II)phosphine oxide complexes have been reported to be products of the reaction of PPh_3 with $CuCl_2$ in acetone[109, 110)]. The compounds $OPPh_3$, $CuCl_2(OPPh_3)_4 \cdot 2H_2O$, $CuCl_2(OPPh_3)_2$, and $Cu_4OCl_6(OPPh_3)_4$ were reported to be formed at a 1 : 1 $CuCl_2 : PPh_3$ ratio, whereas $CuCl(PPh_3)_3$ and $CuCl(PPh_3)_2$ were products at a 1 : 4 $CuCl_2 : PPh_3$ ratio.

There have been a large number of structural investigations of Cu(I)monophosphine complexes but relatively few detailed reports of diphosphine complexes. For monodentate phosphines a wide variety of different structures have been found for complexes with Cu : P ratios of 4 : 1, 3 : 1, 2 : 1, 3 : 2 and 1 : 1[111)]. They have been prepared either by reduction of the appropriate Cu(II) salt with excess tertiary phosphine, or by direct reaction of the appropriate Cu(I) salt (generally halide) with PR_3. It can be seen from Fig. 22 that they can have monomeric, dimeric or tetrameric structures depending on the stoichiometry and the nature of the additional ligands (or anions) present. The copper atom is almost always tetrahedrally coordinated, (unless steric contraints between bulky phosphines prevent four coordination), and, in common with Ag(I)phosphine complexes, this is often achieved via halide bridges. Tetrakis Cu(I)phosphine complexes $[Cu(PR_3)_4]X$ are known where X is a non-coordinating anion, for example $[Cu(PPh_3)_4]ClO_4$[97)], but rarely when X is halide. Instead, halide binds with the exclusion of the fourth phosphine ligand producing complexes of the type $CuCl(PR_3)_3$[126)].

Investigations of Cu(I)diphosphine complexes have been restricted to the ligands dppm and dppe, and in general the complexes have been prepared by direct reaction of the phosphine with Cu(I)halide. Complexes with P : Cu ratios of 3 : 1, 2 : 1, 3 : 2, 4 : 3 and

Fig. 22. Some structures of copper(I) complexes with tertiary phosphine ligands

P : Cu Ratio	Monodentate Ligands	Example	Ref.	Bidentate Ligands	Example	Ref.
4 : 1	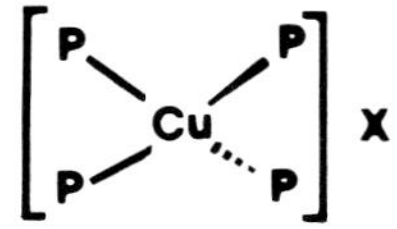	$[Cu(PPh_3)_4]ClO_4$ $[Cu(PMe_3)_4]Cl$	97 99	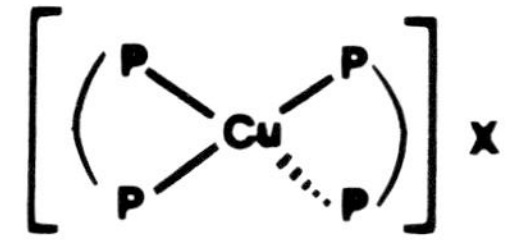	$[Cu(dppe)_2]X$ $X = CF_3CO_2^-$ $X = [Cu(mesityl)_2]$ $X = NO_3^-$	 112, 113 123 115
3 : 1	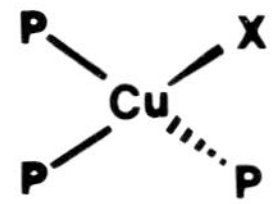	$[Cu(PPh_3)_3Cl]$ $[Cu(PMePh_2)_3NO_3]$	111 115		$[(CuX)_2(dppe)_3]$ $X = Cl^-$ $X = OAc^-$ $X = NO_3^-$ $X = N_3^-$	 114 124 115 125
2 : 1	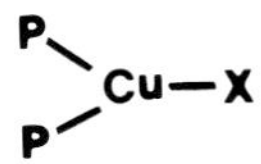 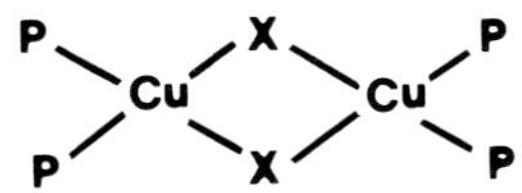 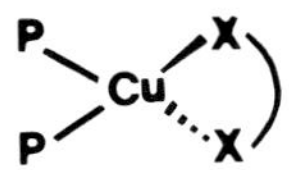	$Cu(PPh_3)_2Br \cdot \frac{1}{2}C_6H_6$ $[(Me_3P)_2CuCl]_2$ $[(Ph_3P)_2CuBH_4]$	116 99 111	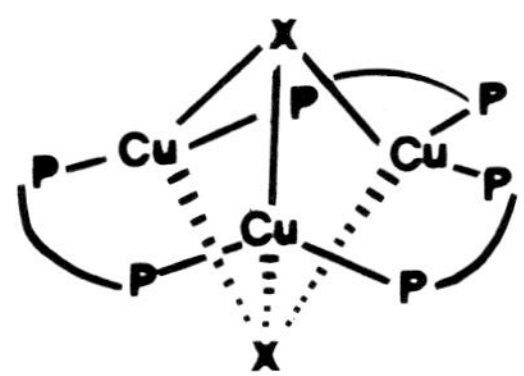	$[Cu_3Cl_2(dppm)_3]Cl$	117
3 : 2	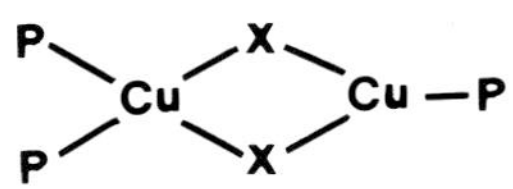	$(Ph_3P)_3Cu_2Cl_2 \cdot C_6H_6$	111			
4 : 3				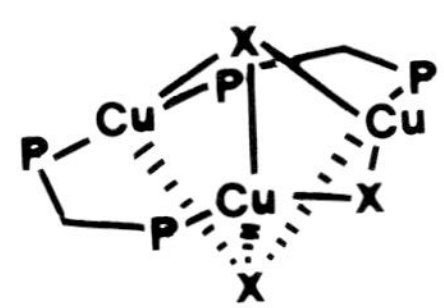	$[(CuI)_3(dppm)_2]$	118

1 : 1	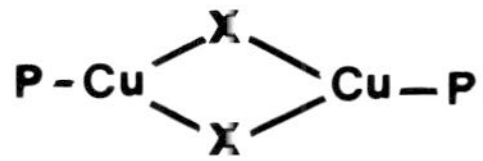	$[(cyc)_3PCuCl]_2$	119
	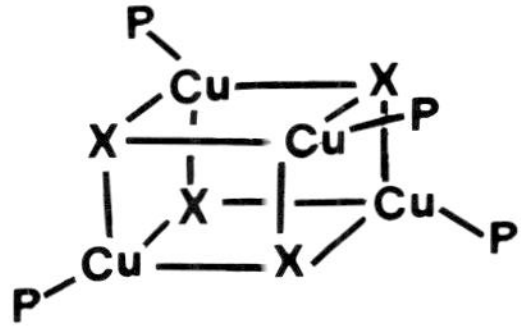	$[Ph_3PCuCl]_4$	120
	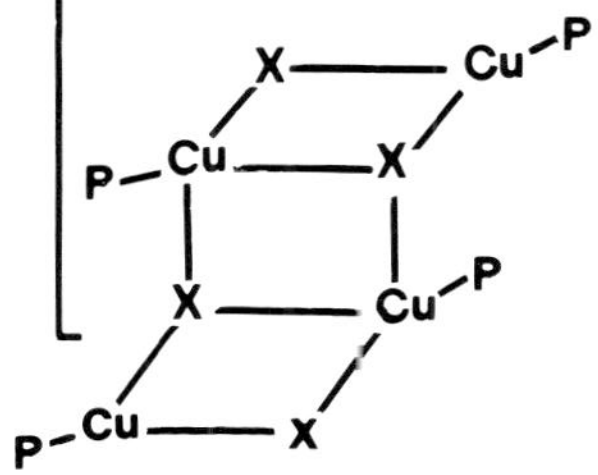	$[Ph_3PCuBr]_4$	120
	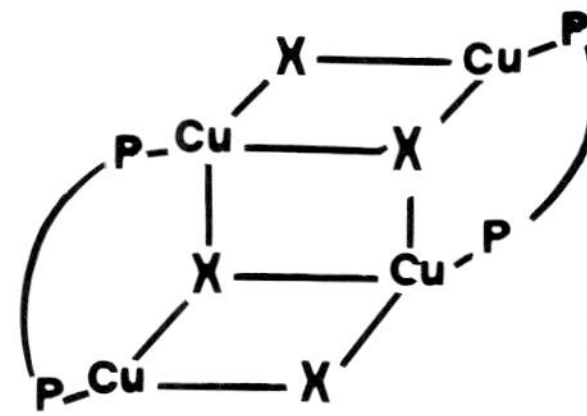	$[(CuX)_2(dppm)_2]$ $X = Cl^-, Br^-, I^-$	121 122

1 : 1 have been isolated[127] (depending on the amount of phosphine added) but few have been structurally characterised. These are mostly polynuclear compounds containing bridging dppm and halide (Fig. 22). The 1 : 1 CuCl : dppe species has not been characterised but molecular weight measurements indicate that it has a trimeric structure $[CuCl(dppe)]_3$[127]. The 2 : 3 CuCl : dppe complex has a dinuclear structure containing both bridging and chelated diphosphine[114] (Fig. 22, Table 7). Again, chloride appears to bind in preference to the fourth phosphine ligand: the bis-chelated cation $[Cu(dppe)_2]^+$ had been isolated with CF_3COO[112, 113], $[Cu(mesityl)_2]$[123] and NO_3[115] as counter anions (Table 7).

It will not be an easy task to identify whether a particular Cu(I)diphosphine complex plays a key role in the cytotoxic and antitumour activity of diphosphine ligands. In addition to the variety of possible Cu(I)diphosphine species that could form it is also probable that these species will undergo extensive dissociation in solution. Solution studies of monodentate phosphine complexes of Cu(I)halides have shown that the complexes undergo complicated dissociative equilibria involving species of the form $L_m(CuX)_n$ with a variety of m : n stoichiometries. The extent of the dissociation depends on the solvent, temperature, concentration and the bulkiness of the phosphine[126, 128].

^{31}P NMR studies[129] have shown that the dppe complex $(CuCl)_2(dppe)_3$ dissociates in solution; both dppe and $[Cu(dppe)_2]^+$ were involved in the equilibria. In a biological system the position of equilibria involving Cl^- would be expected to change greatly on passage of a complex from outside to inside cells where the Cl^- concentrations decrease from ca. 104 mM to ca. 3 mM. There is also the possibility that Cu(II) phosphine-oxide complexes could be formed. Anderson and coworkers[115] reported that the Cu(I) complex $(CuNO_3)_2(dppe)_3$ rapidly converted into a green Cu(II)bisphosphine oxide complex in chlorinated solvents. The crystal structure of a $CuCl_2$ complex of $Ph_2P(O)(CH_2)_2P(O)Ph_2$ ($dppeO_2$) has been determined[130]. The complex consists of infinite chains of $CuCl_2$ units linked by O-coordinated bridging $dppeO_2$ ligands. The complex $CuCl_2 \cdot dppeO_2$ has been prepared either by reaction of $CuCl_2$ with dppe in the presence of O_2[115, 130] or by the direct reaction of $CuCl_2$ with $dppeO_2$[131].

In view of the potent cytotoxicity and antitumour activities of tetrahedral Au(I) and Ag(I) bis-chelated diphosphine complexes the chemistry and biological activity of analogous Cu(I) complexes[129, 132] were also investigated. $[Cu(dppe)_2]Cl$ could not be prepared by reaction of CuCl with an excess of dppe, but instead the binuclear complex $(CuCl)_2(dppe)_3$ was isolated[129]. In contrast, bis-chelated complexes $[Cu(R_2P(CH_2)_nPR'_2)_2]Cl$ were readily isolated for the ligands, eppe (n = 2, R = Ph, R′ = Et)[132], dppp (n = 3, R = R′ = Ph) and *cis* -$Ph_2PCH = CHPPh_2$[129]. These are rare examples of tetrakis (tertiary phosphine) complexes of Cu(I)chloride. Moreover, both $[Cu(eppe)_2]Cl$[132] and $[Cu(dppey)_2]Cl$ were found to be remarkably stable in solution with no evidence of significant dissociation or equilibria with coordinately-unsaturated species. $[Cu(dppp)_2]Cl$ did appear to dissociate in chloroform solution[129].

3.6.2 Antitumour Activity of Cu(I) Diphosphine Complexes

$(CuCl)_2(dppe)_3$, and the bis-chelated complexes $[Cu(P\text{–}P)_2]Cl$ where P–P is dppey and dppp exhibited good activity against ip. P388 leukaemia in mice (Table 9)[129]. Thc activity was comparable to that of the free ligands, and in parallel with the bis-chelated Au(I)diphosphine complexes, they were at least 20-fold more potent. Preliminary results of their activities against other ip. tumours B16 melanoma and M5076 reticulum cell sarcoma, suggest that they may exhibit a similar spectrum of antitumour activity to the Au(I) analogues. In parallel with the analogous Au(I) complexes, the effect of exchanging phenyl substituents for ethyls was to reduce activity. $[Cu(eppe)_2]Cl$ exhibited only marginal activity against ip. P388 leukaemia[129]. The complex $[Cu(dppey)_2]Cl$ (where dppey is *cis*-$Ph_2PCH = CHPPh_2$), which has high antitumour activity and a well-defined structure in solution, would make a suitable probe for investigating the effect of Cu(I)bisphosphines on critical cellular processes. Although $[Cu(eppe)_2]Cl$ is less active as an antitumour agent *in vivo* it has high-water solubility which may makes it more suitable for some model studies.

3.7 Mechanism of Cytotoxic Action

Little is known about the mechanisms by which some phosphines and metal phosphine complexes are able to kill cells. However, studies to date indicate that the phosphine itself is the cytotoxic agent. In the case of auranofin and other monophosphine Au(I) complexes, the role of the metal may be largely to protect the phosphine and deliver it to cellular targets. Auranofin is potently cytotoxic to cells *in vitro,* but its activity *in vivo* is limited[54]. This may be because Au(I) is very reactive towards biologically important ligands, particularly thiols[85, 86], so that it binds to proteins and is prevented from reaching the critical targets.

Tetrahedral bis-chelated diphosphine complexes of Au(I) have a broader spectrum of antitumour activity. The chemistry of these complexes is very different to that of linear 2-coordinate Au(I) complexes. The chelation of ligands and high-coordination number introduce considerable activation barriers to ligand-exchange so that Au(I) is not reactive towards thiols[83]. In addition, the metal is able to offer a greater degree of protection to the phosphine compared to that of auranofin by virtue to the chelate effect. Taken together these factors are likely to increase the chances of delivery of the toxic phosphine to critical targets. Indeed, ^{31}P NMR studies[84] have shown that the $[Au(dppe)_2]^+$ cation remains essentially intact in human blood plasma (Fig. 14) and is readily transferred to red cells where it appears to bind *intact* within the membrane (Fig. 15).

The unusual biological profile of bis-chelated Au(I)diphosphine complexes is likely to result from a combination of their high lipophilicity which allows them to penetrate cells, and their high kinetic and thermodynamic stability which prevents unwanted side-reactions *in vivo.* However, in addition there must be sufficient lability in the Au-P bond so that the phosphine itself is reactive at the target site[133]. This fine tuning of the balance of kinetic and thermodynamic stability is also obtained with the analogous tetrahedral bis-chelated Ag(I) and Cu(I) diphosphine complexes. They exhibit similarly enhanced stabilities with respect to complexes with monophosphine ligands. ^1H and ^{13}C NMR studies have shown that the Au(I), Ag(I) and Cu(I) complexes $[M(eppe)_2]^+$ undergo an intramolecular inversion process in solution (Fig. 23). The solvent-dependence of the rates of inversion (higher in $CDCl_3$ than D_2O) is consistent with a mechanism involving facile breaking of at least one M–P bond[132]. The kinetic stability of the complexes decreases in the order Cu(I) > Au(I) > Ag(I).

The high kinetic stability of bis-chelated Ag(I) diphosphines protects the Ag(I) ion from precipitation as AgCl in physiological fluids. For these complexes, the cytotoxic and

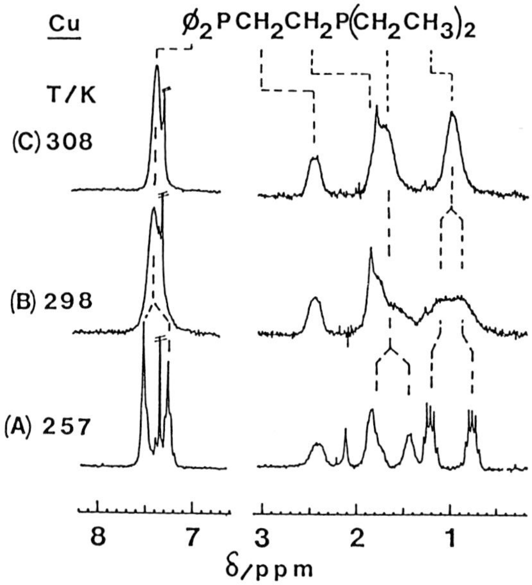

Fig. 23. Temperature dependence of the 200 MHz ^{1}H NMR spectrum of the eppe complex $[Cu(Ph_2P(CH_2)_2PEt_2)_2]Cl$, showing the coalescence of the peaks for non-equivalent Ph and Et groups at the high temperature due to rapid inversion of the tetrahedral complex (from Ref. 132)

antitumour activities may be partly due to delivery of toxic Ag^+ to a cell. Similarly, for the Cu(I) diphosphine complexes the delivery of Cu(I) into a cell could be important. Indeed, it is possible that a Cu(I)diphosphine species is the ultimate cytotoxic product of the metabolism of the Au(I) (and Ag(I)?) complexes and the ligands themselves. $[Au(dppe)_2]Cl$ reacts in aqueous solution with Cu(II) to give a Cu(I)dppe complex as one of the products[83)]. The cytotoxic potency of dppe *in vitro* and its toxicity *in vivo* are significantly increased when incubated in the presence of non-cytotoxic concentrations of Cu(II) salts whereas Mg(II), Fe(II), Co(II) and Cd(II) have no effect[90)]. For the series of diphosphines $Ph_2P(CH_2)_nPPh_2$ the cytotoxic potency is increased by $CuCl_2$ when the P–P bridge contains 2, 3, 4 or 5 carbon atoms but not for n = 6. *Cis*-$Ph_2PCH = CHPPh_2$ is more potent in the presence of $CuCl_2$ but the potencies of ligands containing *trans* CH = CH or C ≡ C bridges are not affected. The antitumour activity of the ligands alone shows a similar depedence on the length of the P–P bridge and is highest for those capable of forming 5- or 6-membered chelate rings (Fig. 19) indicating that chelation of metal ions *in vivo* may be important. The increased potency could occur because complexation protects the ligand from oxidation and may promote its uptake into cells. In addition, the delivery of Cu(I) into cells and its redistribution within cells could play a key role in the cytotoxic mechanism.

Copper has been implicated in the anticancer activity of a number of potential chelating agents, such as thiosemicarbazones[134)], 2,9-dimethyl 1,10-phenanthroline[135)], 1,10-phenanthroline[136–138)] and 4′(9-acridinylamino)methanesulphon-m-anisidide[139)]. For the latter two agents there is evidencc that DNA strand-cleavage results from a Cu(II) dependent production of oxygen free radicals. Chelation of other metals that are critical to cell function could occur. For example, zinc is now thought to play a crucial role in the natural regulation of genes[140]. It is noticeable that many nucleic-binding and gene-regulatory proteins appear to contain $(Cys)_4$, $(His)(Cys)_3$, and $(His)_2(Cys)_2$ sites[141]. What would be the functional consequences in a cell if Au(I) or Cu(I) were delivered to these sites. Diphosphines may be able to remove metal ions from metalloproteins and metalloenzymes, or act as inhibitors by binding to metal ions in active sites. Although a few studies have been carried out with haem proteins (see Sect. 4.5), there are many other possibilities still to be explored.

It is interesting to compare the cytotoxic metal phosphine complexes with Pt(II) amine complexes which are also cytotoxic. *Cis*-$[PtCl_2(NH_3)_2]$ (cisplatin) is an anticancer drug in widespread clinical use[142, 143]. The activity is attributed to Pt(II) binding to DNA bases. The NH_3 ligands are normally thought to remain bound to platinum, and even if they are displaced they are unlikely to be cytotoxic. For the metal phosphine complexes there is a completely different situation. Direct binding of the metal to DNA is unlikely to be the important cytotoxic event particularly for the gold(I) complexes. Au(I) (d^{10}) is a very much softer ion than Pt(II) and binds only weakly to N-ligands[3]. Attack on DNA could, however, be mediated by a protein. The Pt(II) complex $[PtCl_2(dppe)]$ does not exhibit anticancer activity[89]. Presumably *trans* labilisation of Cl by P will mean that the complex will be too reactive to be targetted to DNA, and the dppe ligand may be bound too tightly to be reactive.

Although there are some similarities in the chemistry of NH_3 and PH_3 there are many important differences. For example, NH_3 is a much stronger base (pK_a 9.21) than PH_3 (pK_a − 14). Comparison of the oxidation state diagrams for N and P (Fig. 1) demonstrates that NH_4^+, in contrast to PH_3, is not a strong reducing agent, and will not be readily oxidised *in vivo*.

There may be many targets within cells where reactions with reactive phosphines could be destructive. In the next sections we will consider some possible reactions of phosphines in biological systems. It is interesting to note that PH_3 itself is highly toxic and is widely used as a pesticide.

3.8 Insecticidal Activity of PH_3

PH_3 (phosphine) is in worldwide use as a fumigant to control insect pests in a range of stored agricultural products. It is likely to become increasingly used in the future with the decreasing utilitsation of chlorinated hydrocarbon insecticides. The major advantage of PH_3 is that it forms non-toxic decomposition and metabolism products (Fig. 1). (The detoxification of organophosphines (PR_3) by oxidation is similarly a point of consideration in the use of organophosphines as pharmaceuticals.)

PH_3 is generated *in situ* by the action of moisture on Mg or Al phosphide. In many countries, zinc phosphide is used as a rodenticide. It is usually mixed with some form of bait. Since Zn_3P_2 is hydrolysed only in acidic conditions, it decomposes to PH_3 only in the gut. The gas is toxic to insects, mites and all vertebrates including humans[146, 147].

---

3 It is possible to characterise P–Au–N linkages by ^{31}P and ^{15}N NMR in solution and in the solid state by the observation of two-bond ^{31}P–^{15}N couplings[144, 145]. The Et_3PAuN(imido) complexes that we have studied (for the imides phthalimide, saccharin, 5,5-diphenylhydantoin and riboflavin) themselves exhibit good anti-inflammatory activity. The imido ligand is readily displaced by thiols[144].

A few studies have been carried out to elucidate the mechanism of action of PH_3. These have centred on the inhibition of cytochrome oxidase[148, 149] and catalase[149, 150]. The reduction of catalase levels in insects appears to be only an indirect effect since PH_3 does not inhibit catalase "in vitro"[149]. Phosphine concentrations down to 50 μM inhibit mitochondrial respiration[151]. It is notable that PH_3 loses its insecticidal potential in the absence of O_2 or in atmospheres of very low O_2[152], and also, that some strains of certain insect species are resistant to phosphine[146]. The mechanism of resistance is not known but may be related to the ability of these insects to exclude the gas actively (using metabolic energy)[153]. There is some evidence (N. R. Price, personal communication) that although PH_3 does not react directly with haemoglobin in the absence of oxygen, oxyhaemoglobin is converted through Fe^{3+}-containing compounds to a verdichromogen-like substance. Concentrations of PH_3 between 3–19 μM are reached in susceptible insect strains in stored products[154].

In vitro, oxidised preparations of cytochrome oxidase and cytochrome c are reduced by PH_3. The observed changes in the absorption spectra are similar to those produced by dithionite, suggesting reduction at the haem site[154]. The possibility that PH_3 can bind directly to Fe at these haem sites must also be considered (see Sect. 4.5).

4 Reactions of Phosphines of Relevance to Biology

4.1 Phosphines as Deoxygenation Agents

The reactions of tertiary phosphines are frequently driven by formation of the thermodynamically-stable phosphoryl (P=O) bond. Deoxygenation reactions are common in phosphine chemistry; many oxygen-containing compounds, for instance diacyl peroxides, peresters, alkyl hydroperoxides, ozonides, amine oxides and sulphoxides, react with phosphines[1, 13, 14].

Although many tertiary phosphines are readily oxidised in air there have been remarkably few investigations of the oxidation of phosphines by O_2. The limited data available indicate that autoxidation generally proceeds via a radical pathway. Buckler[155] investigated the reaction of Bu_3P with O_2 in solution. He observed four products: Bu_3PO (42%), $Bu_2P(O)OBu$ (49%), $BuP(O)(OBu)_2$ (6%) and $P(O)(OBu)_3$ (3%). Floyd and Boozer[156] studied the kinetics of the reaction and concluded that the process requires free-radical initiation. Formation of both phosphine-oxide and phosphinate products involves the initial formation of an intermediate phosphoranyl radical which decomposes via either an α-scission or β-scission pathway. Arylphosphines appear to react via a slightly different pathway. The reaction of Ph_3P with O_2 is *slow* and Ph_3PO is the only product. The high stability of the intermediate phosphoranyl radical $Ph_3\dot{P}O_2R$ is thought to account for the lower reactivity of $RO_2^{\cdot}$ with Ph_3P[157].

There appear to have been no detailed investigations of the autoxidation of bidentate tertiary phosphine ligands, despite their common utilisation as ligands in organometallic chemistry. We have observed that phenyldiphosphine ligands slowly oxidise in solution to give only the mono- and bis-phosphine oxides whereas $Et_2P(CH_2)_2PEt_2$ (depe) oxidises very much faster to give a mixture of phosphine oxide, phosphinite and phosphonate products[158]. The nature of the products formed depended on the type of solvent, but the formation of mixed products is consistent with a radical chain mechanism.

Although most organic compounds are thermodynamically unstable with respect to oxidation by dioxygen, there is a kinetic constraint because the ground state of molecular dioxygen has two unpaired electrons[159]. Therefore reactions of O_2 can be very slow unless it reacts with atoms or molecules containing unpaired electrons. It is interesting to note that the autoxidation of phosphines proceeds readily without the addition of a radical initiator to the solution. This suggests that trace impurities in the solvents may act as initiators.

The reduction products of dioxygen, e.g. $O_2^{\cdot-}$, H_2O_2 and $OH^{\cdot}$, are all strong oxidants and do not suffer from the same kinetic constraints as O_2. Little seems to be known chemically about oxidation of PR_3 with $OH^{\cdot}$ or $O_2^{\cdot-}$ which are potentially available in a biological system. It is notable that $O_2^{\cdot-}$ is considerably more reactive in aprotic solvents than in aqueous media. It is a good nucleophile and reductant and reacts with many compounds. Its reactions with biological molecules such as polyunsaturated fatty acids and α-tocopherol in *aprotic* solvents have been investigated because certain membrane or enzyme microenvironments might be sufficiently aprotic to promote such reactions[160]. These are regions where lipophilic phosphines would be expected to localise. There are several examples from organic chemistry of reactions of tertiary phosphines with a variety of peroxides (hydroperoxides, peroxy-acids, diaroyl peroxides, dialkyl- and diaryl-peroxides). In almost all cases the mechanism involves nucleophilic attack by P on peroxidic oxygen. This was demonstrated for the reaction of Ph_3P with benzoyl peroxide (PhCO–OO–COPh) by labelling the carbonyl groups with ^{18}O[161]. In aprotic solvents 5-coordinate phosphoranes have been isolated from the reaction of dialkyl peroxides with tertiary phosphines[162, 163]. The mechanism involves a concerted biphilic insertion of PR_3 into the peroxo bond. In the presence of H_2O the phosphorane intermediates normally hydrolyse to give the phosphine-oxide and appropriate alcohol[164]. An interesting example of this, from a biological point of view, is the reaction of a prostaglandin endoperoxide model compound with Ph_3P[165]:

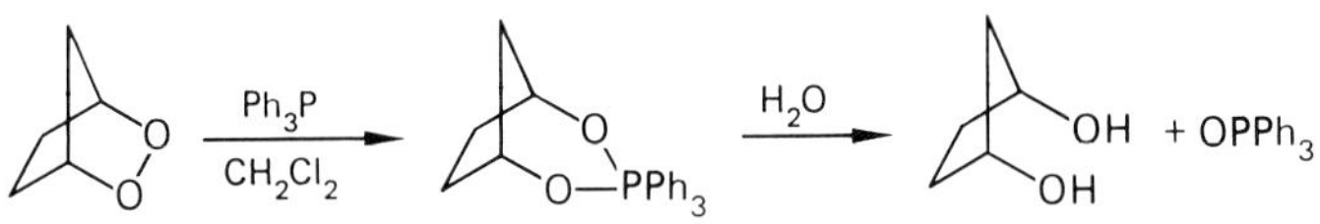

Under conditions where alkoxy radicals can be generated, phosphines will also react with dialkylperoxides via a radical mechanism. For instance Buckler[155] observed that $^nBu_3P$ reacted with di-t-butyl peroxide at 130 °C to give an 80% yield of $^nBu_2PO^tBu$ and only 20% $^nBu_3PO$. This indicates that α-scission of the intermediate phosphoranyl radical $[^nBu_3\dot{P}O^tBu]$ predominates over β-scission despite formation of the strong P=O bond.

Hydroperoxides are very rapidly reduced by both alkyl and aryl tertiary phosphines. The rate of reaction of ^tBuOOH with a series of phosphines decreases in the order[166] $^nBu_3P > Et_3P > Ph_3P$. The reaction is not inhibited by free radical traps[167], and the mechanism involves nucleophilic attack at the peroxide linkage.

In addition to the reactive reduced forms of dioxygen discussed above, biological systems contain metalloenzymes which are able to activate kinetically-inert dioxygen. These enzymes generally contain Fe, Cu or Mo at the active site and they have a variety of different roles. For instance they can act as O_2 carriers, (e.g. haemoglobin) or produce activated forms of oxygen for O-atom insertion reactions (e.g. cytochrome P 450, tyrosin-

Table 12. Some sources of activated oxygen in enzymes and their reactivity with phosphines (where known)

Enzyme	Postulated activated oxygen intermediate	Comment
Oxidases		
Molybdenum-oxo-transferases[a] e.g. xanthine dehydrogenase sulphite oxidase	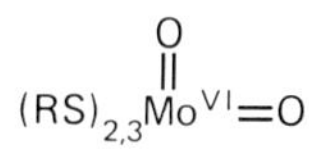	O derived from H_2O in the enzyme system. O atom transfer to PR_3 (PPh_3, PPh_2Et, $PPhEt_2$, PEt_3) shown in model system[b]: $MoO_2(S_2CNEt_2)_2 + PR_3 \rightarrow MoO(S_2CNEt_2)_2 + OPR_3$
Flavoproteins[c] e.g. glucose oxidase, some monamine oxidases, amino acid oxidases	Flavin C(4a)OOH 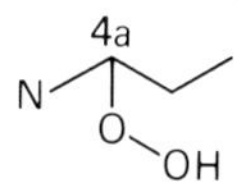	Convert $O_2 \rightarrow H_2O_2$ without O_2^- as an intermediate. Reaction with PR_3 not known
Hydroperoxidases e.g. peroxidase, catalase	$Fe^{IV}O^{2-}$ (Compound II) $Fe^{V}O^{2-}$ (Compound I)	$H_2O_2(ROOH) \rightarrow 2H_2O(H_2O + ROH)$ $2H_2O_2 \rightarrow 2H_2O + O_2$ Model compound $PFe^{III}O{-}O{-}Fe^{III}P$ (P is a porphyrin dianion) catalytically oxidises PPh_3[d].
Myeloperoxidase	?	$H_2O_2 + X^- \rightarrow 2HOX$ (X = Cl, Br, SCN) Phagocytosis in neutrophils, monocytes[e]. Note phosphines react with organic hypochlorites: $PR_3 + ClOR \rightarrow OPR_3 + RCl$
Glutathione peroxidase	Se = 0? (selenocysteine)	Red cells, liver, kidney $ROOH + 2GSH \rightarrow GSSG + ROH + H_2O$
Cytochrome oxidase[f]	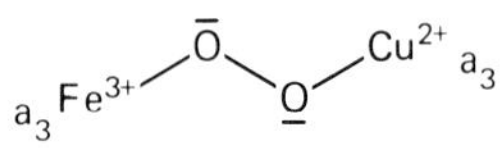	Inner mitochondrial membrane $O_2 + 4H^+ + 4e^- \rightarrow 2H_2O$ PH_3 inhibits cyt oxidase (see section 3.8)
Ascorbate oxidase, Ceruloplasmin	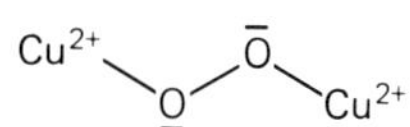	$O_2 \rightarrow H_2O$
Ribonucleotide reductase	$Fe^{3+}{-}O{-}Fe^{3+}$	ribonucleotide → deoxyribonucleotide
Oxygenases		
Monooxygenases Cytochrome P450[g]	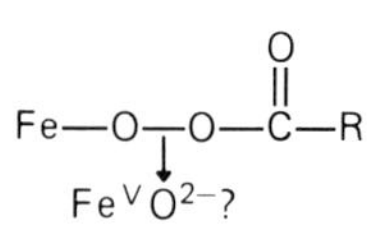	The model compound $PFe^{III}{-}OOFe^{III}P$ catalytically oxides PPh_3 (see peroxidase)
Phenol o-monooxygenase[h] (tyrosinase) Dopamine-β-monooxygenase	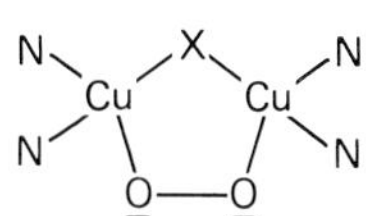	$RH + O_2 + 2H^+ \xrightarrow{2e^-} ROH + H_2O$ The diCu(I) complex 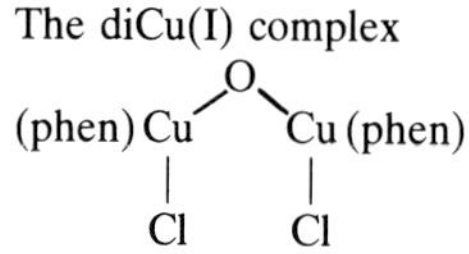 transfers O to PPh_3[i]

Table 12 (continued)

Enzyme	Postulated activated oxygen intermediate	Comment
Dioxygenases		
Tryptophan 2,3 dioxygenase, Indolamine 2,3 dioxygenase	haem Fe-oxo?	
Catechuate dioxygenases	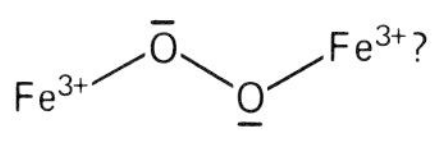	see peroxidase
O_2 Carriers		
Haemoglobin	$Fe^{2+} - O_2$, $Fe^{3+} - O_2^-$	Potential source of O_2^-

[a] Ref. 168; [b]Ref. 169; [c]Ref. 170; [d]Ref. 171; [e]Ref. 172; [f]Ref. 173; [g]Ref. 174; [h]Ref. 175; [i]Ref. 176

ase) or control the reactivity of oxygen species produced during the reduction of O_2 (e.g. cytochrome oxidase). Tertiary phosphines are likely to react readily with these activated oxygen species.

There are some examples in the literature where tertiary phosphines have been used as substrates to follow O-transfer in synthetic models of these metalloenzymes. In Table 12 we show examples of enzymes which contain activated oxygen intermediates with could be potential targets for tertiary phosphines. We have also included details of the reactivity of model compounds with phosphines where known. Note that phosphines could interfere with electron transport in other ways, for example by binding to Fe at the haem site (see Sect. 4.5) or by acting as an electron acceptor or donor (see Sect. 4.3).

4.2 Reactions of Phosphines and Au(I) Phosphine Complexes with Disulphides

The reactions of phosphines with disulphides have been studied extensively due to the importance of these reactions in organic syntheses. Under appropriate conditions tertiary phosphines can act as desulphurizing agents (producing $R_3P{=}S$ and the thioether) or reduce disulphides to thiols. The reaction of disulphides with phosphines has also been used as the dehydration step in the syntheses of peptides and nucleotides. The subject has been reviewed by Mukaiyama and Takei[177)].

The majority of the reactions between disulphides and PR_3 compounds are thought to proceed via ionic intermediates. The first step involves nucleophilic attack by PR_3 on the disulphide bond forming a phosphonium salt intermediate. In aprotic solvents (e.g. benzene) this decomposes to the phosphine sulphide:

$$R_3P \; + \; R'S{-}SR' \rightarrow \left[\begin{array}{c} R_3P^{+}SR' \\ {}^{-}SR' \end{array} \right] \rightarrow R_3P{=}S + R'SR'.$$

The reaction depends on the nature of both the phosphine and disulphide and the rate is influenced by the solvent polarity and the pK_a of the thiol corresponding to the displaced thiolate anion. PPh_3 does not react with either alkyl or aryl disulphides but does react with acyl or aroyl disulphides in boiling benzene. On the other hand, $P(NEt_2)_3$ is a very efficient desulphurizing agent and reacts readily with many disulphides at room temperature. It is notable that $P(NEt_2)_3$ has been used to desulphurize biologically-important disulphides including cystine and α-lipoic acid derivatives. It is unlikely that desulphurization reactions will play an important role in the biological chemistry of phosphines when H_2O is present because then reactions with disulphides are likely to take a different course (vide infra). However this type of reactivity may have to be considered for more lipophilic phosphines with disulphides in hydrophobic regions of proteins.

In aqueous solvents the intermediate phosphonium salt hydrolyses to form the phosphine-oxide and thiol:

$$R_3P \quad + \quad R'S\text{–}SR' \rightleftharpoons R_3P^+SR' + RS^-$$
$$R_3P^+SR' \quad + \quad H_2O \rightarrow R_3PO + R'SH + H^+.$$

The initial attack of PR_3 on the disulphide bond is the rate-limiting step. Hydrolysis of the phosphonium salt is rapid and irreversible. The rate of reaction is enhanced by either dilute acid or dilute base[178]: at high pH hydrolysis of the phosphoium salt is rapid but at low pH the thiolate anion becomes protonated, thus suppressing the reverse reaction in the rate-limiting step.

In general, the rate of reduction of disulphides by phosphines in aqueous solvents decrease in the order[177] $PR_3 > PR_2Ph > PRPh_2 > PPh_3$ (where R = alkyl).

PPh_3 rapidly reduces diaryldisulphides to thiols in aqueous methanol at room temperature but dialkyldisulphides are reduced to only a small extent under these conditions. PBu_3 is an efficient reducing agent for dialkyldisulphides at room temperature.

In a biological system the reductive-cleavage of disulphide (cystine) bonds in proteins and peptides may be a very important reaction of alkylphosphines. PBu_3 and the water-soluble phosphines $P(CH_2OH)_3$ and $P(CH_2COOH)_3$ have been used as alternatives to sulphydryl compounds (e.g. 2-mercaptoethanol) to cleave disulphide (cystine) bonds selectively in several proteins, under mild conditions (see Table 13).

The higher reactivity of alkyl- compared to aryl-phosphines towards disulphide bonds is an important point to consider when contrasting the biological activities of phosphines. The formation of the phosphine-oxide by reductive cleavage of disulphides may provide a means of reducing the toxicity of a phosphine in a biological system. It is notable that diphenyldiphosphines exhibit potent antitumour activity whereas dialkydiphosphines are inactive (see Sect. 3.4.1). We have observed contrasting reactivity for the tetrahedral Au(I)diphosphine complexes $[Au(Ph_2P(CH_2)_2PR_2)_2]Cl$ when R = Ph (dppe) or Et (eppe). The eppe complex slowly cleaves disulphide bonds of bovine serum albumin with release of the phosphine-oxide whereas the dppe complex appears to be less reactive in serum[87]. This may account for the reduced antitumour activity of $[Au(eppe)_2]Cl$ with respect to $[Au(dppe)_2]Cl$ (Table 9). The reactivity of metal-coordinated phosphines towards disulphides in proteins is likely to be related to the lipophilicity of the complex and the kinetic lability of the M–P bond. The water soluble complex $[Au(PEt_3)_2]Cl$ causes whole blood samples to solidify in a few hours by denaturation of albumin whereas $ClAuPEt_3$ does not cleave disulphide bonds in plasma proteins[185].

Finally, it is worth noting that in the presence of free-radical initiators, PR_3 compounds can react with alkyldisulphides and thiols via a radical-chain mechanism[186]. The key step is the formation of an intermediate thiophosphoranyl radical which decomposes via a β-scission pathway to the phosphine sulphide:

Table 13. Reduction of protein and peptide disulphide bonds by phosphines

PR_3	Protein/Peptide	Comment	Ref.
$P(CH_2OH)_3$	Keratin		179
P^nBu_3	wool	Potent and specific cleavage of cystine.	180
$P(CH_2COOH)_3$, $P(CH_2OH)_3$	human γ-globulin	Reduced protein similar to that from treatment with 2-mercaptoethanol.	181
P^nBu_3	Insulin, human serum albumin, bovine ribonuclease [Lys]-vasopressin	Fully reduced with 5–20% molar excess of PR_3. Reduction rapid (< 40 min), 25 °C, slightly alkaline pH. Only cystine residues modified.	182
P^nBu_3	Papain	Activates SH-dependent enzyme.	183
P^nBu_3	glutathione (GSSG)		184
M(PR₃)			
$[Au(PEt_3)_2]Cl$	human serum albumin (human plasma)	$Au(PEt_3)_2^+ + RSSR + H_2O \rightarrow 2\,RSH + OPEt_3 + AuPEt_3^+$ $Au(PEt_3)^+ + RSH \rightarrow RSAuPEt_3$ Whole blood samples were solidified in a few hours by denaturation of albumin.	185
$[Au(eppe)_2]Cl$	bovine serum albumin	Slowly cleaved disulphide bonds of albumin with release of the phosphine-oxide.	87

$$RS^{\cdot} + PR_3 \rightarrow RSPR_3^{\cdot}$$
$$RSPR_3^{\cdot} \rightarrow R^{\cdot} + S{=}PR_3$$

The thiyl radical is regenerated via the propogation steps:

$$R^{\cdot} + RSSR \rightarrow RSR + RS^{\cdot}$$

or

$$R^{\cdot} + RSH \rightarrow RH + RS^{\cdot}.$$

The order of reactivity of PR_3 compounds towards alkylthiyl radicals decreases in the order[187] $P^nBu_3 > P(OEt)_3 > PPh_3 > P(OPh)_3$. This type of radical reactivity is discussed further in the next section.

4.3 Radical Reactions of Phosphines

Radical reactions may play an important role in the biological chemistry of phosphines. As discussed in Sects. 4.1 and 4.2, under certain conditions phosphines can react with dialkyl peroxides, disulphides and thiols by radical pathways rather than ionic mechanism. The autoxidation of phosphines also appears to involve a radical mechanism. For all of these examples the intermediate species is a phosphoranyl radical $R_4P^{\cdot}$ which contains

9 electrons in the P valence shell. However there are three additional types of phosphorus radical species that may be relevant to the biological chemistry of phosphines. These are phosphonium radical anions ($R_3P^{\cdot-}$), which also have 9 valence electrons, and the two 7-electron species, phosphinium radical cations ($R_3P^{\cdot+}$) and phosphino radicals ($R_2P^{\cdot}$).

The chemistry of phosphorus radicals has been studied extensively and the subject has been reviewed[1, 188)]. However, there appear to have been no investigations of the radical chemistry of phosphines under conditions relevant to biology. In Table 14 we give examples of the types of radical reactions that phosphines could potentially undergo in a biological system.

The ability of phosphines to act as either one-electron acceptors or one-electron donors may be crucial in the mechanism of their cytotoxicity and allow them to interfere with electron transport processes. Phosphinium radical cations have been generated by oxidation of phosphines at a mercury anode[189)] or by γ-irradiation of tertiary phosphines on silica[190)] and in sulphuric acid[191)]. The oxidation potentials appear to be within the range accessible to biological systems[189, 192)]. Aryl phosphines are more easily oxidised than alkyl phosphines: $E_{1/2}$ (polarography in CH_3CN, anodic oxidation) PPh_3 + 50 mV, PEt_3 − 415 mV[192)]. Phosphine dimer cation radicals $(R_3PPR_3)^{\cdot+}$ have been identified by ESR spectroscopy as products generated during electrochemical oxidation of PR_3 compounds in solution; the initially-formed phosphinium radical cation reacts rapidly with a further molecule of phosphine[193)]. Phosphonium radical anions are the least well characterised of all phosphorus radicals. Electrolytic reduction of PPh_3 in CH_3CN, at a dropping mercury cathode, produced biphenyl and diphenylphosphonic acid. The radical anion was presumed to be an intermediate[194)]:

$$Ph_3P \xrightarrow[CH_3CN]{+e} [PPh_3]^{\cdot-} \xrightarrow{H_2O} Ph\text{–}Ph + Ph_2P(O)H.$$

Trialkylphosphines were not reduced polarographically[195)] but $[PPhMe_2]^{\cdot-}$ was produced by electrolytic reduction of $PPhMe_2$ and characterised by ESR spectroscopy[196)]. The unpaired electron appeared to be largely delocalised over the phenyl ring. This may be another important point of consideration when comparing the cytotoxicity of aryl and alkyl phosphines. The ability of aryl phosphines to delocalise the unpaired electron is likely to have a pronounced effect on the stability and reactivity of the radical species. It is unlikely that there are any strong enough reductants available in biological systems. The polarographic half-wave potentials of various mono-, di- and triphenylphosphines (DMF calomel reference electrode) range from −2.2 to −2.7 V[195)].

Although phosphino radicals ($R_2P^{\cdot}$) have been generated by photolysis of tertiary phosphines such as PPh_3[197)], in a biological system reactions of PR_3 compounds via phosphino radicals are likely to be more important for PH_3 and primary and secondary phosphines. For these, phosphino radicals can be generated by radical-initiated hydrogen atom abstraction:

$$R_2PH \xrightarrow{R^{\cdot}} R_2P^{\cdot} + RH.$$

Once formed, phosphino radicals either abstract hydrogen, or add rapidly to an unsaturated double bond (Table 14). The free radical addition of PH_3 to alkenes can be used as a method for preparing tertiary phosphines[198)].

Table 14. Some radical species derived from phosphines

Radical	Generation	Reactivity
Phosphino $R_2P^{\cdot}$	$PHR_2 \xrightarrow{R^{\cdot}} {}^{\cdot}PR_2 + RH$ Particularly for PH_3, PRH_2 and PR_2H	Could react by $H^{\cdot}$ abstraction or by rapid addition to double bond: $R_2P^{\cdot} + CH_2 = CHR' \rightarrow R_2PCH_2 - {}^{\cdot}CHR'$ $R_2PCH_2 - {}^{\cdot}CHR' + R_2PH \rightarrow R_2PCH_2CH_2R' + R_2P^{\cdot}$ etc.
Phosphinum radical cation $R_3P^{\cdot +}$	PR_3 + suitable 1-electron acceptors	Could interfere with electron transport pathways; note the reaction[205]: 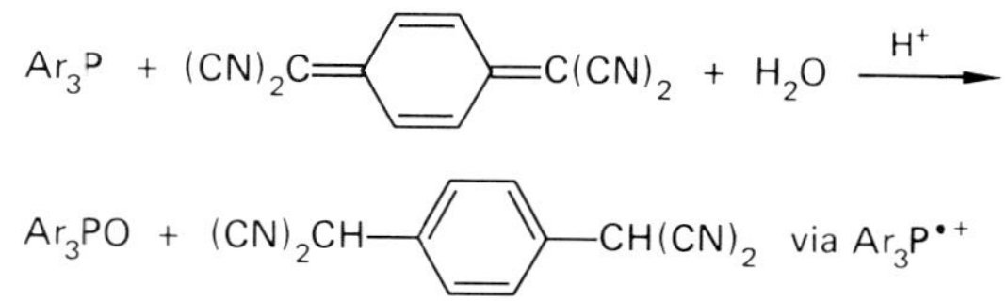
Phosphoranyl $R_4P^{\cdot}$	$R_3P + {}^{\cdot}OR' \rightarrow [R_3\dot{P}OR']$ $({}^{\cdot}OOR') \rightarrow [R_3\dot{P}OOR']$ Intermediate in the reaction of PR_3 with O_2, peroxides, or other activated O species.	$[R_3\dot{P}OR'] \xrightarrow{\alpha\text{-scission}} R_2POR + R^{\cdot} \rightarrow$ products β-scission $\rightarrow R_3PO + R'^{\cdot}$ The major product may be phosphine-oxide or phosphinite species depending on R (see text).
	$R_3P + {}^{\cdot}SR' \rightarrow [R_3\dot{P}SR']$ thiyl radicals, e.g. from GSSG	$[R_3\dot{P}SR'] \rightarrow R_3P = S + R^{\cdot}$ β-scission predominates? (see text)
Phosphonium radical anion $R_3P^{\cdot -}$	R_3P + 1-electron donor Probably no strong enough reductants available in biological systems (see text).	

Phosphoranyl radicals ($R_4P^{\cdot}$) are the most widely studied type of phosphorus radical[199, 200]. These may be important intermediates of the reactions of phosphines with radical species that are potentially available in biology, e.g. thiyl ($RS^{\cdot}$), alkoxy ($RO^{\cdot}$) or alkylperoxy ($RO_2^{\cdot}$) radicals. The reaction of alkoxy[201] or alkylthiyl[202] radicals with PR_3 compounds take place at close to diffusion controlled rates (ca. $10^8 - 10^9 M^{-1}s^{-1}$), although $^tBuOO^{\cdot}$ reacts very much slower than $BuO^{\cdot}$ with R_3P[199]. An important feature of the chemistry of phosphoranyl radicals $X_3P^{\cdot}$–A–B is that they can undergo two types of fragmentation involving cleavage of the P–A bond (α-scission) or A–B bond (β-scission). For reactions of phosphines with oxy or thiyl radicals the two pathways lead to either substitution or oxidation products:

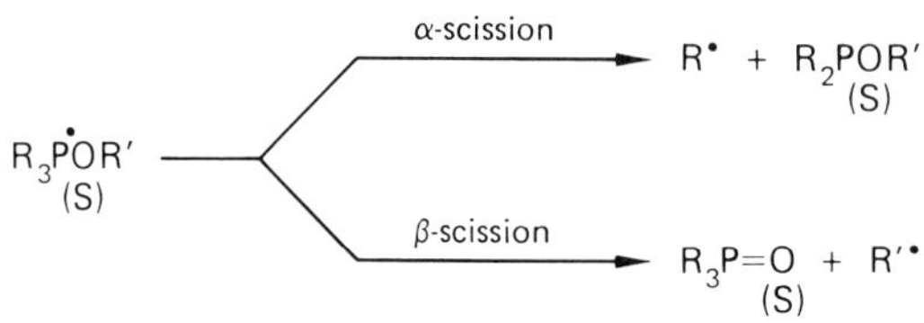

As discussed in Sect. 4.1 the reaction of P^nBu_3 with $^tBuOOBu^t$ at 130 °C gives an 80% yield of the substitution product $^nBu_2PO^tBu$ although β-scission would be favoured by ca. 100 kJ mol^{-1} on thermodynamic grounds[199]. There appear to be several factors that govern the competition between α and β-scission. For instance, it has been suggested that the phosphoranyl radical $^tBuOP^{\cdot}(X)(OEt)_2$ would undergo α-scission at 335 K if $\bar{D}$(P–X) in $\dot{P}X_3$ is < 314 kJ mol^{-1}, but mainly β-scission with ^tBu–O cleavage if the P–X bond is stronger[202].

The majority of phosphoranyl radicals appear to possess trigonal-bipyramidal structures. It has been suggested that there is a site selectivity so that α-scission takes place preferentially from apical positions whereas β-scission occurs most readily from an equatorial position[199]. The presence of aromatic substituents on P has the effect of increasing the rate of β-scission relative to α-scission. For example, $[Ph(^nPr)MePO^tBu]^{\cdot}$ undergoes exclusively β-scission[203], whereas for $^nPr_3P^{\cdot}O^tBu$, α-scission is favoured. This may be a consequence of a delocalisition of the unpaired electron onto the benzene rings[199, 201]. The dominance of the β-scission pathway for arylphosphines also accounts for the observation that Ph_3PO is the only product of the autoxidation of Ph_3P whereas a mixture of phosphine-oxide and phosphinite products are found for alkyl phosphines (see Sect. 4.1). These differences may be reflected in a different type of biological reactivity of aryl- and alkyl-phosphines. Thiophosphoranyl radicals have not been studied in such detail but the available data suggest that β-scission predominates so that $R_3P{=}S$ is the major (or exclusive) product.

Little is known about the reactions (as opposed to the fragmentation) of phosphoranyl radicals. However, it has been shown that phosphoranyl radicals react very rapidly with molecular oxygen to form phosphoranylperoxyl radicals[199, 204]

$$R_4P^{\cdot} + O_2 \rightarrow R_4POO^{\cdot}.$$

4.4 Formation of Phosphonium Salts

One of the most useful reactions of phosphines in organic synthesis is the Wittig reaction[2, 13, 14]. The first step involves formation of a phosphonium salt by nucleophilic attack of P on an alkyl halide:

$$R_3P + R'X \rightarrow [R_3PR']^+X^-.$$

Quaternary phosphonium salts are generally stable crystalline solids which have high solubility in polar solvents. They are relatively stable towards dealkylation but hydrolyze in the presence of hydroxide ion to the phosphine oxides[2]:

$$[R_3PR']^+X^- + OH^- \rightarrow R_3PO + R'H + X^-.$$

If the phosphonium salt contains an available α-proton then this can be removed in the presence of a suitable base to give a phosphorus ylid. The reactivity of ylids towards carbonyl compounds is the basis of the Wittig olefin snythesis:

$$[R_3PCHR^1R^2]^+X^- \xrightarrow{B^-} R_3P{=}CR^1R^2 + BH + X^-$$

$$R_3P{=}CR^1R^2 + O{=}CR^3R^4 \longrightarrow R_3PO + R^1R^2C{=}CR^3R^4.$$

The phosphine is generally Ph_3P and the strength of the required base depends on the pK_a of the phosphonium salt and the nature of the groups R^2 and R^3. If these groups are strongly electron withdrawing the ylid may be stable in the presence of H_2O, but in general they are rapidly hydrolysed to phosphine oxides[2]. Therefore, in a biological system the conditions are unlikely to be suitable for phosphorus ylid chemistry, but it is possible that phosphonium salts could be formed from phosphines.

With the exception of C–I bonds in thyroxines, no compounds analogous to alkyl halides are found in animals. However, as discussed earlier, methylation pathways are known in mammalian arsenic biochemistry, and there may be potential sources of CH_3^+ which could react with phosphines.

Once formed phosphonium salts could have a significant effect on cellular processes, particularly on mitochondrial function. Ph_3PMe^+ has been used as a probe to measure membrane potentials, for instance in thyroid cells[206], lymphocytes[207, 208], *Escherichia coli*[209, 210] and human granulocytes[211]. The lipid-soluble cation distributes itself across the membrane in accordance with the membrane potential[212]. Monovalent lipophilic cations which have significant membrane permeability are concentrated in mitochondria. For instance the laser dye Rhodamine 123 has been used as a specific probe for the localisation of mitochondria in living cells[213]. It has also been shown to have *in vivo* antitumour activity[214].

It is possible that the tetrahedral, bis-chelated Cu(I), Ag(I) or Au(I) diphosphine complexes (especially those with phenyl substituents) are recognised by cells in a similar manner to the phenyl phosphonium salts discussed above: as lipophilic cations which distribute according to membrane potentials. Indeed, Mirabelli and coworkers (unpublished results) have found preliminary evidence for extensive specific damage to mitochondria by $[Au(dppe)_2]^+$ in hepatocytes. It is also notable that positively-charged

Tc(III) diphosphine complexes are taken up by tissues with high mitochondrial activity, e.g. heart, which enable radio-imaging studies (see Sect. 5).

It is interesting to note that on the National Cancer Institute's files there are 70 compounds which contain the substructure P^+–C and also have shown activity against the primary screen P388 leukaemia. It might be expected that if mitochondrial effects were important then the degree of cytotoxicity would be related to the degree of lipophilicity of the phosphonium salts. Thus aryl-substituents may produce a greater cytotoxic effect than alkyl-substituents.

4.5 Binding of Phosphines to Haems

There have been several reports of the interaction of phosphine ligands with iron porphyrin complexes and haem proteins. Many of these studies were stimulated by the need to interpret the characteristic "hyperporphyrin" (split-Soret) UV-visible absorption spectra of COFe(II)cytochrome-P450: a highly red-shifted Soret band at ca. 450 nm and a second Soret band at ca. 360 nm. The rationale for these studies was that phosphines and CO are both good π-acceptor ligands. Indeed it is found that phosphines (along with NO, nitrosoalkanes and isocyanides[215]) do generate similar "hyper" spectra due to the formation of a R_3PFe(II)Scys system. The origin of the red-shifted band is attributed to an orbital mixing mechanism between the Soret π-π* porphyrin transitions and charge-transfer transitions from an axial π system. The electronic properties of the ligand *trans* to cysteine are considered to be of critical importance in establishing the correct conditions for orbital mixing to occur[215].

Table 15 summarises some of the data on the binding of phosphines to haems. It can be seen that phosphines bind to both ferric and ferrous porphyrins, the former slightly more strongly than the latter. In both cases the products are low-spin, showing the strong-field nature of the phosphine ligand.

It is noticeable that the affinity of cyt P450 for phosphines in liver microsomes (dissociation constants in the μM range) may be relevant to the pharmacology of phosphines. There may be a selectivity of phosphines for reactive haem sites which will depend on factors including the bulkiness and lipophilicity of the phosphine, and the architecture of the haem sites. Apparently even bulky phosphines such as $PPh_2(OEt)$ can bind to cytochrome P-450[219], whereas binding constants for chloroperoxidase adducts of phosphines are much lower; PMe_2Ph and PEt_2Ph have very low affinities for chloroperoxidase but the less bulky and more hydrophilic ligand $(HOCH_2)_2PMe$ forms a 1:1 adduct with both ferrous and ferric forms of the enzyme[217].

There is little information on the reactivity of haem-phosphine complexes. There is a hint that some complexes of ferrous-P450 are reactive towards O_2[218], but little suggestion that ferric-haems are reduced by tertiary phosphines. In contrast, PH_3 does appear to reduce both Fe(III) cytochrome c and Fe(III) cytochrome oxidase (see Sect. 3.8). As discussed in Sect. 2.1 there is a clear difference between the oxidative pathways of PH_3 and PR_3. It seems likely that PH_3 can bind directly to Fe in haem sites (Fe(II) complexes of PH_3 are known[17]), and this should be investigated.

Et_3PAuCl and Et_3PAuNO_3 can induce reversible low-spin to high-spin state changes of Fe(III) cytochrome c and cytochrome b_5[220]. They also induce autoxidation of oxy-Fe(II) myoglobin and haemoglobin again giving high-spin Fe(III) products. The effects

Table 15. Some literature data on tertiary phosphines and phosphites binding to haems

Haem[a]	PR_3	Comment	Ref.
Fe(II)TPP	PEt_3, P^nBu_3, $P(OMe)_3$, $P(OEt)_3$, $P(O^nBu)_3$	Low-spin $Fe(II)TPP(PR_3)_2$ λ_{max} 335–360 nm, 445–457 nm (hyperporphyrin spectra).	216
Fe(II)CP Fe(III)CP	$PMe(CH_2OH)_2$	Low-spin $Fe(II)CP(PR_3)$ K_{diss} 7.7 mM, λ_{max} 458 nm low-spin $Fe(III)CP(PR_3)$ K_{diss} 2mM, λ_{max} 376, 450 nm $g_1 = 2.59$, $g_2 = 2.29$, $g_3 = 1.82$. Hyperporphyrin spectra	217
Fe(III)Mb Fe(III)HRP	$PMe(CH_2OH)_2$	λ_{max} 427 nm (his)N–Fe(III)–PR_3 (single Soret-band)	217
CAM-cyt P450	$PPhMe_2$	Low spin $Fe(II)PR_3$ λ_{max} 345, 460 nm (hyperporphyrin)	215
		$g_1 = 2.51$, $g_2 = 2.28$, $g_3 = 1.86$.	217
Liver Microsomal cyt P450			
Fe(II) P450 Fe(III) P450	$PPhEt_2$	$Fe(II)PR_3$ λ_{max} 459 nm, K_{diss} 10 μM $Fe(III)PR_3$ λ_{max} 375, 455 nm, K_{diss} 50 μM	218
Fe(II) P450	$P(OR)_3(R = C_4H_7$, C_2H_4Cl, Me, Et, Ph) $PPh_2(OEt)$, $PPh(OMe)_2$	$Fe(II)PR_3$ All give hyperporphyrin spectra	219
	$P(OCH_2C_6H_5)(OEt)_2$	$Fe(II)PR_3$, K_{diss} 4.2 and 70 μM.	

[a] TPP = tetraphenylporphine; CP = chloroperoxidase; HRP = horseradish peroxidase – the 5th ligand is probably cysteine thiolate; Mb = myoglobin; CAM = camphor

may be related to the ability of Et_3PAu^+ to penetrate hydrophobic haem pockets and bind to active site histidine residues. However there was no evidence of direct binding of the phosphine to iron[220)].

More experiments need to be carried out on the effects of PR_3 ligands on haem sites which are turning over O_2. Cytochrome P-450 activates molecular oxygen for insertion into substrate molecules. Clearly a phosphine can act as an O-atom acceptor (see Sect. 4.1). It is interesting that $(Ph_3P)_2Fe(II)P$ (where P is a porphyrin dianion) reacts with dioxygen to form a peroxy-bridged dimer PFeOOFeP. This does not bind Ph_3P but catalytically oxidises it to Ph_3PO[171)]. This system has been proposed as a model for peroxidase (Compound II) (see Table 12).

4.6 Proton Affinities of Phosphines

Differences in the availability of the P lone pair for a series of phosphines are likely to have profound effects on their biological chemistry. For instance, the question arises as to whether a particular phosphine is likely to become protonated *in vivo*. The pK_a's of hydrophobic phosphines can not be measured in H_2O. Consequently most of the experimental data have been obtained by titration in non-aqueous media (usually nitromethane). Conversion to aqueous pK_a values is usually carried out using a relationship between half neutralisation potentials in CH_3NO_2 and $pK_a(H_2O)$ that exists for amines[221]. For monodentate tertiary phosphines, introduction of phenyl-substituents lowers the pK_a (Table 2). There appear to have been no similar determinations of the pK_a's of diphosphines. It seems likely that phenyldiphosphines (e.g. dppe) are weakly basic ($pK_a < 3$) and therefore will not be significantly protonated at pH 7. Preliminary experimental data measured by R. Norman in our laboratory suggest that this is the case[158].

However, solvation effects can have a major influence on the pK_a. In the gas phase PPh_3 is apparently a stronger base than Me_3P[30, 32], the opposite of that found from pK_a measurements in solution (see Table 2). This has been attributed to the poor solvation of the bulky PPh_3H^+. There is a good correlation between the gas-phase proton-affinities and the lone pair ionisation energies which can be measured by photoelectron spectroscopy. The lone pair ionisation energies follow the sequence $PH_3 > PMe_3 > PMe_2Ph > PMePh_2 > PPh_3$ (see Table 2).

In a biological system solvation effects will clearly be important. It is difficult to predict whether pK_a's or gas-phase proton-affinities will be more reliable guides to the degree of protonation of phosphines *in vivo*. There are a variety of compartments of greatly differing polarities that are potentially available. Lipophilic phosphines would be expected to become localised in non-aqueous environments, e.g. membranes. It is possible that a phenyldiphosphine could act as a H^+ shuttle across a membrane. The only way to probe this may be to investigate whether proton translocation occurs in a model system.

5 Technetium Phosphine Complexes: Myocardial Imaging

Tertiary phosphines can stabilise Tc in low oxidation states. The isotope ^{99m}Tc has very favourable nuclear properties for diagnostic nuclear medicine ($T_{1/2}$ 6 h, 140 keV γ-emission)[222]. The positively charged Tc(III) complex *trans*-$[Tc(dmpe)_2Cl_2]^+$ (where dmpe is $Me_2P(CH_2)_2PMe_2$) shows significant uptake into the heart in several animal species including man so that it can be used for heart imaging[223]. There is also significant uptake into the lung, liver, spleen and kidney. In rats, 4.1% of the injected dose is found in the heart 5 min after injection (1.3% at 90 min)[224]. Unfortunately, the high accumulation in the liver obscures the myocardial apex and prevents acquisition of clinically useful images[225].

The positive charge on the complex is probably critical for myocardial cell uptake. Indeed, this was the rationale of Deutsch and coworkers[223] who noted that cationic species such as ammonium salts and alkali metal ions accumulate in normal heart muscle and have themselves been used in diagnostic nuclear medicine for gamma camera imaging of the heart. The lipophilicity of the complex is probably also important – whether it shows selective uptake into mitochondria, distributing according to membrane potentials, does not appear to have been studied (see Sect. 4.4). The more lipophilic complexes are excreted primarily via the hepatobiliary system whereas the more hydrophilic complexes are excreted through the kidneys.

Trans-$[Tc(dmpe)_2Cl_2]^+$ is highly water soluble and can be administered by injection in saline. However, the ligand itself is highly unstable in air. How kinetically and thermodynamically stable is the Tc complex? It is thought to be relatively stable towards substitution *in vivo* since the complex can apparently be recovered intact from tissue (heart, liver and bile) soon after injection[224].

The complex itself is an intermediate obtained during the reduction of Tc(IV) or Tc(VII) with excess diphosphine:

$$Tc(IV)(Hal)_6^{2-} \quad \text{or} \quad TcO_4^- + \text{excess dmpe} \rightarrow [Tc(III)(dmpe)_2X_2]^+.$$

For medical use the pertechnetate route (using ^{99m}Tc) is preferred. Chemistry can be done with the β-emitter ^{99}Tc. The exact conditions of the reaction are critical in determining its course. Besides *trans*-$[Tc(III)(dmpe)_2Cl_2]^+$ (which has been crystallised and shown to have *trans* octahedral geometry with Cl^- ligands occupying the axial coordination positions[223]), *trans*-$[Tc(V)(dmpe)_2(O)_2]^+$ and $[Tc(I)(dmpe)_3]^+$ can be formed, the latter being the thermodynamically stable product in the presence of excess dmpe[226]. This complex is cleared so slowly from the blood that myocardial images can be obtained only 6–10 h after injection[227].

Reduction of the cationic Tc(III) complex to a neutral Tc(II) complex may play a role in determining its biodistribution[222, 224]. The Tc(III)/Tc(II) redox potentials for bis ditertiary phosphine complexes are within the range of those within biological systems. The $E^{0\prime}$ (cyclic voltammetry) value for *trans*-$[Tc(III)(dmpe)_2Cl_2]^+$ is – 208 mV. The substitution of alkyl groups by phenyl groups tends to make Tc(III) easier to reduce to Tc(II): *trans*-$[Tc(III)(dppe)_2Cl_2]^+$ has an $E^{0\prime}$ value of – 40 mV[222].

The Re complex *trans*-$[Re(dmpe)_2Cl_2]^+$ is more difficult to reduce (by 190 mV) than the Tc analogue. This may account for the lower liver and higher heart uptake of the ^{186}Re complex compared to the ^{99m}Tc complex despite their similar structures[224]. It would be interesting to compare the kinetics of their ligand substitution reactions.

Is the depe complex *trans*-$[^{99m}Tc(depe)_2Cl_2]^+$ metabolised by a different pathway to the dmpe analogue? It does not accumulate in the heart[228]. In similarity to $[^{99m}Tc(dmpe)_3]^+$ the myocardial uptake of the complex $[^{99m}Tc(pom - pom)_3]^+$, where pom is bis(dimethoxyphosphino)ethane, is obscured for several hours by the high blood background, slow lung clearance and high liver uptake[229]. It would be interesting to investigate the rates of ring-opening, ligand substitution and diphosphine oxidation of all these complexes under biologically relevant conditions. An improved rational design of these heart imaging agents may then emerge.

It seems unlikely that the physiological consequences resulting from the administration of Tc phosphine complexes will be significant. The amounts used are minute (nmol).

6 Conclusions

The main focus of attention in this article has been the cytotoxicity and antitumour activity of phosphines and metal phosphine complexes. Activity is likely to stem from the strong reducing properties of phosphines. In natural biological systems phosphorus is present only as P(V): phosphate chemistry. Reduced phosphorus is rarely (if ever) detected.

PH_3 itself is cytotoxic and used as a fumigant against pests in stored products. It inhibits mitochondrial respiration but the details of the process are unclear. The oxidation pathways for PH_3 differ significantly from those of tertiary phosphines PR_3 and for these the available pathways depend on the substituents R. Pathways involving radicals usually lead to β-scission when R = aryl but α-scission when R = alkyl. The autoxidation of alkyl phosphines is usually rapid compared to aryl phosphines. Similar differences exist in the reduction of disulphide bonds; the alkyl derivatives are more reactive.

Much work remains to be done on the pathways of phosphine oxidation in biological systems. It will not be easy. The choice of solvent is a problem. Biological systems contain both aqueous (e.g. plasma, cytoplasm) and non-aqueous (e.g. membranes, lipoproteins) compartments. These differences could change the course of phosphine reactions. For example sulphur abstraction from a disulphide would lead to $R_3P = S$ in a non-aqueous medium but R_3PO in an aqueous one. There are many systems in biology which handle activated dioxygen, oxygen atoms and related species. Many of these are potential sites of phosphine attack. Few have been studied as yet.

The ease of oxidation of phosphines means that they are difficult to test reliably for biological activity, but oxidation also provides cells with an inbuilt resistance mechanism. Shoulders are often seen on cell survival curves. The donation of an oxygen atom to a phosphine can be a controlled, non-destructive 2-electron process. In cells one-electron acceptors are also available e.g. Cu(II), Fe(III), peroxy radicals. The ensuing radical chemistry could be very destructive especially it occurs in membranes. Membrane bound enzymes handling electrons (e.g. cytochrome oxidase, 4 one-electron processes are involved in the reduction of dioxygen to water) may be particular targets.

Metal ions can protect phosphines against oxidation until they reach intracellular target sites. Evidently, a fine balance between kinetic and thermodynamic stability of the M–P bonds needs to be achieved; Pt(II)dppe complexes appear to be inactive, Cu(I), Ag(I) and Au(I) dppe complexes are active. Complexes such as $[Au(dppe)_2]Cl$ possess enough kinetic lability in the Au–P bonds to react via a ring-opening mechanism.

The key step will now be to establish whether complexes of this type, which exhibit a reasonable spectrum of antitumour activity in animal models, attack DNA, and if so, by what mechanism. Perhaps copper could play an important role in this. Cu(II) potentiates the cytotoxicity of dppe and $[Au(dppe)_2]Cl$ is reactive towards Cu(II) ions. There is a potential wealth of Cu(I)phosphine chemistry (particularly aqueous) involving chelate ring opening, halide (Cl^-) competition and free radical reactions which could be explored. The mechanism of cytotoxicity of metal diphosphine complexes seems likely to be different from that of cisplatin. This could be a promising sign for possible combination chemotherapy.

It is not known whether phosphines (like arsines) can be methylated *in vivo*. Lipophilic phenyl phosphines might then be good candidates for disrupting membrane potentials, e.g. in mitochondria. Whether they can translocate protons across membranes may have to be the subject of experiment rather than theory. The predictions are that alkyldiphosphines are likely to be protonated at pH 7 whereas phenyldiphosphines will not. However, the measurement of pK_a's in non-aqueous solutions is difficult and the solvent will play a crucial role in determining proton affinities. The distribution of positively-charged lipophilic metal phosphine complexes may also be responsive to membrane potentials. The consequences of this remain to be seen.

The recent introduction of the first metal phosphine complex into clinical use, the triethylphosphine Au(I) complex auranofin ("Ridaura"), as an antiarthritic drug has provided a stimulus for further exploration of the biological chemistry of phosphines and their metal complexes. The diverse chemistry of phosphines provides many potentially attractive features for drug-design. In the case of auranofin the role of the phosphine is both to stabilise a low oxidation state (gold(I)) and introduce lipophilicity (oral uptake). It might be possible to use phosphines to deliver other metals into cells, for example for their antimicrobial properties (Ag(I) diphosphine complexes) or tracer properties (^{99m}Tc as a γ-camera imaging agent). The carrier (phosphine) can be oxidised to a relatively non-toxic product (phosphine-oxide), so trapping the metal ion inside the cell. However, as discussed above, some oxidative processes could be destructive.

There may be further scope for exploring metal-ligand synergy in the design of metal-phosphine complexes as chemotherapeutic agents. The metal, its oxidation state and the phosphine substituents could all be altered to control the reactivity of a metal-phosphine complex in a biological system. In the case of homogeneous hydrogenation catalysts and models for nitrogen fixation (Mo and W phosphine complexes) not only does the phosphine stabilise multiple oxidation states of the metal (thus controlling the reactivity of the metal) but also dictates the course of the reaction (dissociation of bulky ligands, monodentate versus chelating ligands).

There is a wealth of knowledge available on the reactivity of metal-phosphine complexes but little has been obtained under conditions relevant to biological activity (i.e. in the presence of H_2O or O_2). Progress in the design of metal-phosphine complexes as drugs will depend also upon elucidating the coordination chemistry of these complexes in intact cells and bio-fluids. This is a difficult task. Practical approaches are needed to this problem of *in vivo* speciation. Some progress can be made with multinuclear NMR but others need to be developed.

7 New Data Added at Proof Stage

7.1 Metal Bis(Diphosphine) Complexes

Recently, Timmer et al.[230)] have synthesized a series of complexes of the type $[M(dppe)_2]^{n+}nX^-$ (M = Fe(II), Fe(III), Co(II), Rh(I), Rh(III), Ir(I), Ir(III), Ni(II), Pd(II), X = Cl, Br, NO_3, ClO_4, CF_3SO_3) and tested them for antitumour activity (Table 16). All of the complexes tested exhibited marginal to good activity against i.p. P388 leukaemia in mice, but were generally less potent (dose >50 μmol/kg) and less active than dppe alone. Only $[Rh(dppe)_2]X$ (X = CF_3SO_3 and ClO_4) and $Pd(dppe)_2I_2$ were significantly more potent than dppe, and the latter showed only marginal activity (ILS 39%). The Pd(II) complexes $[Pd(dppe)_2](NO_3)_2$ and $[Pd(dppe)_2Br]Br$ were potently

Table 16. Cytotoxicity and antitumour data for bis(diphosphine) and related complexes reported by Timmer et al[230)]

Compound	*IC_{50}[a] (μM)*		*ILS%[b] (dose, μmol/kg)*		
	B16	HCT116	P388	B16	L1210
Pd(II)					
$[Pd(dppe)_2]Cl_2$	59	–[c]	44(62)	77(12)	0(2)
$[Pd(dppe)_2](NO_3)_2$	4	19	–	–	0(0.8)
$[Pd(dppey)_2]Cl_2$	29	14	–	–	–
$[Pd(dppe)_2Br]Br$	0.7	6	–	–	–
$[Pd(dppe)_2I_2]$	–	–	39(4)	–	–
$[Pd(dppe)_2](ClO_4)_2$	–	–	78(181)	104(54)	–
Ni(II)					
$[Ni(dppe)_2](NO_3)_2$	4	1	39(51)	83(30)	0(82)
$[Ni(dppe)_2](ClO_4)_2$	–	–	140(24)	–	–
Ir(I)					
$[Ir(dppe)_2]ClO_4$	–	–	67(184)	50(18)	–
Ir(III)					
$[Ir(dppe)_2O_2]ClO_4$	–	–	70(178)	–	–
Rh(I)					
$[Rh(dppe)_2]CF_3SO_3$	–	–	53(4)	38(3)	–
$[Rh(dppe)_2]ClO_4$	–	–	50(12)	31(12)	33(160)
Rh(III)					
$[Rh(dppe)_2Cl_2]ClO_4$	–	–	67(93)	71(28)	14(75)
$[Rh(appe)_2Cl_2]ClO_4$	–	–	36(173)	–	–
$[Rh(dppe)_2Cl_2]Cl$	–	–	33(50)	–	–
Co(II)					
$[Co(dppe)_2NO_3]NO_3$	421	>510	61(102)	71(20)	0(102)
$[Co(dppe)_2Br]Br$	–	–	67(98)	–	0(197)
$[Co(appe)_2](ClO_4)_2$	–	–	30(175)	–	–
$[Co(dppe)_2](ClO_4)_2$	–	–	44(95)	50(19)	42(114)
Fe(II)					
$[Fe(dppey)_2Cl_2]$	492	416	50(217)	–	–
Fe(III)					
$[Fe(dppe)_2Cl_2]FeCl_4$	–	–	60(89)	–	–

[a] Concentration that inhibits growth of B16–F10 murine melanoma and HCT-116 human colon carcinoma cells by 50%; [b] Maximum percentage increase in lifespan with respect to non-treated control animals at the indicated dose. Compounds were administered i.p. on day 1 only for P388 and L1210, and days 1–9 for B16 melanoma; [c] Not reported. Key: dppe, $Ph_2P(CH_2)_2PPh_2$; dppey $Ph_2PCH = CHPPh_2$; appe, $Ph_2P(CH_2)_2AsPh_2$

cytotoxic in *vitro* (IC_{50} 4 and 0.7 μM against B16–F10 murine melanoma, 19 and 6 μM against HCT-116 human colon carcinoma) but antitumour activity *in vivo* was not reported. $[Ni(dppe)_2](NO_3)_2$ was potently cytotoxic in *vitro* (IC_{50}: B16, 4 μM, HCT-116, 1 μM) and active against P338 leukaemia (ILS 39% at 51 μmol/kg) and B16 melanoma (ILS 83% at 30 μmol/kg). Several other complexes (notably $[Pd(dppe)_2]X_2$,X = Cl and ClO_4) exhibited also high activity against i.p. B16 melanoma, but only 2 of the 7 complexes tested were active against i.p. L1210 leukaemia in mice, although dppe itself is active in this tumour model. The high doses of many of the complexes required for the demonstration of antitumour activity suggest that they are readily decomposed *in vivo* or in the testing media. Our own work on $[Ni(dppe)_2]^{2+}$ (P. S. Jarrett and P. J. Sadler, unpublished) shows that this may be the case. Again, the need to characterise complexes fully under the conditions used for antitumour testing is emphasized.

pK_a's of Diphosphines

These measurements have now been completed[158]. In summary, the difference between the first and second pK_a's decreases as the length of the carbon chain between the phosphorus centres increases. Unsaturation in the chain lowers pK_a's, and $-PEt_2$ centres are more basic than $-PPh_2$ centres:

diphosphine,	$pK_a(1)$,	$pK_a(2)$
dppe	0.99	3.86
eppe	1.48	8.04
depe	5.11	8.41

Acknowledgements. We are very grateful to Drs. C. K. Mirabelli, R. K. Johnson, S. T. Crooke, D. T. Hill, B. M. Sutton and colleagues at SK & F Laboratories (Philadelphia) for many stimulating discussions throughout the course of this work. We also thank Drs. M. Nasr and V. Narayanan (National Cancer Institute, Bethesda) for supplying anticancer testing data, Dr. N. R. Price (MAFF, Slough UK) for discussions on PH_3, and Professor E. A. Deutsch for supplying data on Tc complexes, and Heather Robbins for excellent technical assistance. We thank SK & F Laboratories, the SERC and the MRC for their support for this work. We are also grateful to Professor R. J. P. Williams and Dr. D. M. P. Mingos (Oxford University), and Professor J. M. Pratt (Surrey University) for their critical comments on this paper.

8 References

1. a. Organophosphorus Chemistry (eds. Hutchinson, D. W., Walker, B. J.), Royal Society of Chemistry Specialist Periodical Reports, vol. 15 (1985), vol. 16 (1986)
   b. Puddephatt, R. J.: Chem. Soc. Rev. *12,* 99 (1983)
   c. Hall, K. P., Mingos, D. M. P.: Prog. Inorg. Chem. *32,* 237 (1984)
   d. Garrou, P. E.: Chem. Rev. *85,* 171 (1985).
   e. Homogeneous Catalysis with Metal Phosphines (ed. Pignolet, L. H.), Plenum Press, New York 1983
2. Emsley, J., Hall, D.: The Chemistry of Phosphorus, Harper & Row, London, 1976, and references sited therein
3. Latimer, W. M.: The Oxidation States of The Elements and Their Potentials in Aqueous Solutions, 2nd edn., Prentice Hall, N.J. 1952
4. Thomson, A. J. in: Metalloproteins (Part 1) (ed. Harrison, P.), Macmillan, London 1985, p. 79
5. Mellor, J. W.: A Comprehensive Treatise on Inorganic and Theoretical Chemistry, vol. 8, Longmans, London 1940, pp. 802–822
6. Partington, J. R.: A Textbook of Inorganic Chemistry, 6th edn. Macmillan, London 1953, pp. 571–573
7. Tsubota, G.: Soil pl. Fd. Tokyo *5,* 10 (1959)
8. a. Skinner, F. A. in: The Ecology of Soil Bacteria (eds. Gray, T. R. G., Parkinson, D.), Liverpool University Press 1968, pp. 573–592
   b. Burford, J. R., Bremner, J. M.: Soil Biol. Biochem. *4,* 489 (1972)
   c. Mills, A. A.: Chem. Brit. *16,* 69 (1980)
9. Irgolic, K. J. in: Frontiers in Bioinorganic Chemistry (ed. Xavier, A. V.), VCH Publishers, Weinheim 1986, p. 399

10. Cheng, C. N., Focht, D. D.: Envir. Health. Persp. *19,* 147 (1977)
11. McBride, B. C., Wolfe, R. S.: Biochem. *10,* 4312 (1971)
12. Friedman, O. M., Myles, A., Colvin, M.: Advances in Cancer Chemotherapy (ed. Rosowsky, A.), Marcel Dekker, New York 1979, p. 143.
13. Hudson, R. F.: Organo-Phosphorus Chemistry, Academic Press, London 1965
14. Cadogan, J. I. G., Mackie, R. K.: Chem. Soc. Rev. *3,* 87–138 (1974)
15. Hartley, S. B., Holmes, W. S. Jacques, J. K., Mole, M. F., McCoubrey, J. C.: Quart. Rev. *17,* 204 (1963)
16. Fluck, E.: Fortschr. Chem. Forsch. *35,* 1–64 (1973)
17. Chow, K. K., Levason, W., McAuliffe, C. A. in: Transition Metal Complexes of P, As, Sb and Bi Ligands (ed. McAuliffe, C. A.), MacMillan, London 1973 (and references cited therein)
18. Stelzer, O. in: Topics in Phosphorus Chemistry, vol. 9 (eds. Griffith, E. J., Grayson, M.), John Wiley, New York 1977, p. 1
19. Green, J. C., King, D. I., Eland, J. H. D.: Chem. Commun. 1121 (1970)
20. Collman, J. P., Hegedus, L. S.: Principles and Applications of Organotransition Metal Chemistry, University Science Books, Mill Valley, CA, 1980, pp. 54–60 and pp. 316–348, and references cited therein
21. Mason, R., Meek, D. W.: Angew. Chem. Int. Ed. Engl. *17,* 183 (1978)
22. Pidcock, A.: in Transition Metal Complexes of P, As, Sb and Bi Ligands (ed. McAuliffe, C. A.), Macmillan, London 1973, p. 1
23. Tolman, C. A.: Chem. Rev. *77,* 313 (1977)
24. Stelzer, O., Unger, E.: Chem. Ber. *108,* 1246 (1975)
25. Maier, J. P., Turner, D. W.: J. Chem. Soc. *68,* 711 (1972)
26. Weston, R. E., Bigeleisen, J.: J. Amer. Chem. Soc. *76,* 3074 (1954)
27. Staley, R. H., Beauchamp, J. L.: ibid. *96,* 6252 (1974)
28. Behan, J., Johnstone, R. A. W., Puddephatt, R. J.: J. Chem. Soc. Chem. Comm. 444 (1978)
29. Lappert, M. F., Pedley, J. B., Wilkins, B. T., Stelzer, O., Unger, E.: J. Chem. Soc. Dalton Trans. 1207 (1975)
30. Puddephatt, R. J., Bancroft, G. M., Chan, T.: Inorg. Chim. Acta *73,* 83 (1983)
31. Henderson, W. A., Streuli, C. A.: J. Amer. Chem. Soc. *82,* 5791 (1960)
32. Ikuta, S., Kebarle, P., Bancroft, G. M., Chan, T., Puddephatt, R J.: ibid. *104,* 5899 (1982)
33. Allman, T., Goel, R. G.: Can. J. Chem. *60,* 716 (1982)
34. Schafer, W., Schweig, A.: Angew. Chem. *84,* 898 (1972)
35. Debies, T. B., Rabalais, J. W.: Inorg. Chem. *13,* 308 (1974)
36. Weiner, M. A., Lattmann, M., Grim, S. O.: J. Org. Chem. *40,* 1292 (1975)
37. Mann, B. E., Musco, A.: J. Chem. Soc. Dalton Trans. 1673 (1975)
38. Tolman, C. A., Gerlach, D. H., Jesson, J. P., Schunn, R. A.: J. Organomet. Chem. *68,* C23 (1974)
39. Jardine, F. H.: Prog. Inorg. Chem. *28,* 63 (1981)
40. Montelactici, S., van der Ent, A., Osborn, J. A., Wilkinson, G.: J. Chem. Soc. A 1054 (1968)
41. Stiefel, E. I.: Prog. Inorg. Chem. *22,* 160 (1970)
42. a. Chatt, J., Pearman, A. J., Richards, R. L.: Nature *253,* 39 (1975)
    b. Chatt, J., Pearman, A. J., Richards, R. L.: J. Chem. Soc. Dalton Trans. 1852 (1977)
43. a. Pickett, C. J., Talarmin, J.: Nature *317,* 652 (1985)
    b. Al-Salih, T. I., Pickett, C. J.: J. Chem. Soc. Dalton 1255 (1985)
44. Sadler, P. J.: Struct. Bond. *29,* 170 (1976)
45. Berners-Price, S. J., Sadler, P. J. in: Frontiers in Bioinorganic Chemistry (ed. Xavier, A. V.), VCH Publishers, Weinheim 1986, p. 376 and references cited therein
46. Sutton, B. M., McGusty, E., Walz, D. T., DiMartino, M. J.: J. Med. Chem. *15,* 1095 (1972)
47. Blodgett, R. C. Jr., Heuer, M. A., Pietrusko, R. G.: Seminars in Arthritis and Rheumatism *13,* 25 (1984)
48. Chaffman, M., Brogden, R. N., Heel, R. C., Speight, T. M., Avery, G. S.: Drugs *27,* 378 (1984)
49. Intoccia, A. P., Flanagan, T. L., Walz, D. T., Gutzait, L., Swagdis, J. E., Flagiello, J. Jr.: J. Rheumatol. *9* (suppl. 8) 90 (1982)
50. Simon, T. M., Kunishima, D. H., Vibert, G. J., Lorber, A.: Cancer Philad. *44,* 1965 (1979)
51. Simon, T. M., Kunishima, D. H., Vibert, G. J., Lorber, A.: Ibid. *41,* 94 (1981)

52. Sadler, P. J., Nasr, M., Narayanan, V. L. in: Platinum Coordination Complexes in Cancer Chemotherapy (eds. Hacker, M. P., Douple, E. B., Krakhoff, I. H.) Martinus Nijhoff Pub., Boston 1984, pp. 290–304 (and references cited therein)
53. Shaw, C. F. III, Beery, A., Stocco, G. C.: Inorg. Chim. Acta *123,* 213 (1986)
54. Mirabelli, C. K., Johnson, R. K., Sung, C.-M., Faucette, L., Muirhead, K., Crooke, S. T.: Cancer Res. *45,* 32 (1985)
55. Mirabelli, C. K., Johnson, R. K., Hill, D. T., Faucette, L. F., Girard, G R., Kuo, G. Y., Sung, C-M., Crooke, S. T.: J. Med. Chem. *29,* 218 (1986)
56. Puddephatt, R. J.: The Chemistry of Gold, Elsevier, Amsterdam 1978
57. Nyholm, R. S.: Nature *168,* 705 (1951)
58. Harris, C. M., Nyholm, R. S., Stephenson, N. A.: Rev. Trav. Chim. *75,* 687 (1956)
59. Harris, C. M., Nyholm, R. S.: J. Chem. Soc. 63 (1957)
60. Cochran, W., Hart, F. A., Mann, F. G.: ibid 2816 (1957)
61. Davis, M., Mann, F. G.: ibid. 3791 (1964)
62. McAuliffe, C. A., Parish, R. V., Randall, P. D.: J. Chem. Soc. Dalton Trans. 1730 (1979)
63. Ludwig, W., Meyer, W.: Helv. Chim. Acta *65,* 934 (1982)
64. Al-Baker, S., Hill, W. E., McAuliffe, C. A.: J. Chem. Soc. Dalton Trans. 2655 (1985)
65. Carty, A. J., Efraty, A.: Inorg. Chem. *8,* 543 (1969)
66. Schmidbaur, H., Wohlleben, A., Wagner, F., Orama, O., Huttner, G: Chem. Ber. *110,* 1748 (1977)
67. Bates, P. A., Waters, J. M.: Inorg. Chim. Acta *98,* 125 (1985)
68. Eggleston, D. S., Chodosh, D. F., Girard, G. R., Hill, D. T.: ibid. *108,* 2211 (1985)
69. Cooper, M. K., Mitchell, L. E., Henrick, K., McPartlin, M., Scott, A.: ibid. *84,* L 9 (1984)
70. Jones, P. G.: Acta Cryst. Sect. B *36,* 2775 (1980)
71. Schmidbaur, H., Wohlleben, A., Schubert, U., Frank, A., Huttner, G.: Chem. Ber. *110,* 2751 (1977)
72. Schmidbaur, H., Franke, R.: ibid. *105,* 2985 (1972)
73. Parish, R. V., Rush, J. D.: Chem. Phys. Lett. *63,* 37 (1979)
74. Jones, P. G.: J. Chem. Soc. Chem. Comm. 1031 (1980)
75. Elder, R. C., Zeiher, E. H. K., Onady, M., Whittle, R. R.: ibid. 900 (1981)
76. Mays, M. J., Vergnano, P. A.: : J. Chem. Soc. Dalton Trans. 1112 (1979)
77. Colburn, C. B., Hill, W. E., McAuliffe, C. A., Parish, R. V.: J. Chem. Soc. Chem. Comm. 218 (1979)
78. Parish, R. V., Al-Sa'ady, A. K. H., McAuliffe, C. A., Moss, K., Fields, R.: J. Chem. Soc. Dalton Trans. 491 (1984)
79. Berners-Price, S. J., Mazid, M. A., Sadler, P. J.: ibid. 969 (1984)
80. Bates, P. A., Waters, J. M.: Inorg. Chim. Acta *81,* 151 (1984)
81. Berners-Price, S. J., Sadler, P. J.: Inorg. Chem. *25,* 3822 (1986)
82. Berners-Price, S. J., Jarrett, P. S., Sadler, P. J.: Inorg. Chem. *26,* 3074 (1987)
83. Berners-Price, S. J., Mirabelli, C. K., Johnson, R. K., Mattern, M. R., McCabe, F. L., Faucette, L. F., Sung, C-M., Mong, S-M., Sadler, P.J., Crooke, S. T.: Cancer Research *46,* 5486 (1986)
84. Berners-Price, S. J., Sadler, P. J.: Abstract from the 192nd A.C.S. Meeting, Anaheim, CA, September 7–12, 1986 and J. Inorg. Biochem. in press
85. Coffer, M. T., Shaw, C. F. III, Eidsness, M. K., Watkins, J. W. III, Elder, R. C.: Inorg. Chem. *25,* 333 (1986)
86. Razi, M. T., Otiko, G., Sadler, P. J.: A.C.S. Symp. Ser. *209,* 371 (1983)
87. Berners-Price, S. J., Mirabelli, C. K., Johnson, R. K., Sadler, P. J. et al.: manuscript in preparation (for J. Med. Chem.)
88. Struck, R. F., Shealey, Y. F.: J. Med. Chem. *9,* 414 (1966)
89. a. Johnson, R. K., Mirabelli, C. K., Faucette, L. F., McCabe, F. L., Sutton, B. M., Bryan, D. L., Girard, G. R., Hill, D. T.: Proc. Amer. Assoc. Cancer Res. *26,* 254 (1985)
    b. Mirabelli, C. K., Hill, D. T., Faucette, L. F., McCabe, F. L., Girard, G. R., Bryan, D. B., Sutton, B. M., Bartus, J. O'L., Crooke, S. T., Johnson, R. K.: J. Med. Chem. in press
90. Snyder, R. M., Mirabelli, C. K., Johnson, R. K., Sung, C-M., Faucette, L. F., McCabe, F. L., Zimmermann, J. P., Whitman, M., Hempel, J. C., Crooke, S. T.: Cancer Res. *46,* 5054 (1986)
91. Mirabeili, C. K., Jensen, B. D., Mattern, M. R., Sung, C-M., Mong, S-M., Hill, D T., Dean, S. W., Schein, P. S., Johnson, R. K., Crooke, S. T.: submitted to Anticancer Drug Design

92. Sadler, P. J.: Inorg. Persp. Biol. Med. *1,* 233 (1978)
93. Churchill, M. R., Deboer, B. G.: Inorg. Chem. *14,* 2502 (1975)
94. Churchill, M. R., Donahue, J., Rotella, F. J.: ibid. *15,* 2752 (1976)
95. Alyea, E. C., Ferguson, G., Somogyvari, A.: ibid. *21,* 1369 (1982)
96. Socol, S. M., Jacobson, R. A., Verkade,J. G.: ibid. *23,* 88 (1984)
97. Cotton, F. A., Goodgame, D. M. L.: J. Chem. Soc. 5267 (1960)
98. Muetterties, E. L., Alegranti, C. W.: J. Amer. Chem. Soc. *94,* 6386 (1972)
99. Schmidbauer, H., Adlkofer, J., Schwirter, K.: Chem. Ber. *105,* 3382 (1972)
100. Sartori, P., Kuhn, V. N.: Chemiker-Zeitung *105,* 87 (1981)
101. Van der Ploeg, A. F. M. J., Van Koten, G., Spek, A. L.: Inorg. Chem. *18,* 1052 (1979)
102. Van der Ploeg, A. F. M. J., Van Koten, G.: Inorg. Chim. Acta *51,* 225 (1981)
103. Ho, D. M., Bau, R.: Inorg. Chem. *22,* 4073 (1983)
104. Berners-Price, S. J., Brevard, C., Pagelot, A., Sadler, P. J.: Ibid. *24,* 4278 (1985)
105. Berners-Price, S. J., Johnson, R. K., Giovenella, A. J., Faucette, L. F., Mirabelli, C. K., Sadler, P. J.: submitted for publication
106. Zelonka, R. A., Baird, M. C.: Can. J. Chem. *50,* 1269 (1971)
107. Zelonka, R. A., Baird, M. C.: Chem. Commun. 780 (1971)
108. Axtell, D. D., Yoke, J. T. Inorg. Chem. *12,* 1265 (1973)
109. Makanova, D., Ondrejovic, G., Gazo, J.: Proc. Conf. Coord. Chem. 3rd, 215–20 (1971)
110. Makanova, D., Ondrejovic, G., Gazo, J.: Chem. Zvesti *27,* 4 (1973)
111. Gill, J. T., Mayerle, J. J., Welcker, P. S., Lewis, D. F., Ucko, D. A., Barton, D. J., Stowens, D., Lippard, S. J.: Inorg. Chem. *15,* 1155 (1976) (and references cited therein)
112. Dines, M. B.: ibid. *11,* 2949 (1972)
113. Camus, A., Marsich, N., Nardin, G., Randaccio, L.: Trans. Met. Chem. *1,* 205 (1976)
114. Albano, V. G., Bellon, P. L., Ciani, G.: J. Chem. Soc. Dalton Trans. 1938 (1972)
115. Anderson, W. A., Carty, A. J., Palenik, G. J., Schreiber, G.: Can. J. Chem. *49,* 761 (1971)
116. Davis, P. H., Belford, R. L., Paul, I. C.: Inorg. Chem. *12,* 213 (1973)
117. Brescani, N., Marsich, N., Nardin, G., Randaccio, L.: Inorg. Chim. Acta *10,* L5 (1974)
118. Nardin, G., Randaccio, L., Zangrando, E.: J. Chem. Soc. Dalton Trans. 2566 (1975)
119. Churchill, M. R., Rotella, F. J.: Inorg. Chem. *18,* 167 (1979)
120. Churchill, M. R., DeBoer, B. G., Mendak, S. J.: ibid. *14,* 2041 (1975)
121. Nardin, G., Marsich, N., Randaccio, L.: J. Amer. Chem. Soc. *95,* 4053 (1973)
122. Nardin, G., Camus, A., Randaccio, L.: Inorg. Chim. Acta *12,* 23 (1975)
123. Leoni, P., Pasquali, M., Ghilardi, C. A.: J. Chem. Soc. Chem. Comm. *240,* (1983)
124. Edwards, D. A., Richard, R.: J. Chem. Soc. Dalton Trans. 637 (1975)
125. Gaughan, A. P., Ziolo, R. F., Dori, Z.: Inorg. Chem. *10,* 2776 (1971)
126. Lippard, S. J., Mayerle, J. J.: ibid. *11,* 753 (1972)
127. Marsich, N., Camus, A., Cebulec, E.: J. Inorg. Nucl. Chem. *34,* 933 (1972)
128. Fife, D. J., Moore, W. M., Morse, K. W.: Inorg. Chem. *23,* 1684 (1984)
129. Berners-Price, S. J., Mirabelli, C. K., Johnson, R. K., Faucette, L. F., McCabe, F. L., Sadler, P. J.: Inorg. Chem. *26,* 3383 (1987)
130. Mathew, M., Palenik, G. J.: Inorg. Chim. Acta *5,* 573 (1971)
131. Brisdon, B. J.: J. Chem. Soc. Dalton Trans. 2247 (1972)
132. Berners-Price, S. J., Brevard, C., Pagelot, A., Sadler, P. J.: Inorg. Chem. *25,* 596 (1986)
133. Berners-Price, S. J., Sadler, P. J.: Chem. Brit. *23,* 541 (1987)
134. Saryan, L. A., Ankel, E., Krishnamurti, C., Petering, D. H., Elford, H.: J. Med. Chem. *22,* 1218 (1979)
135. Mohindru, A., Fisher, J. M., Rabinovitz, M.: Biochem. Pharmaol. *32,* 2637 (1983)
136. Marshall, L. E., Graham, D. R., Reich, K. A., Sigman, D. S.: Biochemistry *20,* 244 (1981)
137. Downey, K. M., Que, B. G., So, A. G.: Biochem. Biophys. Res. Commun. *93,* 264 (1980) and Biochemistry *19,* 5987 (1980).
138. Graham, D. R., Marshall, L. E., Reich, K. A., Sigman, D. S.: J. Amer. Chem. Soc. *102,* 5419 (1980)
139. Wong, A., Huang, C. H., Crooke, S. T.: Biochemistry *23,* 2939 and 2945 (1984)
140. Miller, J., McLachlan, A. D., Klug, A.: EMBO J. *4,* 1609 (1985)
141. Berg, J. M.: Science *232,* 485 (1986)
142. a. Rosenberg, B. in: Cisplatin, current status and new developments (eds. Prestayko, A. W., Crooke, S. T., Carter, S. K.), Academic Press, New York 1980, p. 9 and other articles in this

volume
b. Roberts, J. J., Fraval, H. N. A.: ibid. p. 57
143. Lippard, S. J. in: Platinum coordination complexes in cancer chemotherapy (eds. Hacker, M. P., Douple, E. B., Krakhoff, I. H.), Martinus Nijhoff, Boston 1984, p. 11 and other articles in this volume
144. Berners-Price, S. J., DiMartino, M. J., Hill, D. T., Kuroda, R., Mazid, M. A., Sadler, P. J.: Inorg. Chem. *24,* 3425 (1985)
145. Berners-Price, S. J., Morden, K., Opella, S. J., Sadler, P. J.: Mag. Res. Chem. *24,* 734 (1986)
146. Price, N. R.: MAFF/ADAS Agricultural Science Service R & D Report "Pesticide Science", 1980, pp. 22–27
147. Price, N. R.: J. Stored Prod. Res. *21,* 157 (1985)
148. Cherfuka, W., Kashi, K. P., Bond, E. J.: Pest. Biochem. Physiol. *6,* 65 (1976)
149. Price, N. R., Dance, S. J.: Comp. Biochem. Physiol. *76 C,* 277 (1983)
150. Price, N. R., Mills, K. A., Humphries, L. A.: ibid. *73 C,* 411 (1982)
151. Price, N. R.: Insect Biochem. *10,* 65 (1980)
152. Bond, E. J., Robinson, J. R., Buckland, C. T.: J. Stored Prod. Res. *5,* 289 (1969)
153. Price, N. R.: ibid. *20,* 163 (1984)
154. Kashi, K. P., Cherfuka, W.: Pest. Biochem. Physiol. *6,* 350 (1976)
155. Buckler, S. A.: J. Amer. Chem. Soc. *84,* 3093 (1962)
156. Floyd, M. B., Boozer, C. E.: ibid. *85,* 984 (1963)
157. Ogata, Y. Yamashita, M.: J. Chem. Soc. Perkin (II) Trans. 730 (1972)
158. Berners-Price, S. J., Norman, R. E., Sadler, P. J.: J. Inorg. Biochem. *31,* 197 (1987)
159. Hill, H. A. O. in: New Trends in Bio-Inorganic Chemistry (eds. Williams, R. J. P., Da Silva, J. R. R. F.), Academic Press, London 1978, p. 173
160. Aust, S. D., Morehouse, L. A., Thomas, C. E.: J. Free Rad. Biol. Med. *1,* 3 (1985)
161. Greenbaum, M. A., Denney, D. B., Hoffmann, A. K.: J. Amer. Chem. Soc. *78,* 2563 (1956)
162. Denney, D. B., Denney, D. Z., Chang, B. C., Marsi, K. L.: ibid. *91,* 5243 (1969)
163. Baumstark, A. L., McCloskey, C. J., Williams, T. E., Chrisope, D. R.: J. Org. Chem. *45,* 3593 (1980)
164. Holtz, H. D., Solomon, P. W., Mahan, J. E.: ibid. *38,* 3175 (1973)
165. Clennan, E. L., Heah, P. C.: ibid. *46,* 4105 (1981)
166. Shulman, J. I.: ibid. *42,* 3971 (1977)
167. Hiatt, R., Smythe, R. J., McColeman, C.: Can. J. Chem. *49,* 1707 (1971)
168. Holm, R. H., Berg, J. M.: Pure Appl. Chem. *56,* 1645 (1984)
169. Reynolds, M. S., Berg, J. M., Holm, R. H.: Inorg. Chem. *23,* 3057 (1984)
170. Malstrom, B. G.: Ann. Rev. Biochem. *51,* 21 (1982)
171. Chin, D-H., La Mar, G. N., Balch, A. L.: J. Amer. Chem. Soc. *102,* 5945 (1980)
172. Harrison, J. E., Schultz, J.: J. Biol. Chem. *251,* 1371 (1976)
173. Hatefi, Y.: Ann. Rev. Biochem. *54,* 1015 (1985)
174. Murray, R. I., Fisher, M. T., Debrunner, P. G., Sligar, S. G. in: Metalloproteins Part 1 (ed. Harrison, P.), MacMillan, London 1985, p. 157
175. Lerch, K. in: Metal Ions in Biological Systems vol. 13 (ed. Sigel, H.), Marcel Dekker, N.Y. 1981, p. 143
176. Lapinte, C., Riviere, H., Roselli, A.: J. Chem. Soc., Chem. Comm. 1109 (1981)
177. Mukaiyama, T., Takei, H. in: Topics in Phosphorus Chemistry, vol. 8 (eds. Griffith, E. J., Grayson, M.), John Wiley, N.Y. 1976, pp. 587–645 (and references cited therein)
178. Overman, L. E., O'Connor, E. M.: J. Amer. Chem. Soc. *98,* 771 (1976)
179. Wolfram, L. J.: 3rd int. Wool Text. Res. Conf. Paris, Sect. 2, p. 505 (1965)
180. Maclaren, J. A., Sweetman, G. J.: Aust. J. Chem. *19,* 235 (1966)
181. Levison, M. E., Josephson, A. S., Kirschenbaum, D. M.: Experimentia *25,* 126 (1969)
182. Ruegg, U. T., Rudinger, J.: Methods Enzymol. *47,* 111 (1977)
183. Kirschenbaum, D. M., Bakker, E., Thiel, L., Plotch, S.: Fedn. Proc. Fedn Am. Socs. exp. Biol. *26,* 837 (1969)
184. Kirkpatrick, A., Maclaren, J. A.: Anal. Biochem. *56,* 137 (1973)
185. Malik, N. A., Otiko, G., Sadler, P. J.: J. Inorg. Biochem. *12,* 317 (1980)
186. a. Walling, C., Rabinowitz, R.: J. Amer. Chem. Soc. *81,* 1243 (1959)
b. Walling, C., Basedow, O. H., Savas, E. S.: ibid. *82,* 2181 (1960)
187. Walling, C., Pearson, M. S.: ibid. *86,* 2262 (1964)

188. Bentrude, W. G. in: Free Radicals (Vol. II) (ed. Kochi, J. K.), John Wiley, New York 1973, pp. 595–663 (and references cited therein)
189. Horner, L., Haufe, J.: Chem. Ber. *101,* 2921 (1968)
190. Wong, R. K., Allen, A. O.: J. Phys. Chem. *74,* 774 (1970)
191. Lyons, A. R., Neilson, G. W., Symons, M. C. R.: J. Chem. Soc. Chem. Commun. 507 (1972)
192. Matschiner, H., Krause, L., Krech, F.: Z. Anorg. Allg. Chem. *373,* 1 (1970)
193. Gara, W. G., Roberts, B. P.: J. Chem. Soc. Perkin II Trans. 150 (1978)
194. Santhanam, K. S. V., Bard, A. J.: J. Amer. Chem. Soc. *90,* 1118 (1968)
195. Matschiner, H., Tzschach, A., Steinert, A.: Z. Anorg. Allg. Chem. *373,* 237 (1970)
196. Gerson, F., Plattner, G., Bock, H.: Helv. Chim. Acta *53,* 1629 (1970)
197. Wang, S. K., Sytryk, W., Wan, J. K. S.: Can. J. Chem. *49,* 994 (1971)
198. Sonovsky, G.: Free Radical Reactions in Preparative Organic Chemistry, MacMillan, New York 1964, Chap. 5
199. Roberts, B. P.: Adv. free radical chem. *6,* 225 (1980) (and references cited therein)
200. Bentrude, W. G.: A.C.S. Symp. Ser. *69,* 321 (1978)
201. Griller, D., Ingold, K. U., Patterson, L. K. Scaiano, J. C., Small, R. D. Jr.: J. Amer. Chem. Soc. *101,* 3780 (1979)
202. Bentrude, W. G., Hanse, E. R., Khan, W. A., Min, T. B., Rogers, P. E.: ibid. *95,* 2286 (1973)
203. Bentrude, W. G., Hargis, J. H., Rusek, P. E.: J. Chem. Soc. Chem. Commun. 296 (1969)
204. Howard, J. A., Tait, J. C.: Can. J. Chem. *56,* 2163 (1978)
205. Naan, M. P., Powell, R. L., Hall, C. D.: J. Chem. Soc. B, 1683 (1971)
206. Grollmann, E. F., Lee, G., Ambesi-Impiombato, F. S., Meldolesi, M. F., Aloj, S. M., Coon, H. G., Kaback, H. R., Kohn, L. D.: Proc. Natl. Acad. Sci. USA *74,* 2352 (1977)
207. Holian, S. W., Holian, A., Daniele, R. P.: Fed. Proc. Fed. Am. Soc. Exp. Biol. *36,* 1233 (1977)
208. Lever, J. E.: Biochemistry *16,* 4329 (1977)
209. Szmeleman, S., Adler, J.: Proc. Natl. Acad. Sci. USA *73,* 4387 (1976)
210. Schuldiner, S., Kaback, H. R.: Biochemistry *14,* 5451 (1976)
211. Korchak, H. M., Weissmann, G.: Proc. Natl. Acad. Sci. USA *75,* 3818 (1978)
212. Liberman, E. A., Skulachev, V. P.: Biophys. Acta *216,* 30 (1970)
213. Johnson, L. V., Walsh, M. L., Chen, L. B.: Proc. Natl. Acad. Sci. USA *77,* 990 (1980)
214. Chen, L. B., Lampidis, T. J., Bernal, S. D., Nadakaurkareni, K. K., Summerhayes, I. S. in: Genes and Proteins in Oncogenesis, Academic Press 1983, pp. 369–387
215. Dawson, J. H., Andersson, L. A., Sono, M.: J. Biol. Chem. *258,* 13637 (1983) (and references cited therein)
216. Ohya, T., Morohoshi, H., Sato, M.: Inorg. Chem. *23,* 1303 (1984)
217. Sono, M., Dawson, J. H., Hager, L. P.: ibid. *24,* 4339 (1985)
218. Mansuy, D., Duppel, W., Ruf, H-H., Ullrich, V.: Hoppe-Seyler's Z. Physiol. Chem. *355,* 1341 (1974)
219. Dahl, A. R., Hodgson, E.: Chem. Biol. Interactions *21,* 137 (1978)
220. a. Otiko, G., Sadler, P. J.: FEBS Lett. *116,* 227 (1980)
b. Grootveld, M. C., Otiko, G., Sadler, P. J., Cammack, R.: J. Inorg. Biochem. *27,* 1 (1986)
c. Berners-Price, S. J., Grootveld, M. C., Otiko, G., Robbins, H. R., Sadler, P. J.: Inorg. Chim. Acta *79,* 186 (1983)
221. Streuli, C. A.: Anal. Chem. *32,* 985 (1960)
222. Deutsch, E., Libson, K., Jurisson, S.: Prog. Inorg. Chem. *30,* 75 (1983)
223. Deutsch, E., Bushong, W., Klavan, K. A., Elder, R. C., Sodd, V. J., Fortman, D. L., Lukes, S. J.: Science *214,* 85 (1981)
224. Vanderheyden, J–L., Heeg, M. J., Deutsch, E.: Inorg. Chem. *24,* 1666 (1985)
225. Gerson, M. C., Deutsch, E. A., Nishiyama, H.: Eur. J. Nucl. Med. *8,* 371 (1983)
226. Vanderheyden, J-L., Ketring, A. R., Libson, K., Heeg, M. J., Roecker, L., Motz, P., Whittle, R., Elder, R. C., Deutsch, E.: Inorg. Chem. *23,* 3184 (1984)
227. Gerson, M. C., Deutsch, E. A., Libson, K. F.: Eur. J. Nucl. Med. *9,* 403 (1984)
228. Thakur, M. L., Park, C. H., Fazio, F.: Int. J. Appl. Radiat. Isot. *35,* 507 (1984)
229. Gerundini, P., Savi, A., Gilardi, M. A., Margonato, A., Vicedomini, G., Zecca, L., Hirth, W., Libson, K., Bhatia, J. C., Fazio, F., Deutsch, E. A.: J. Nuc. Med. *27,* 409 (1986)
230. Timmer, K., Meinema, H. A., Schurig, J. E.: Eur. Pat. Appl. 86 201853 8, Pub., July 1987

Transition and Main-Group Metal Cyclopentadienyl Complexes: Preclinical Studies on a Series of Antitumor Agents of Different Structural Type

Petra Köpf-Maier[1] and Hartmut Köpf[2]

1 Institut für Anatomie, Freie Universität Berlin, Königin-Luise-Straße 15, D-1000 Berlin 33
2 Institut für Anorganische und Analytische Chemie, Technische Universität Berlin, Straße des 17. Juni 135, D-1000 Berlin 12

Cyclopentadienyl metal complexes are organometallic compounds which exhibit antiproliferative properties in vivo and in vitro. They are represented by compounds of various structural type. The metallocene diacido complexes $(C_5H_5)_2MX_2$ contain early transition metal atoms such as titanium(IV) and vanadium(IV), the ionic metallicenium salts $[(C_5H_5)_2M]^+X^-$ include medium transition metals, e.g. iron(III), whereas the uncharged decasubstituted metallocenes $(C_5R_5)_2M$ comprise the main group elements tin(II) and germanium(II) as central metal atoms M. A variety of biological data are available about metallocene diacido complexes, especially titanocene dichloride. These substances exhibit antitumor properties against numerous experimental tumors, e.g. Ehrlich ascites tumor, B16 melanoma, colon 38 carcinoma, Lewis lung carcinoma, as well as against various human tumors heterotransplanted to athymic mice. Biological experiments using $(C_5H_5)_2TiCl_2$ and $(C_5H_5)_2VCl_2$ pointed to the nucleic acids as a probable target for metallocene diacido complexes within the cells, revealed a main accumulation of titanium or vanadium in the liver and the intestine and unfolded a pattern of organ toxicity which fundamentally differs from that of other cytostatic drugs. These biological features confirm the cyclopentadienyl metal complexes to be an independent group of non-platinum-group metal antitumor agents being characterized by unusual biological properties. They represent interesting candidates for future biological and clinical investigations.

I. Introduction . . . 105

II. Antitumor Metallocene Complexes . . . 107
 A. Metallocene Diacido Complexes . . . 107
  1. Metallocene Dichlorides . . . 110
  2. Titanocene Diacido Complexes . . . 110
  3. Titanocene Complexes with Modified Cyclopentadienyl Ligands . . . 112
  4. Mono(cyclopentadienyl) Titanium Complexes . . . 113
  5. Polynuclear Cyclopentadienyl Titanium Complexes . . . 114
  6. Ionic Cyclopentadienyl Titanium Complexes . . . 114
 B. Metallicenium Salts . . . 115
  1. Mononuclear Ferricenium Salts . . . 116
  2. Polynuclear Ferricenium Salt . . . 116
 C. Main Group Metallocenes . . . 116

Structure and Bonding 70
© Springer-Verlag Berlin Heidelberg 1988

III. Antiproliferative Properties . . . 118
A. Cytostatic Properties in Vitro . . . 119
B. Antitumor Properties Against Animal Tumors . . . 121
1. Fluid Ehrlich Ascites Tumor . . . 123
2. Fluid Sarcoma 180 . . . 126
3. Leukemias L 1210 and P 388 . . . 126
4. Solid Ehrlich Ascites Tumor . . . 128
5. Solid Sarcoma 180 . . . 130
6. Solid B 16 Melanoma . . . 130
7. Colon 38 Adenocarcinoma . . . 132
8. Lewis Lung Carcinosarcoma . . . 132
9. Mouse Mammary Tumor TA 3 Ha . . . 133
Summary – Animal Antitumor Data . . . 133
C. Antitumor Properties Against Xenografted Human Tumors . . . 134
1. Human Colorectal Carcinomas . . . 134
2. Human Lung Malignancies . . . 138
3. Heterotransplanted Human Tumors . . . 139
Summary – Xenografted Human Tumors . . . 141

IV. Antiviral, Insecticidal and Antiinflammatory Properties . . . 142

V. Structure-Activity Relationships . . . 143
A. Influence of the Central Metal Atoms . . . 143
B. Influence of the Acido Groups . . . 144
C. Influence of the Cyclopentadienyl Ring Ligands . . . 145
D. Influence of Charge Effects . . . 148

VI. Cellular Mode of Action . . . 149
A. Incorporation Studies . . . 149
B. Subcellular Distribution of Central Metal Atoms . . . 150
C. Cytokinetic Investigations . . . 150
D. Cytologic Phenomena in Fibroblasts, Experimental and Human Tumors . . . 152
1. Human Embryonal Fibroblasts . . . 152
2. Ehrlich Ascites Tumor . . . 152
3. Human Colon Adenocarcinoma . . . 156

VII. Model Complexes . . . 162

VIII. Dissociation and Hydrolysis Reactions . . . 165

IX. Organ Distribution and Pharmacokinetics . . . 166
A. Organ Distribution of Titanium . . . 167
B. Organ Distribution of Vanadium . . . 169

X. Toxicologic Properties . . . 169
A. Organ Toxicity . . . 169
1. Kidneys . . . 170
2. Liver . . . 171
3. Endocrine Glands . . . 174
4. Bone Marrow . . . 176
B. Embryotoxicity . . . 177

XI. Summary . . . 179

XII. Abbreviations . . . 181

XIII. References . . . 181

I. Introduction

Cancer diseases are together with angiocardiopathies the main causes of death in most of the civilized countries. In principle, cancer diseases can be treated by surgery, radiation and chemotherapy. Apart from isolated attempts during the last century, the history of a systematic therapy of cancer using medicines started only about fifty years ago. Until the middle of the seventies, organic compounds such as alkylating agents, antimetabolites and vinca rosea alkaloids were the most common cytostatic drugs, generally administered as drug combinations with or without surgery and/or radiation. Towards the end of the 1970's, a newly developed inorganic platinum complex, *cis*-$(NH_3)_2PtCl_2$ (cisplatin) (Fig. 1), was introduced into clinical use and added to the panel of approved cytostatics. This provides a markedly enhanced therapeutic benefit to patients suffering from certain malignancies, especially urogenital malignancies, carcinomas of the head and neck, and, to a less extent, against lung tumors. A broad interest into experimental and clinical chemotherapy has been aroused by this development.

Fig. 1. Structure of the inorganic antitumor drug *cis*-diamminedichloroplatinum(II) (cisplatin)

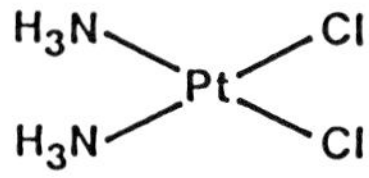

Besides cisplatin, an impressive number of platinum complexes of the second and third generation have been developed during the past years and introduced into early and advanced clinical trials[4, 5]. Typical representatives of second generation platinum complexes which have undergone clinical trials are diammine(cyclobutane-1,1-dicarboxylato)platinum (II) (carboplatin), aquo-1,1-bis(aminomethyl)cyclohexane(sulfato)platinum(II) (spiroplatin) and bis(isopropylamine)-*cis*-dichloro-*trans*-dihydroxoplatinum(IV) (iproplatin) (Fig. 2). Their clinical antitumor spectrum largely resembles that of cisplatin, urogenital tumors being the most sensitive human tumors[5–7].

A

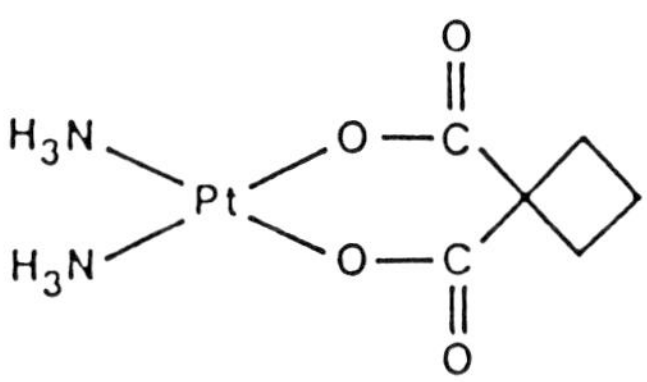

B

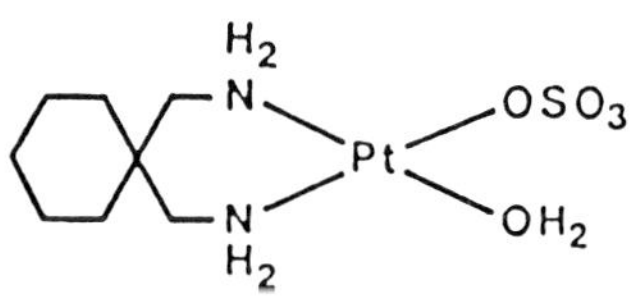

Fig. 2A–C. Structure of the second generation antitumor platinum complexes carboplatin (**A**), spiroplatin (**B**), and iproplatin (**C**)

C

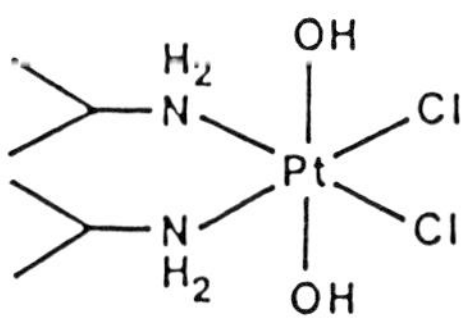

On the other hand, various and structurally different types of non-platinum-group metal complexes have been found to be antitumor agents. They comprise inorganic as well as organometallic compounds, the first often containing organic ligands, but only the latter having direct carbon-to-metal bonds, and either contain main group elements such as gallium[8], germanium[9, 10] and tin[11, 12], or transition metals such as titanium[13–15], iron[16], copper[17, 18], and gold[19, 20] as central atoms. Some examples are listed in Fig. 3.

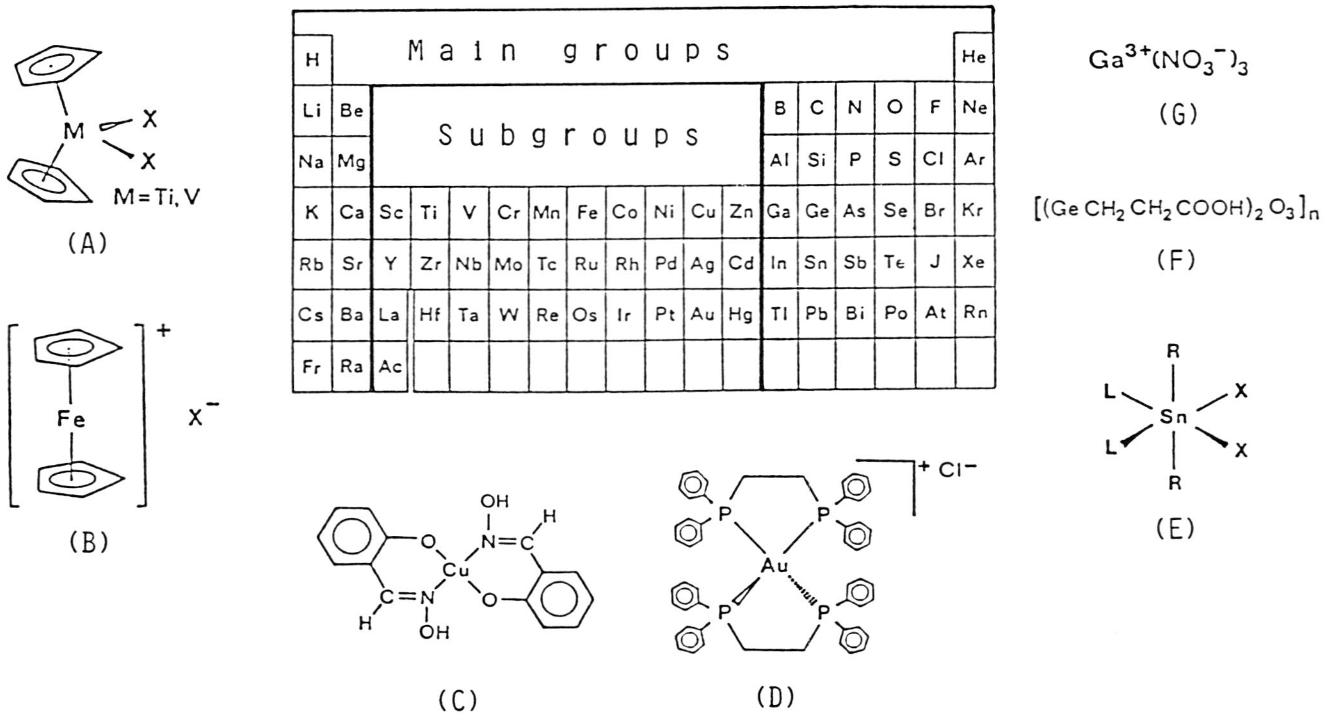

Fig. 3. Examples of non-platinum-group metal antitumor agents and the position of their central metal atoms in the Periodic Table of Elements. *(A)* Bis(η^5-cyclopentadienyl)metal(IV) diacido complexes; *(B)* bis(η^5-cyclopentadienyl)iron(III) complex salts; *(C)* *trans*-bis(salicylaldoximato)-copper(II); *(D)* bis[1,2-bis(diphenylphosphino)ethane]gold(I) chloride; *(E)* diorganodihalotin(IV) complexes with donor molecules (*R* = alkyl or phenyl; *L* = unidentate ligand, e.g. pyridine; or L_2 = bidentate chelating ligand, e.g. 2,2′-bipyridyl or o-phenanthroline); *(F)* bis(carboxyethylgermanium) trioxide ("germanium sesquioxide", Ge-132); *(G)* gallium(III) nitrate

Antiproliferative non-platinum-group metal antitumor agents are generally characterized by spectra of activity against experimental animal tumors which are different from the antitumor spectrum of cytostatic platinum complexes[21, 22]. Moreover, the toxic properties and the cellular mode of action at least of some of the non-platinum complexes fundamentally deviate from that shown for antitumor platinum compounds[21, 22].

The present study deals with organometallic cyclopentadienyl metal complexes and reports upon the structure-activity relationships and the pharmacological properties so far known of these compounds. They are represented by transition metal compounds of two different structural types, the metallocene diacido complexes and the metallicenium salts, the antitumor properties of which were detected in 1979 and 1984[13, 16]. Moreover, pilot experiments performed recently have revealed antiproliferative effectivity of a third structural type of metallocene compounds containing central metal atoms of main group IV[23]. Characteristic examples of these three types are given by the formulae $(\eta^5\text{-}C_5H_5)_2TiCl_2$, $[(\eta^5\text{-}C_5H_5)_2Fe]^+FeCl_4^-$, and $[(\eta^5\text{-}C_5(C_6H_5)_5]_2Sn$.

II. Antitumor Metallocene Complexes

Antitumor cyclopentadienyl metal complexes include (i) neutral bis(η^5-cyclopentadienyl)metal ("metallocene") diacido complexes $(C_5H_5)_2MX_2$ containing early transition metal central atoms in oxidation state +4 such as titanium(IV) and vanadium(IV), (ii) ionic metallicenium salts $[(C_5H_5)_2M]^+X^-$ with medium transition metals in oxidation state +3, e.g. iron(III) and cobalt(III), and (iii) uncharged decasubstituted metallocenes $(C_5R_5)_2M$ with the main group elements tin(II) or germanium(II) as central metals in oxidation state +2. All of them are genuine organometallic compounds containing carbon-to-metal bonds. Most of the antitumor metallocene compounds are air-stable solids at room temperature which generally are thermally stable up to 100–300 °C. They are intensely colored, the colors ranging from yellow, orange and red to green, blue and black. Some chemical properties of antitumor metallocene complexes are summarized in Tables 1–3.

A. Metallocene Diacido Complexes

Metallocene diacido complexes $(C_5H_5)_2MX_2$ are characterized by the following structural features (Fig. 4).

- The geometry of the complexes is that of a distorted tetrahedron.
- The complexes contain two uninegative acido ligands X coordinated to the central metal atom M and arranged in adjacent, "*cis*-like" positions.
- As central metal atoms M, the complexes include early transition metals of subgroups IVa, Va, or VIa of the Periodic Table.
- The sites of the other two ligands are occupied by two anionic cyclopentadienyl rings. These five-membered, aromatic hydrocarbon rings are arranged in an "open sandwich" geometry, the ring planes being tilted to each other, and bound to M by carbon-to-metal bonds.

Table 1. Chemical properties of antitumor η^5-cyclopentadienyl early transition metal complexes

Compound		Molecular weight	Color	Melting point[a] (°C)	Synthesis[b] (Ref.)
Metallocene dichlorides					
I	$(C_5H_5)_2TiCl_2$	249.00	red	~280 w.d.	24, 25
II	$(C_5H_5)_2VCl_2$	252.04	green	250	24
III	$(C_5H_5)_2NbCl_2$	294.00	black	>200	26, 27
IV	$(C_5H_5)_2TaCl_2$	382.05	brown		28, 29
V	$(C_5H_5)_2MoCl_2$	297.04	dark green	270 w.d.	30, 27
VI	$(C_5H_5)_2WCl_2$	384.95	green	250 w.d.	30
Titanocene halides and pseudohalides					
VII	$(C_5H_5)_2TiF_2$	216.09	yellow	250–260 w.d.	24
VIII	$(C_5H_5)_2TiBr_2$	337.91	dark red	307–313 w.d.	24
IX	$(C_5H_5)_2TiI_2$	431.90	dark red	315–318 w.d.	24
X	$(C_5H_5)_2Ti(NCS)_2$	294.25	dark red	305–307 w.d.	31, 32
XI	$(C_5H_5)_2Ti(N_3)_2$	262.13	orange	145–146	33, 34
Titanocene carboxylates					
XII	$(C_5H_5)_2Ti(OCOCF_3)_2$	404.12	orange-red	178–180	35
XIII	$(C_5H_5)_2Ti(OCOCCl_3)_2$	502.85	orange-yellow	173–174	36, 37
XIV	$(C_5H_5)_2Ti(\textit{cis}\text{-}OCOCH{=}CHCOOH)_2$	408.22	orange-red	d. >200	36, 38
Titanocene phenolates, thiophenolates and selenophenolates					
XV	$(C_5H_5)_2Ti(p\text{-}OC_6H_4NO_2)_2$	454.29	orange	232 w.d.	39, 40
XVI	$(C_5H_5)_2Ti(OC_6F_5)_2$	544.20	orange		41
XVII	$(C_5H_5)_2Ti(SC_6F_5)_2$	576.33	orange		41
XVIII	$(C_5H_5)_2TiCl(2,4,6\text{-}OC_6H_2Cl_3)$	409.98	red	200	42
XIX	$(C_5H_5)_2TiCl(o\text{-}SC_6H_4CH_3)$	336.74	red	120–124	43
XX	$(C_5H_5)_2TiCl(o\text{-}SC_6H_4NH_2)$	337.73	black	130–132	43
XXI	$(C_5H_5)_2TiCl(SeC_6H_5)$	369.55	blue-violet	60–80 w.d.	44
Titanocene dithiolene chelate					
XXII	$(C_5H_5)_2Ti[\textit{cis}\text{-}1,2\text{-}S_2C_2(CN)_2]$	318.28	olive-green	d. >280	45–47
Titanocene complexes with substituted η^5-cyclopentadienyl ligands					
XXIII	$(C_2H_5C_5H_4)(C_5H_5)TiCl_2$	277.05	pale red	104–105	48
XXIV	$[Si(CH_3)_3C_5H_4](C_5H_5)TiCl_2$	321.18	brick-red	146–150	48
XXV[c]	$[N(CH_3)_2C_5H_4](C_5H_5)TiCl_2$	292.06	green-black		49
XXVI	$[Si(CH_3)_3C_5H_4]_2TiCl_2$	393.36	copper-red	189–191	48
XXVII	$[Si(CH_3)_2(n\text{-}C_4H_9)C_5H_4]_2TiCl_2$	477.53	light red	66–70	48
XXVIII	$[Ge(CH_3)_3C_5H_4]_2TiCl_2$	482.37	red-brown	189–191	48
XXIX	$[N(CH_3)_2C_5H_4]_2TiCl_2$	333.12	green-black	225 w.d.	49a
XXX	$CH_2(C_5H_4)_2TiCl_2$	261.01	brown-red	305 w.d.	50
XXXI	$CHCH_3(C_5H_4)_2TiCl_2$	275.04	dark brown	d. >220	48
XXXII	$SiHCH_3(C_5H_4)_2TiCl_2$	291.11	red-brown	180–190 w.d.	48
XXXIII	$Si(C_2H_5)_2(C_5H_4)_2TiCl_2$	333.19	purple	190	48
XXXIV	$Ge(CH_3)_2(C_5H_4)_2TiCl_2$	349.64	black-brown	d. >250	48, 51
η^5-Indenyl and η^5-tetrahydroindenyl titanium complexes					
XXXV	$(C_9H_7)_2TiCl_2$	349.12	black-brown	152–153	52
XXXVI	$(C_9H_{11})(C_5H_5)TiCl_2$	303.09	purple	139–140	48
XXXVII	$(C_9H_{11})_2TiCl_2$	357.18	golden-red	214–215	31
Mono(η^5-cyclopentadienyl) titanium complexes					
XXXVIII	$(C_5H_5)TiCl_3$	219.35	yellow	208–211	53, 54
XXXIX	$(C_5H_5)TiCl_2(SC_6H_5)$	293.07	orange-yellow	104–105	43
XL	$(C_5H_5)Ti(NCS)_3$	287.24	brick-red	164–165 w.d.	43

Table 1 (continued)

Compound		Molecular weight	Color	Melting point[a] (°C)	Synthesis[b] (Ref.)
Polynuclear η^5-cyclopentadienyl titanium complexes					
XLI	$[(C_5H_5)_2TiCl]_2(\mu\text{-}O)$	443.09	orange	d. >200	55
XLII	$[(C_5H_5)TiCl(\mu\text{-}O)]_4$	657.79	yellow	d. >200	56, 57
XLIII	$[(C_5H_5)TiCl(SCH_2CH_2NH)]_n$	$(223.58)_n$	purple	d. >130	58
Ionic η^5-cyclopentadienyl titanium complexes					
XLIV	$[(C_5H_5)_2TiCl(NCCH_3)]^+[FeCl_4]^-$	452.20	reddish black	>200	59, 60
XLV	$[(C_5H_5)_2Ti(bipy)]^{2+}[CF_3SO_3]_2^-$	632.38	brown-red		61
XLVI	$[(C_5H_5)_2Ti(phen)]^{2+}[CF_3SO_3]_2^-$	656.40	dark brown-red		61
XLVII	$\{(C_5H_5)_2Ti[o\text{-}S(NHCH_3)C_6H_4]\}^+I^-$	443.20	black	180	62
XLVIII	$[(C_5H_5)Ti(1,2,4\text{-}S_2C_6H_3CH_3)_2]^-[N(C_2H_5)_4]^+$	551.74	dark black-violet	165 w.d.	63

[a] d. = decomposition; w.d. = with decomposition
[b] If available, references on X-ray structure determination are also given
[c] Possibly, XXV is a 1:1 mixture of XXIX with I[49)]

Table 2. Chemical properties of antitumor metallicenium salts

Compound		Molecular weight	Color	Melting point[a] (°C)	Synthesis[b] (Ref.)
Mononuclear ferricenium salts					
XLIX	$[(C_5H_5)_2Fe]^+[FeCl_4]^-$	383.70	dark blue	>300	64, 65
L	$[(C_5H_5)_2Fe]^+[FeBr_4]^-$	561.49	dark brown		66
LI	$[(C_5H_5)_2Fe]_2^+[Cl_3FeOFeCl_3]^{2-}$	712.49	dark blue	>300	67–69
LII	$[(C_5H_5)_2Fe]^+[SbCl_6]^-$	520.45	dark blue		70
LIII	$[(C_5H_5)_2Fe]^+[2,4,6,\text{-}(NO_2)_3C_6H_2O]^-$	414.14	dark green	170–180 w.d.	71–73
LIV	$[(C_5H_5)_2Fe]^+[CCl_3COO]^- \cdot CCl_3COOH$	511.81	dark blue	128–129	74, 75
LV	$[(C_5H_5)_2Fe]^+[CCl_3COO]^- \cdot 2\ CCl_3COOH$	675.19	dark blue	143–144	73–75
Polynuclear ferricenium salt					
LVI	$\{[CH_2C_5H_4(C_5H_4)Fe]^+[FeCl_4]^-\}_n$	$(395.69)_n$	black		76

[a] d. = decomposition; w.d. = with decomposition
[b] If available, references on X-ray structure determination are also given

Table 3. Chemical properties of antitumor main group metallocenes

Compound		Molecular weight	Color	Melting point[a] (°C)	Synthesis (Ref.)
Decasubstituted germanocene					
LVII	$[(C_6H_5)_5C_5]_2Ge$	963.76	yellow	>260 w.d.	77
Decasubstituted stannocenes					
LVIII	$[(C_6H_5)_5C_5]_2Sn$	1009.77	bright yellow	>300 w.d.	78
LIX	$[(C_6H_5CH_2)_5C_5]_2Sn$	1150.04	yellow	95 w.d.	79

[a] w.d. = with decomposition

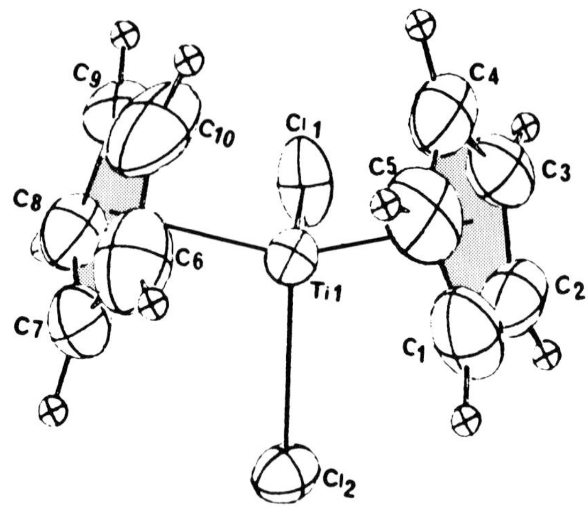

Fig. 4. Molecular structure of titanocene dichloride I. Modified according to Ref. 25

- Normally, the complexes are neutral, uncharged molecules. (It is, however, possible to synthesize charged derivatives by using neutral donor ligands L instead of X, or by introducing excess charges into the acido ligands X; cf. Chap. 6).

Some chemical properties of metallocene diacido and related complexes presented in this chapter are listed in Table 1.

1. Metallocene Dichlorides

The position of the central metal atom M in metallocene dichlorides $(C_5H_5)_2MCl_2$ can be occupied by different early transition metals of subgroup IVa (Ti, Zr, Hf), Va (V, Nb, Ta) or VIa (Mo, W).

Of the eight existing metallocene dichlorides, the six complexes I–VI with M = Ti[24, 25]; V[24], Nb[26, 27], Ta[28, 29]; Mo[27, 30], W[30] exhibited antitumor activity. This was more pronounced in the case of I–III and V than that for IV and VI.

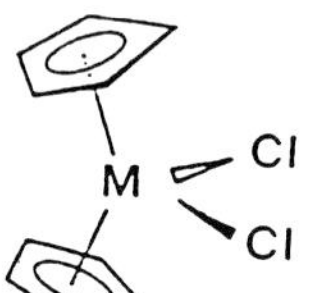

I: M = Ti
II: M = V
III: M = Nb
IV: M = Ta
V: M = Mo
VI: M = W

2. Titanocene Diacido Complexes

Within the group of titanocene diacido complexes, antitumor activity was shown for numerous symmetrically coordinated derivatives $(C_5H_5)_2TiX_2$, containing various acido ligands X. Some representatives of the unsymmetrical type $(C_5H_5)_2TiXY$ with different acido ligands X and Y have also been investigated.

Titanocene Dihalides and Pseudohalides. When both acido ligands X in the titanocene system $(C_5H_5)_2TiX_2$ were represented either by the halide ligands F, Cl, Br, I[24], or by pseudohalide ligands, e.g. NCS[31, 32] or N_3[33, 34], pronounced antitumor activity was found for I (X = Cl) and VII–XI against fluid Ehrlich ascites tumor.

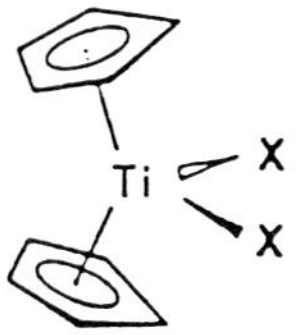

VII: X = F
VIII: X = Br
IX: X = I
X: X = NCS
XI: X = N_3

Titanocene Carboxylates. Besides halide and pseudohalide ligands, oxygen-coordinated carboxylato groups can be introduced at the position of the acido ligands X in $(C_5H_5)_2TiX_2$. Typical titanocene carboxylates exhibiting antitumor activity against various experimental tumor systems are titanocene bis(trifluoroacetate)[35] (XII), titanocene bis(trichloroacetate)[36, 37] (XIII), and titanocene bis(hydrogenmaleinate)[36, 38] (XIV).

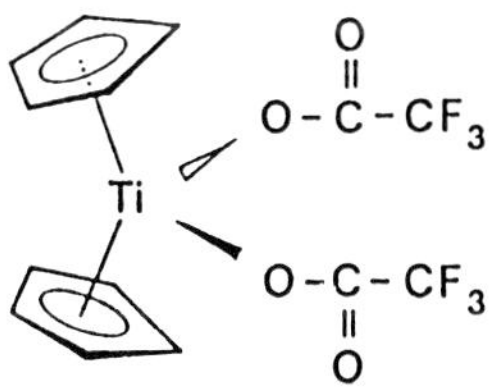

XII

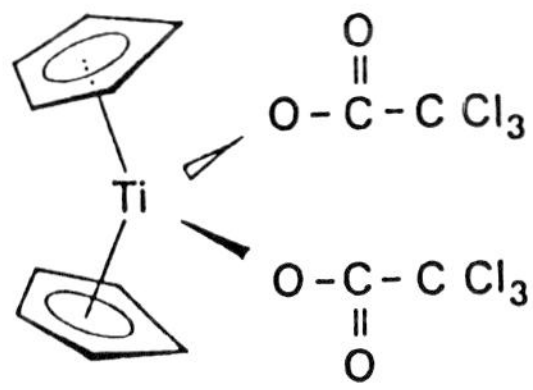

XIII

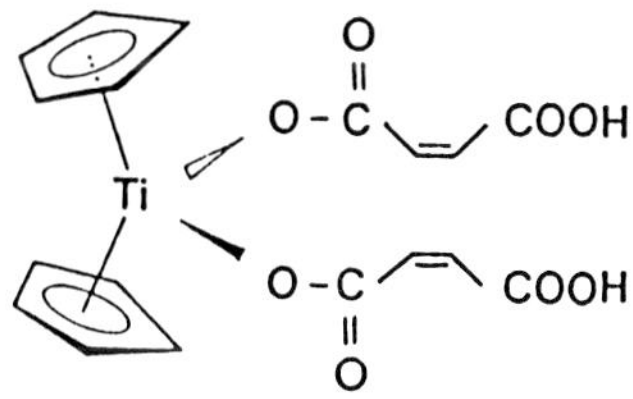

XIV

Titanocene Phenolates, Thiophenolates and Selenophenolates. Other representatives of antitumor titanocene diacido complexes are the chalcogen-coordinated phenolate, thiophenolate and selenophenolate derivatives XV–XXI. The bis(p-nitrophenolate) XV[39, 40], the bis(pentafluorophenolate) XVI[41], and the bis(pentafluorothiophenolate)

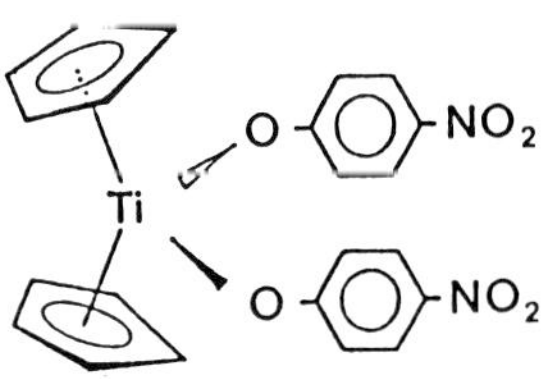

XV

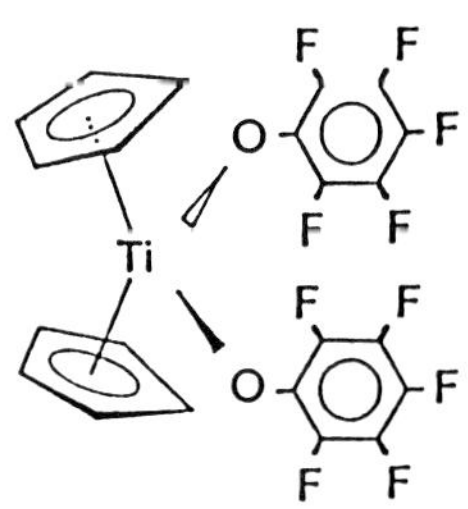

XVI

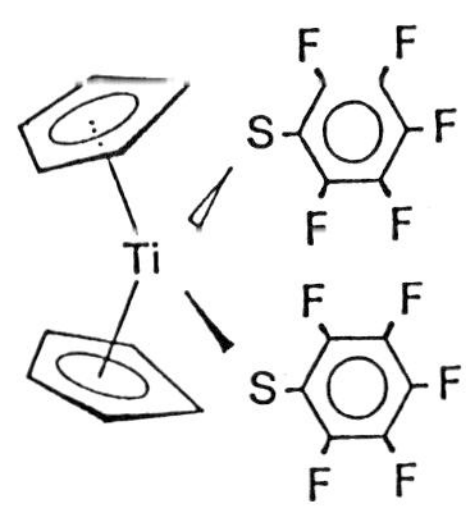

XVII

XVII[41] belong to the symmetrical $(C_5H_5)_2TiX_2$ type, whereas the monochloro complexes XVIII–XXI are unsymmetrically ligated $(C_5H_5)_2TiXY$ compounds with X = Cl and Y = OR, SR, or SeR ligands[42–44].

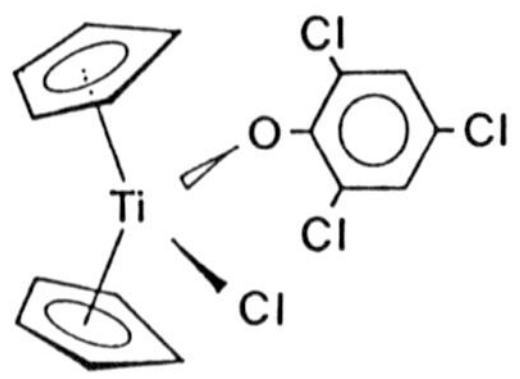

XVIII

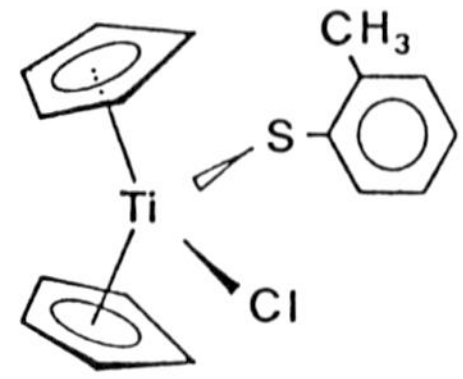

XIX

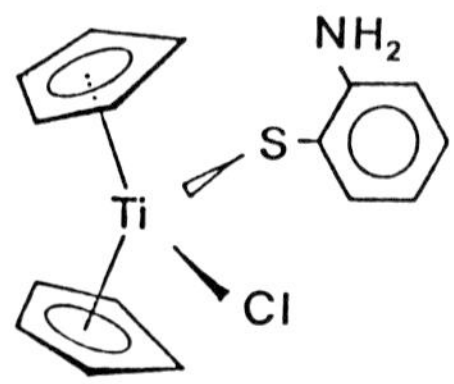

XX

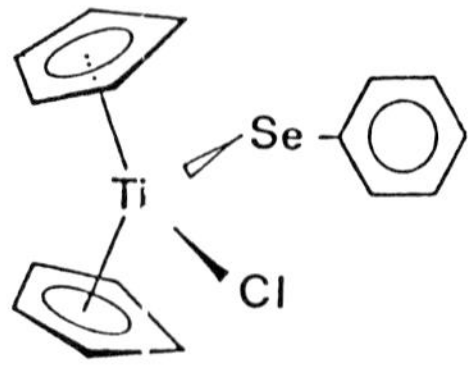

XXI

Compounds XV–XXI exhibited antitumor activity against Ehrlich ascites tumor.

Titanocene Dithiolene Chelate. The complex XXII is the only titanocene dithiolene derivative which showed growth-inhibiting activity against Ehrlich ascites tumor. In this case, the sites of the acido ligands X in $(C_5H_5)_2TiX_2$ are occupied by two sulfur atoms belonging to the bifunctional maleonitriledithiolate chelate ligand. Thus, a five-membered TiS_2C_2 chelate is present in XXII[45–47].

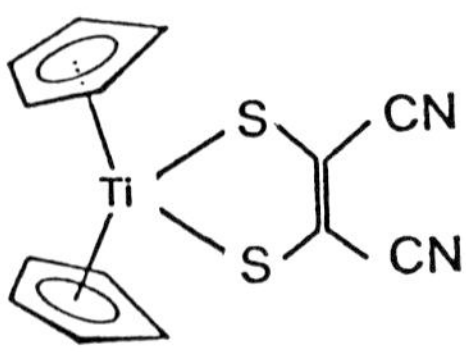

XXII

3. *Titanocene Complexes with Modified Cyclopentadienyl Ligands*

The η^5-bonded cyclopentadienyl ring ligands within the titanocene dichloride system can be widely modified.

Titanocene Complexes with Substituted Cyclopentadienyl Ligands. One or more hydrogen atoms of one or both cyclopentadienyl rings in $(C_5H_5)_2TiCl_2$ can be replaced by other monofunctional residues R in substitution modes ranging from monosubstitution and 1,2- or 1,1′-disubstitution up to decasubstitution, and including phane-like bridging of both rings by a bifunctional group Z. The following titanocene complexes with substituted cyclopentadienyl ring ligands showed antitumor activity, the strength of which was clearly dependent on the degree of modification:

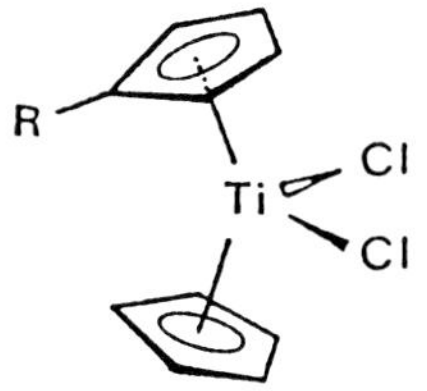

XXIII: R = C_2H_5
XXIV: R = $Si(CH_3)_3$
XXV: R = $N(CH_3)_2$[1]

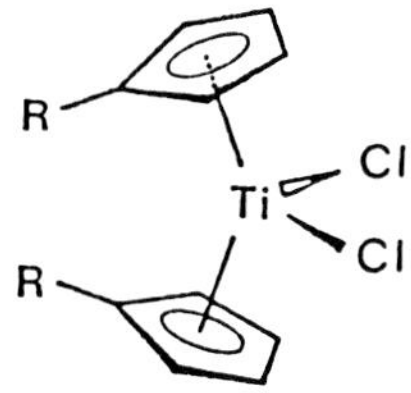

XXVI: R = $Si(CH_3)_3$
XXVII: R = $Si(CH_3)_2n\text{-}C_4H_9$
XXVIII: R = $Ge(CH_3)_3$
XXIX: R = $N(CH_3)_2$

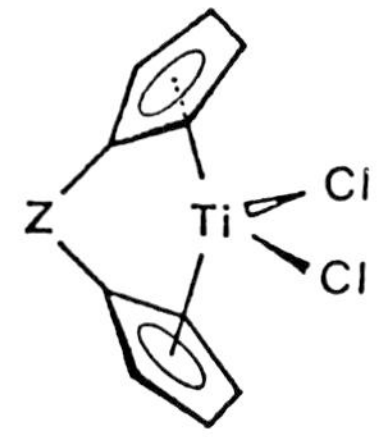

XXX: Z = CH_2
XXXI: Z = $CHCH_3$
XXXII: Z = $SiHCH_3$
XXXIII: Z = $Si(C_2H_5)_2$
XXXIV: Z = $Ge(CH_3)_2$

In the monosubstituted complexes XXIII–XXV one hydrogen atom of one cyclopentadienyl ring is formally substituted by an ethyl, trimethylsilyl or dimethylamino residue[48, 49]. Similarly, in the 1,1′-disubstituted complexes XXVI–XXIX both ring ligands are symmetrically modified by trialkylsilyl, trimethylgermyl, or dimethylamino groups[48, 49]. The 1,1′-bridged compounds XXX–XXXIV contain alkylene, alkylsilylene, or dimethylgermylene as bridging group connecting both ring ligands[48, 50, 51].

Titanocene Analogs with Indenyl or Tetrahydroindenyl Ligands. In continuation of modification of the cyclopentadienyl ligands in $(C_5H_5)_2TiCl_2$, one or both of them can be exchanged for other related, η^5-bound cyclic systems incorporating five-membered aromatic rings, such as the indenyl C_9H_7[52] and tetrahydroindenyl C_9H_{11}[31, 48] ligands. The complexes XXXV–XXXVII again revealed antitumor activity dependent on the degree of modification.

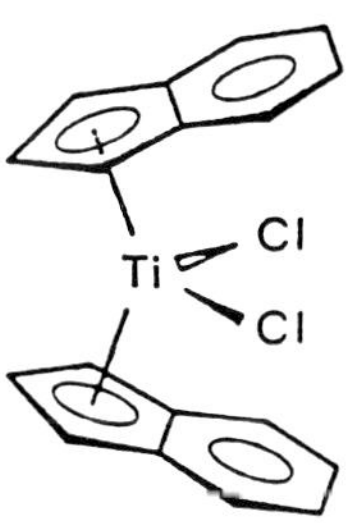

XXXV

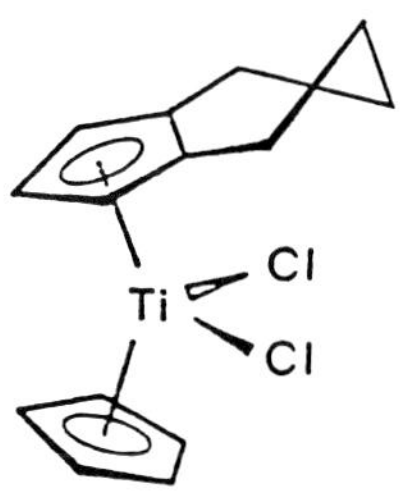

XXXVI

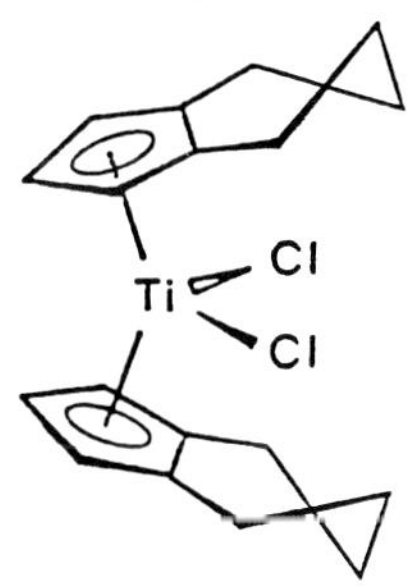

XXXVII

4. *Mono(cyclopentadienyl) Titanium Complexes*

A further extension of the modification of the titanocene system is the total replacement of one cyclopentadienyl ring in $(C_5H_5)_2TiX_2$ by an additional acido ligand Y which may be identical with the other two acido ligands X or different from them[43, 53, 54]. With XXXVIII–XL, weak antiproliferative activity was retained.

1 Possibly, XXV is a 1 : 1 mixture of XXIX with I[49].

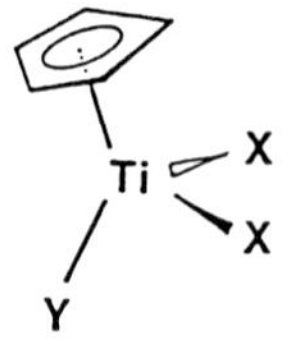

XXXVIII: X = Y = Cl
XXXIX: X = Cl; Y = SC_6H_5
XL: X = Y = NCS

5. Polynuclear Cyclopentadienyl Titanium Complexes

The oxo-bridged dinulcear and tetranuclear complexes XLI[55] and XLII[56, 57] were tested for antitumor properties as typical representatives of products which are formed by hydrolytic reactions after dissolution of $(C_5H_5)_2TiCl_2$ or $(C_5H_5)TiCl_3$ in water. As an example of polynuclear cyclopentadienyl titanium compounds, the polymeric mono(cyclopentadienyl) complex XLIII[58] was investigated.

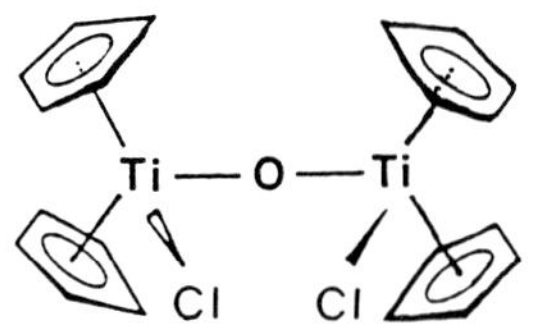

XLI

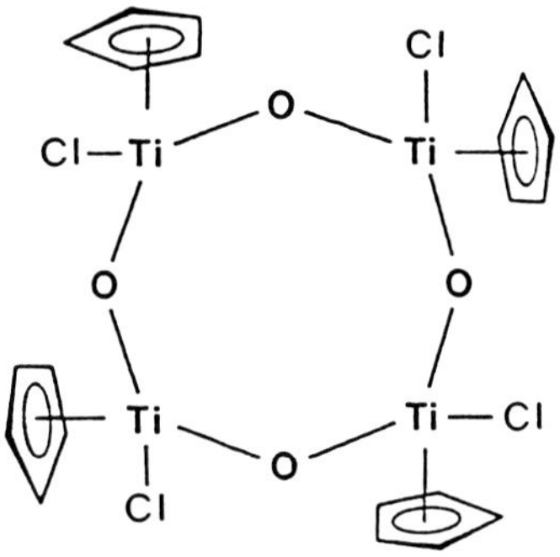

XLII

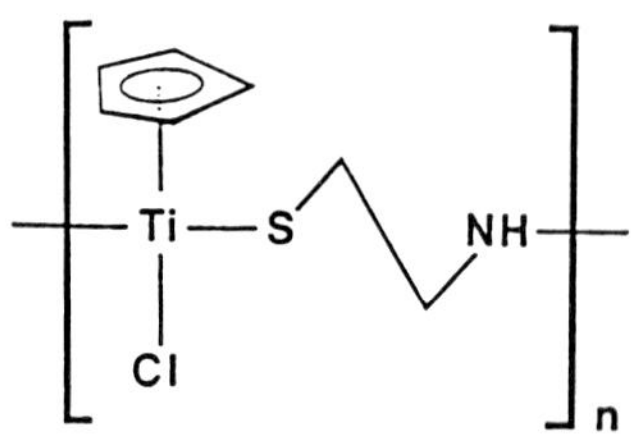

XLIII

6. Ionic Cyclopentadienyl Titanium Complexes

Mainly during the past years, titanocene complexes have been synthesized which fundamentally differ by their ionic character from the neutral titanocene compounds mentioned in the foregoing chapters. Most of them correspond to the general formula $[(C_5H_5)_2TiXL]^+Y^-$ or $[(C_5H_5)_2TiL_2]^{2+}(Y^-)_2$, where X is an anion or ligand anionic donor site, and L is a donor molecule or ligand neutral donor site, and contain the intact bis(cyclopentadienyl)titanium(IV) unit forming together with the covalently bound X and L or two L ligands the cationic complex moiety in tetrahedral configuration. Various anions Y^- are bonded via electrostatic forces. Ionic titanocene complexes are generally characterized by improved water solubility in comparison to I. Typical representatives of ionic titanocene complexes which have shown tumor-inhibiting activity are the complexes XLIV–XLVII[59–62].

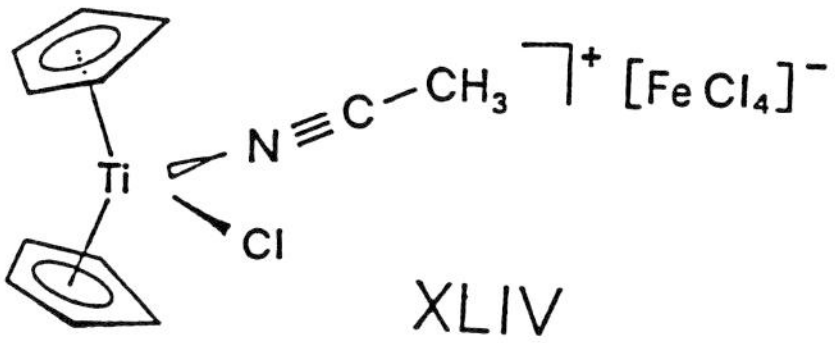

XLIV

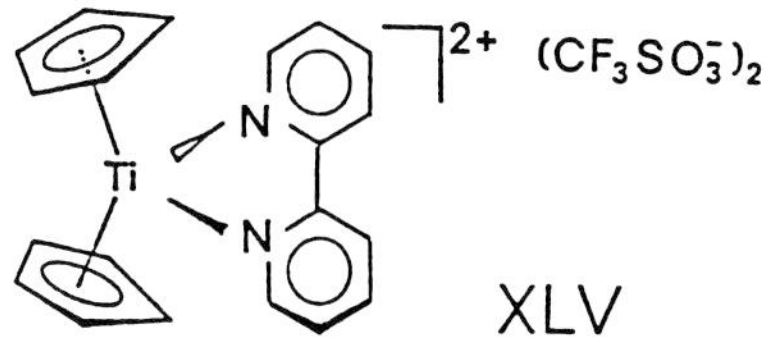

XLV

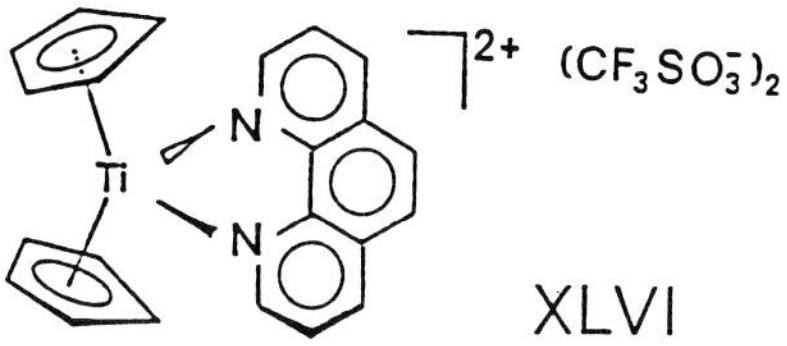

XLVI

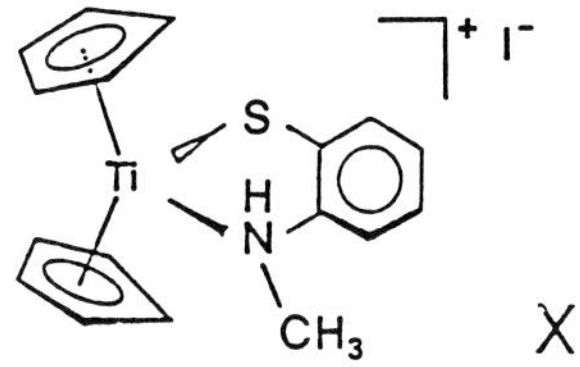

XLVII

Another ionic cyclopentadienyl titanium derivative tested for antitumor properties, the five-coordinate mono(cyclopentadienyl)titanium(IV) bis(dithiolene) chelate XLVIII[63)], belongs to the type $M^+[(C_5H_5)TiX_4]^-$ in which the cyclopentadienyl titanium moiety represents the anionic part of the complex salt.

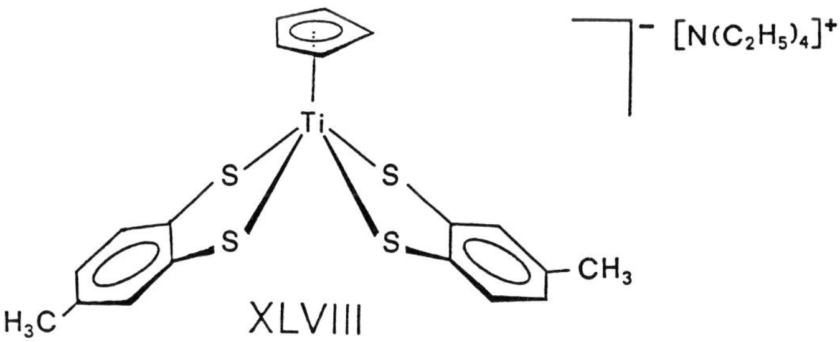

XLVIII

B. Metallicenium Salts

The metallicenium salts $[(C_5H_5)_2M]^+X^-$ are another group of antitumor metallocene compounds. They are characterized by improved water solubility and differ from metallocene diacido complexes with respect to the following properties:

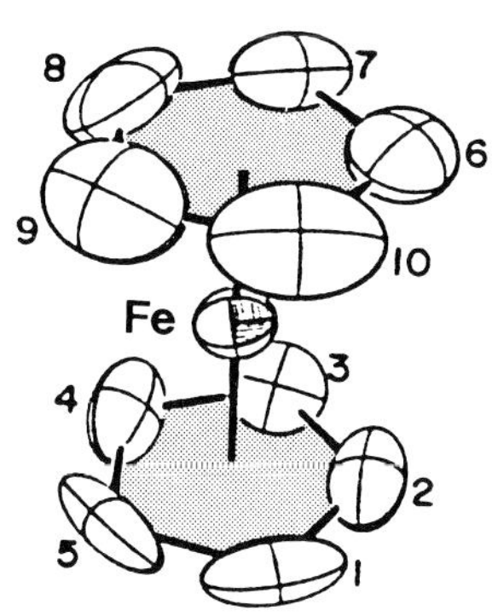

Fig. 5. Molecular structure of a ferricenium cation $[(C_5H_5)_2Fe]^+$. Modified according to Ref. 73. The H atoms are omitted

- Metallicenium salts are ionic organometallic complexes which are composed by metallicenium cations $[(C_5H_5)_2M]^+$ (Fig. 5) and anions X^-. These are not linked together by covalent bonds, but by electrostatic forces in a salt-like crystal lattice.
- Metallicenium salts contain the electron-rich medium transition metals iron or cobalt as central metal atoms M which are, in contrast to the early transition metals in the metallocene diacido complexes, coordinatively saturated when bound to two η^5-cyclopentadienyl ligands.
- The cyclopentadienyl ring planes are parallel to each other forming the usual "sandwich" arrangement (Fig. 5), in contrast to the tilted "open sandwich" geometry in metallocene diacido complexes.

Some chemical properties of typical antitumor ferricenium complexes are summarized in Table 2.

1. Mononuclear Ferricenium Salts

The ferricenium compounds XLIX–LV[64–71] containing diverse anions X^- were found able to reduce the growth of Ehrlich ascites tumor and other experimental tumor systems.

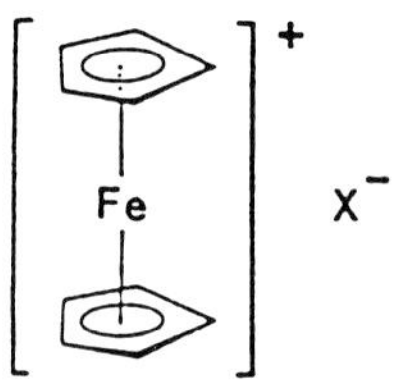

XLIX: $X^- = [FeCl_4]^-$
L: $X^- = [FeBr_4]^-$
LI: $X^- = \frac{1}{2}\,[Cl_3FeOFeCl_3]^{2-}$
LII: $X^- = [SbCl_6]^-$
LIII: $X^- = [2,4,6\text{-}(NO_2)_3C_6H_2O]^-$
LIV: $X^- = [CCl_3COO]^- \cdot 2\ CCl_3COOH$
LV: $X^- = [CCl_3COO]^- \cdot CCl_3COOH$

2. Polynuclear Ferricenium Salt

Because of its analogy to XLIX which includes the same anion, the complex LVI was tested as representative of polymeric ferricenium complexes.

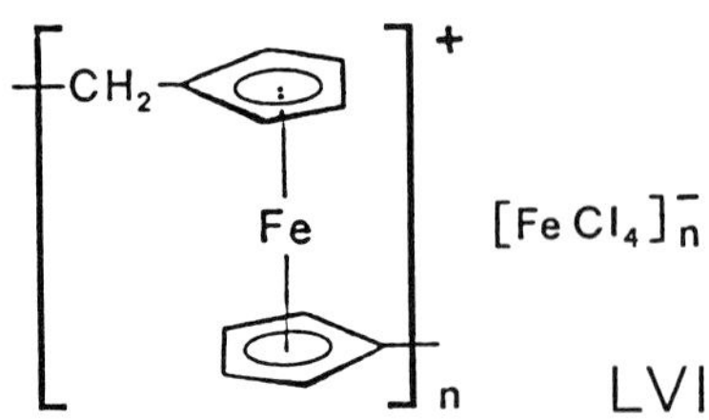

LVI

C. Main Group Metallocenes

Pilot experiments performed during the past months revealed antitumor activity for a third type of metallocene compounds represented by neutral stannocene and germanocene derivatives decasubstituted at the cyclopentadienyl rings. They differ from metallocene diacido complexes as well as from metallicenium salts by the following structural features:

- Neutral group IV metallocenes are uncharged complexes which neither contain acido ligands bound covalently nor counterions linked by electrostatic forces.
- As central metal atoms they include main group elements of group IV such as germanium and tin.
- All ten hydrogen atoms of the two cyclopentadienyl ring ligands are substituted by phenyl (C_6H_5) or benzyl ($C_6H_4CH_2$) residues. In the decaphenylmetallocenes, i.e. bis(η^5-pentaphenylcyclopentadienyl)metal(II) compounds, the cyclopentadienyl rings are forced by steric reasons into the parallel "sandwich" arrangement (Fig. 6a). In the case of the decabenzylmetallocenes, the cyclopentadienyl ring planes are tilted (Fig. 6b) resulting in an "open sandwich" structure similar to that of, e.g., the unsubstituted stannocene $(C_5H_5)_2Sn$.

a

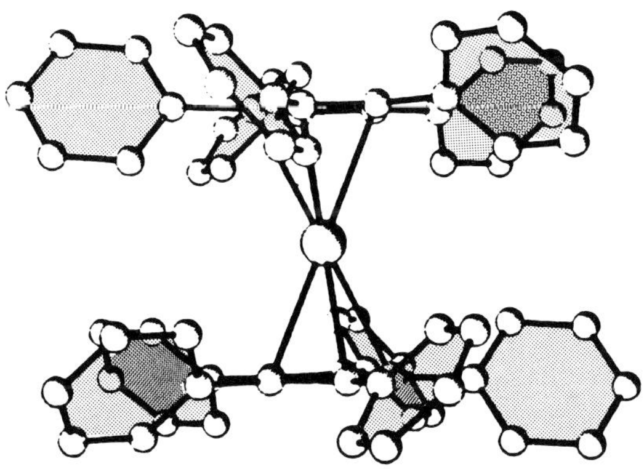

b

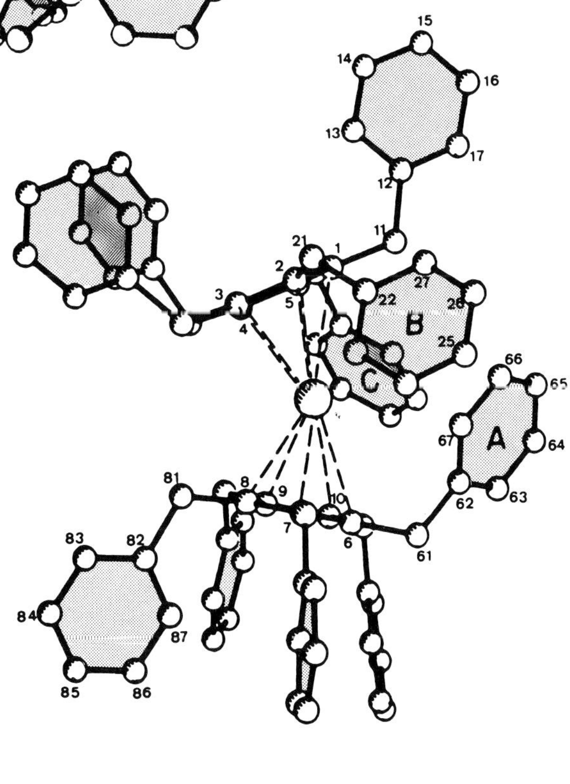

Fig. 6a, b. Molecular structures of decaphenylstannocene LVIII (**a**) and decabenzylstannocene LIX (**b**). Modified according to Refs. 78, 79

Main group metallocenes which were found to exhibit antitumor properties are decaphenylgermanocene LVII[73] as well as decaphenylstannocene LVIII[74] and decabenzylstannocene LIX[75]. All three compounds (Table 3) are air-stable solids at room temperature. The X-ray structures of LVIII and LIX are illustrated in Fig. 6.

III. Antiproliferative Properties

In vitro, the antiproliferative activity of some of the compounds listed in Tables 1–3 was investigated using various strains of cultured tumor and normal cells.

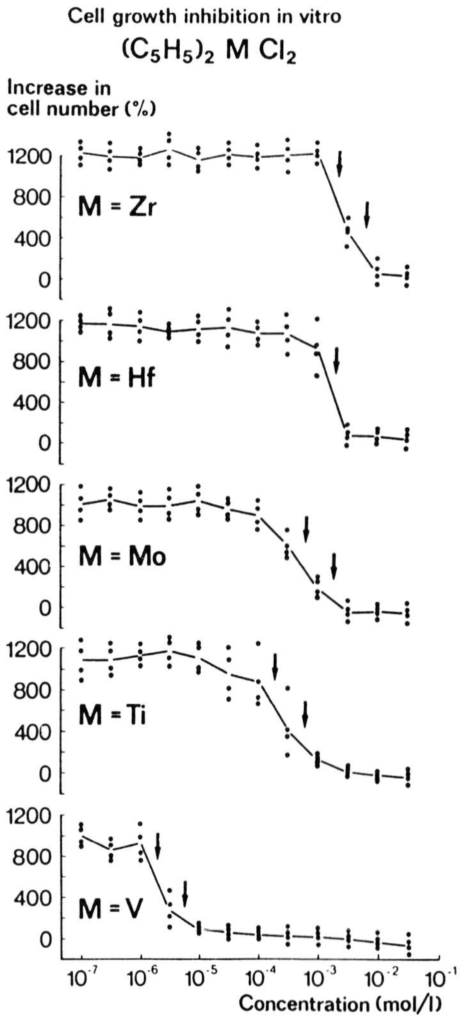

Fig. 7. Effect of a 90-min treatment of in vitro cultured Ehrlich ascites tumor cells with various concentrations of metallocene dichlorides on the increase in cell number, determined 72 h after removal of agents. *Arrows* indicate highly significant differences ($\alpha < 0.1\%$) between proliferation rates of neighboring groups

A. Cytostatic Properties in Vitro

The antiproliferative properties of the metallocene dichlorides $(C_5H_5)_2MCl_2$ with M = Ti (I), Zr, Hf, V (II) and Mo (V) and the ferricenium salts $[(C_5H_5)_2Fe]^+X^-$ XLIX, LI, LIII–LV were studied in vitro using cultured animal Ehrlich ascites tumor cells. The results obtained are illustrated in Figs. 7 and 8. Though the compounds I, II, V, LIII, and LV effected pronounced antitumor activity in vivo against Ehrlich ascites tumor, all inducing an optimum cure rate of 100% (cf. Chap. III.B.), the growth inhibitions observed in vitro were of distinctly different strength[80, 81]:

– Vanadocene dichloride (II) represented that metallocene compound which showed best activity in vitro. It effected highly significant reduction in cell proliferation at a

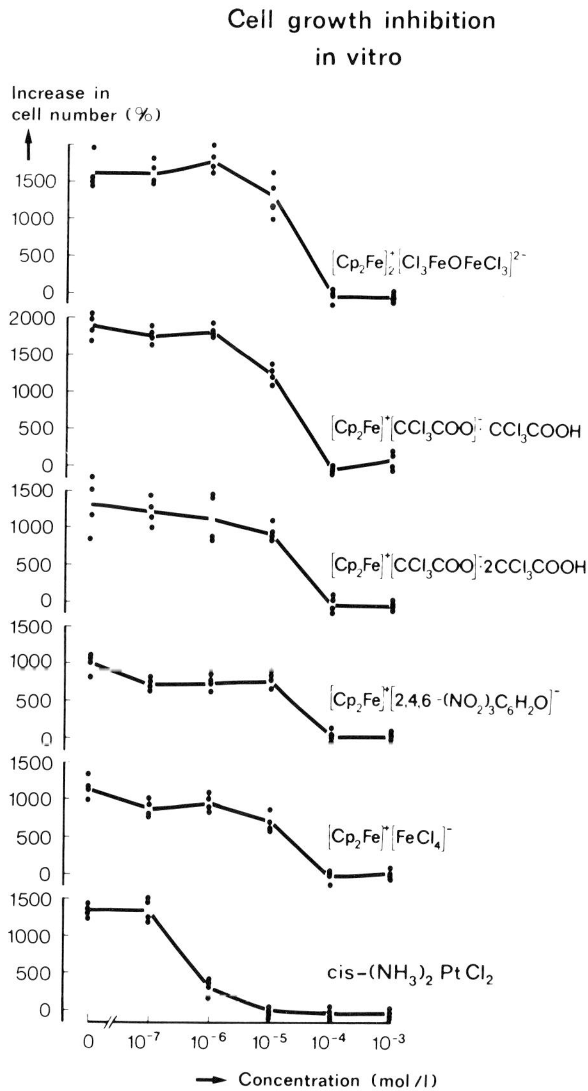

Fig. 8. Effect of continuous exposure of in vitro cultured Ehrlich ascites tumor cells at various concentrations of ferricenium salts on the increase in cell number determined 72 h after drug addition

concentration as low as 5×10^{-6} mol/l; at this concentration, cellular proliferation was inhibited by more than 50% compared with untreated control populations (Fig. 7).

- In the case of the ferricenium salts XLIX, LI, LIII–LV, tenfold higher concentrations ranging between 10^{-5} and 10^{-4} mol/l were needed to cause a 50% reduction of proliferation (Fig. 8).
- For titanocene dichloride (I) and molybdenocene dichloride (V), concentrations of 5×10^{-4} and 10^{-3} mol/l, respectively, i.e., 100-fold higher concentrations than for II, were required to induce equivalent effects in vitro (Fig. 7).
- Zirconocene and hafnocene dichlorides, which were inactive against Ehrlich ascites tumor in vivo, inhibited cellular growth only at concentrations of 5×10^{-3} mol/l and higher (Fig. 7).

Using other cell lines cultured in vitro it was shown that there does apparently not exist cell specifity of the antiproliferative action of metallocene compounds. The growth of animal and human tumor cells as well as of normal, not transformed cells was suppressed by the same concentrations of I and II as mentioned above. The cell lines which were investigated and the ID_{50} values obtained are summarized in Table 4.

Table 4. Antiproliferative activity of $(C_5H_5)_2TiCl_2$ (I) and $(C_5H_5)_2VCl_2$ (II) in vitro

Cell line	Compound	ID_{50}[a] (mol/l)	Ref.
Ehrlich ascites tumor cells	I	4×10^{-4}	80
	II	3×10^{-6}	80
Human KB tumor cells	I	3×10^{-4}	85
	II	3×10^{-6}	85
Human HeLa tumor cells	I	4×10^{-4}	86
	II	4×10^{-6}	86
Human epidermoid (HEP-2) tumor cells	I	[b]	
	II	2×10^{-6}	87, 88
Human embryonic fibroblasts	I	5×10^{-4}	89, 90
	II	3×10^{-6}	89, 90

[a] ID_{50} = concentration effecting 50% inhibition of cellular proliferation
[b] not determined

The differences concerning the cell growth-inhibiting potencies of antitumor metallocene and metallicenium complexes are surprising, considering the pattern of in vivo activity (cf. Chap. III.B.). Either additional mechanisms[82] fortifying the antitumor activity in vivo in the case of some metallocene compounds, e.g. of I, or hydrolytic degradation[36, 57, 83, 84] occurring especially at low concentration levels in the aqueous culture medium, may be responsible factors. In this connection it is worth mentioning that, indeed, the hydrolytic stability of the M-(C_5H_5) bond of metallocene dichlorides in unbuffered aqueous KNO_3 solution was found to be dependent on the metal atom present decreasing according to the order $(C_5H_5)_2VCl_2 > (C_5H_5)_2TiCl_2 \gg (C_5H_5)_2ZrCl_2$[84].

B. Antitumor Properties Against Animal Tumors

All η^5-cyclopentadienyl metal complexes listed in Tables 1–3 were investigated against fluid Ehrlich ascites tumor, and antitumor activity of varying strength was detected (Tables 5–7). Some of those compounds, which were characterized by pronounced tumor-inhibiting activity against Ehrlich ascites tumor, were additionally tested against other fluid and solid animal tumors.

Table 5. Pharmacological data of antitumor η^5-cyclopentadienyl early transition metal complexes in CF1 mice bearing fluid Ehrlich ascites tumor

Compound	Optimum dose range (mg/kg)	Optimum cure rate (%)	T.I.[a]	LD_{50} (mg/kg)	LD_{100} (mg/kg)
Metallocene dichlorides					
I	40– 60	100	3.3	100	140
II	80– 90	100	1.4	110	130
III	20– 25	100	3.5	35	45
IV	80–170	25	–	200	250
V	75–100	100	2.9	175	275
VI	100–400	13	–	500	650
Titanocene halides and pseudohalides					
VII	40	100	2.0	60	80
VIII	40– 80	100	4.5	135	180
IX	60–100	100	2.6	145	180
X	80	100	3.4	135	200
XI	50– 70	100	2.4	95	130
Titanocene carboxylates					
XII	80–120	100	2.0	160	200
XIII	100–360	100	5.5	440	500
XIV	60–120	100	3.8	170	230
Titanocene phenolates, thiophenolates and selenophenolates					
XV	180–240	10	–	260	360
XVI	340–360	100	1.5	480	540
XVII	120–180	100	2.6	260	320
XVIII	60–140	50	–	200	300
XIX	30– 60	25	–	60	80
XX	50–100	13	–	100	120
XXI	80–100	80	–	150	200
Titanocene dithiolene chelate					
XXII	80–120	85	–	240	260
Titanocene complexes with substituted η^5-cyclopentadienyl ligands					
XXIII	50– 70	60	–	100	120
XXIV	40– 70	80	–	100	120
XXV	40– 60	30	–	70	90
XXVI	280–300	20	–	360	460
XXVII	320–360	20	–	420	500
XXVIII	240–320	28	–	400	460
XXIX	80–160	10	–	200	260
XXX	220–260	13	–	360	440
XXXI	200–260	10	–	320	380

Table 5 (continued)

Compound	Optimum dose range (mg/kg)	Optimum cure rate (%)	T.I.[a]	LD_{50} (mg/kg)	LD_{100} (mg/kg)
XXXII	160–220	30	–	300	400
XXXIII	140–200	15	–	220	300
XXXIV	220–260	13	–	280	360
η^5-Indenyl and η^5-tetrahydroindenyl titanium complexes					
XXXV	280–320	13	–	360	400
XXXVI	80–120	73	–	120	150
XXXVII	260–300	13	–	320	150
Mono(η^5-cyclopentadienyl) titanium complexes					
XXXVIII	50–130	31	–	130	150
XXXIX	60–100	24	–	100	120
XL	70–110	24	–	120	140
Polynuclear η^5-cyclopentadienyl titanium complexes					
XLI	40–280	54	–	300	380
XLII	160–480	38	–	480	560
XLIII	220–260	70	–	320	400
Ionic η^5-cyclopentadienyl titanium complexes					
XLIV	80–140	75	–	180	220
XLV	200–220	100	1.6	260	300
XLVI	140–220	67	–	240	340
XLVII	200–320	100	1.7	380	500
XLVIII	40	67	–	60	120

[a] T.I. = therapeutic index (defined as the relation LD_{50}/ED_{90}; when optimum cure rate $<90\%$, no value is given)

Table 6. Pharmacological data of antitumor metallicenium salts in CF1 mice bearing fluid Ehrlich ascites tumor

Compound	Optimum dose range (mg/kg)	Optimum cure rate (%)	T.I.[a]	LD_{50} (mg/kg)	LD_{100} (mg/kg)
Mononuclear ferricenium salts					
XLIX	180–200	67	–	240	300
L	140–180	67	–	260	320
LI	180	83	–	290	320
LII	40–120	90	3.5	140	180
LIII	220–240	100	1.7	340	420
LIV	120–160	83	–	210	260
LV	220–300	100	2.0	400	480
Polynuclear ferricenium salt					
LVI	200–340	15	–	440	580

[a] T.I. = therapeutic index (defined as the relation LD_{50}/ED_{90}; when optimum cure rate $<90\%$, no value is given)

Table 7. Pharmacological data[a] of antitumor main group metallocenes

Compound	Optimum dose range (mg/kg)	Optimum cure rate (%)	LD_{50} (mg/kg)	LD_{100} (mg/kg)
Decasubstituted germanocene				
LVII	140–340	60–100	> 340	> 340
Decasubstituted stannocenes				
LVIII	220–400	60–100	> 460	> 460
LIX	120–260	40– 60	> 260	> 260

[a] All values given are preliminary values which are not exactly determined

1. Fluid Ehrlich Ascites Tumor

In the group of *metallocene dichlorides,* a graduated strength of antitumor activity was shown against fluid Ehrlich ascites tumor. Whereas the complexes I, II, III and V, containing the first and second row transition metals Ti, V, Nb and Mo, caused marked tumor inhibition with distinct dose-activity relationships (Fig. 9) and a cure rate of 100% at optimum doses[13, 14, 91–94)], the metallocene dichlorides IV and VI with the higher homologues Ta and W only sporadically induced the survival of treated tumor-bearing mice[93)]. In the case of $(C_5H_5)_2ZrCl_2$ and $(C_5H_5)_2HfCl_2$, no tumor-inhibiting properties could be observed against fluid Ehrlich ascites tumor[14, 93)].

Numerous titanocene derivatives $(C_5H_5)_2TiX_2$ were tested which had been modified at the position of the acido ligands X. Investigating the antitumor activity of *titanocene dihalides and bis(pseudohalides)* VII–XI[95)] revealed compounds which were equally potent as I, the optimum cure rate amounting to 100%. In the case of VIII with X = Br, the therapeutic range was enlarged leading to an increase of the therapeutic index (T.I.) to 4.5 (Fig. 9), compared with 3.3 after application of I.

Optimum cure rate of 100% was also registered after treatment with the *titanocene carboxylates* XII–XIV[37)], whereby broad therapeutic ranges and therapeutic indices of 5.5 and 3.8 became manifest after administration of the trichloroacetate and hydrogenmaleinate derivatives XIII and XIV (Fig. 9).

Other tumor-inhibiting titanocenes are represented by the *titanocene phenolate, thiophenolate, selenophenolate and dithiolene complexes* XV–XXII[86)]. Best antitumor activity and an optimum cure rate of 100% were observed in the case of the pentafluorochalcogenophenolato complexes XVI and XVII, followed by the dithiolene chelate XXII and the selenophenolate XXI which induced cure rates of 85 and 80%, respectively. On the other hand, no tumor-inhibiting properties against fluid Ehrlich ascites tumor could be detected for other titanocene chalcogeno derivatives, e.g. for $(C_5H_5)_2Ti(SR)_2$[96)] or $(C_5H_5)_2TiS_5$[97)].

When titanocene complexes were modified at the η^5-bound organic cyclopentadienyl rings, a reduction in the strength of antitumor potency against fluid Ehrlich ascites tumor was the general finding, whereby the degree of reduction was clearly dependent on the degree of modification[48)]:

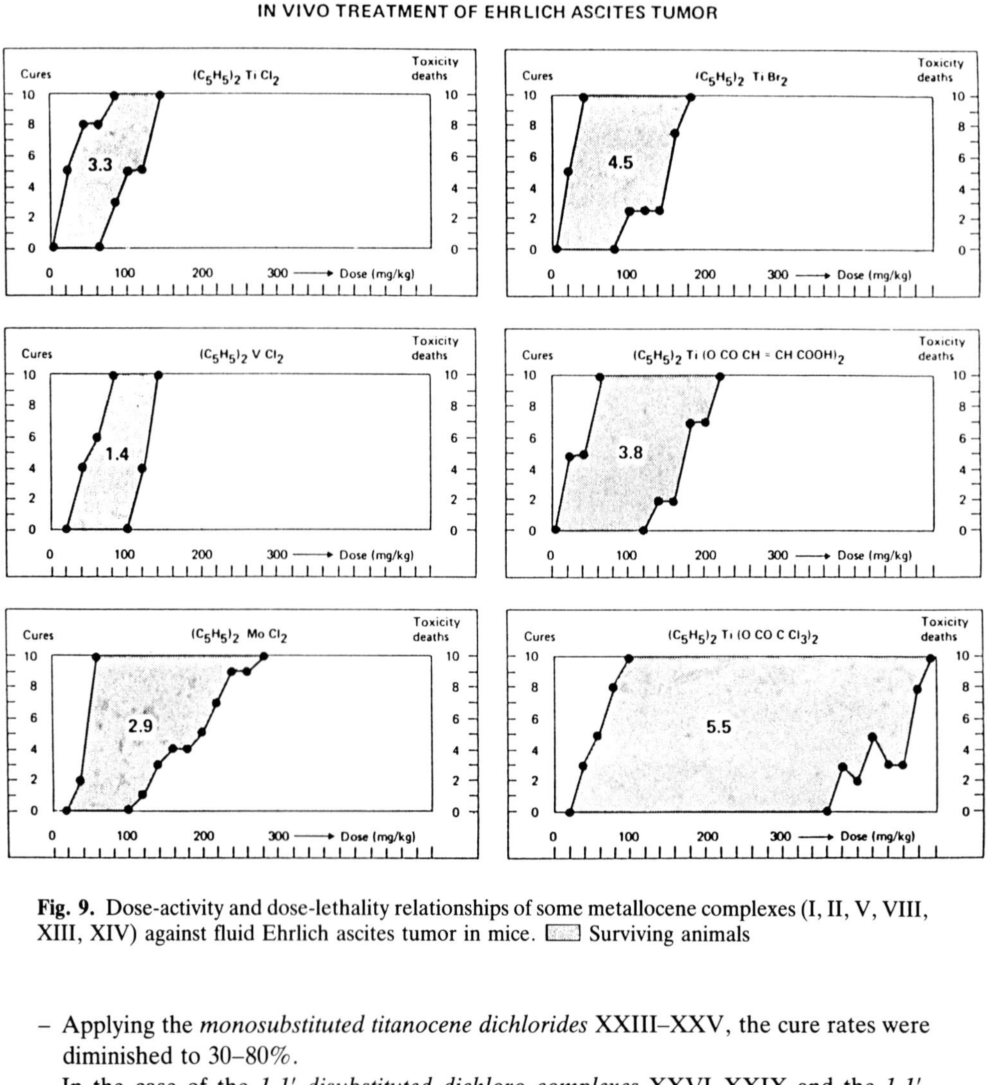

Fig. 9. Dose-activity and dose-lethality relationships of some metallocene complexes (I, II, V, VIII, XIII, XIV) against fluid Ehrlich ascites tumor in mice. ▒ Surviving animals

- Applying the *monosubstituted titanocene dichlorides* XXIII–XXV, the cure rates were diminished to 30–80%.
- In the case of the *1,1'-disubstituted dichloro complexes* XXVI–XXIX and the *1,1'-bridged derivatives* XXX–XXXIV, only sporadic cures without a strong dose-activity relationship occurred and 10–30% of the treated animals survived.
- When four or five substituents were introduced into each of both cyclopentadienyl rings in titanocene or vanadocene complexes, the antitumor activity against Ehrlich ascites tumor was totally lost. This was shown for $[(CH_3)_5C_5]_2TiCl_2$[98] (LXIa) as well as for $[(CH_3)_5C_5]_2V(NCS)_2$ and $[(CH_3)_4C_5H]_2V(NCS)_2$[99].

After treatment with the *indenyl and tetrahydroindenyl titanium complexes* XXXV–XXXVII[48] and the *mono(cyclopentadienyl) titanium derivatives* XXXVIII–XL[43], phenomena comparable with those observed after application of substituted titanocene derivatives occurred. When only one cyclopentadienyl ring was exchanged by the equally η^5-bound tetrahydroindenyl ligand (XXXVI), the optimum cure rate amounted to 73% whereas, in the case of the bis(indenyl) and bis(tetrahydroindenyl) derivatives XXXV and XXXVII, a sporadic cure rate of only 13% was registered. On replacing one of both cyclopentadienyl rings by an additional acido ligand (XXXVIII–XL), the cure rates were again markedly reduced in comparison to the parent compound I and only 24–32% of the animals treated with optimum doses survived.

Regarding the compounds XLI–XLIII as representatives of *di-, tetra- and polynuclear cyclopentadienyl titanium complexes,* the optimum cure rates ranged between 38 and 70%, whereby as long as bis(cyclopentadienyl) titanium moieties were present, the antitumor activity was still more pronounced than in the case of XLII containing four mono-(cyclopentadienyl) titanium units. The oxobridged complexes XLI and XLII are typical examples of products formed by hydrolytic reactions after dissolution of I and XXXVIII, respectively, in water. The clearly reduced strength of their antitumor potency in comparison to the parent compounds underlines that hydrolysis does not seem to be a step producing the intrinsically active species.

Ionic cyclopentadienyl titanium complexes which were tested for antitumor properties comprised the bis(cyclopentadienyl) compounds XLIV–XLVII and the mono(cyclopentadienyl) derivative XLVIII. They contained N atoms as ligand neutral donor sites and/or S atoms as ligand anionic donor sites. All ionic cyclopentadienyl titanium complexes listed in Tables 1 and 5 exhibited antitumor effectivity according to a strong dose-activity relationship[86, 98, 100]. The cure rates which were induced ranged between 67 and 100%. Because of the improved solubility in water and the pronounced antitumor activity against fluid Ehrlich ascites tumor, ionic cyclopentadienyl titanium complexes are interesting candidates for further experimental investigations.

Another type of antitumor cyclopentadienyl metal complexes are *ferricenium salts.* Because of their ionic character, many of them are highly soluble in water. For the

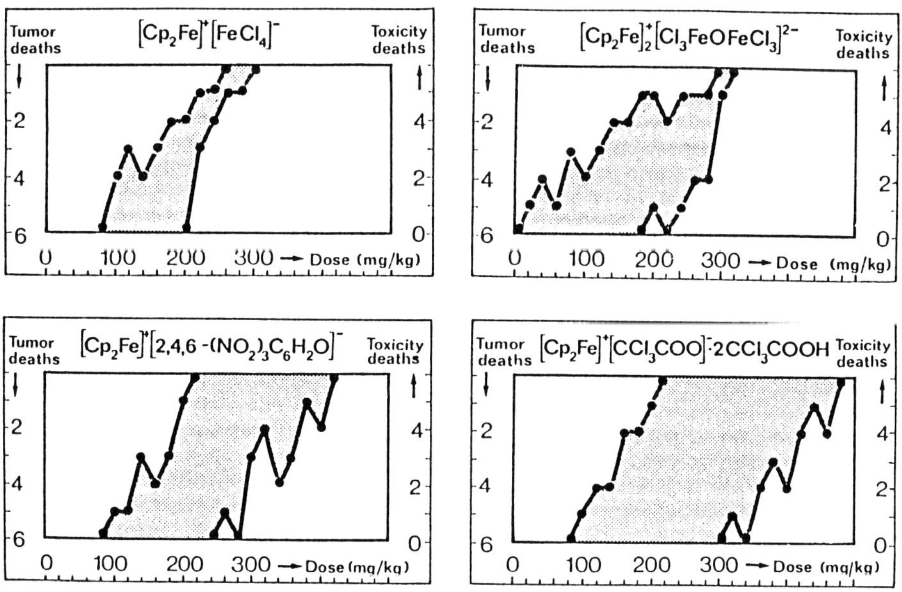

Fig. 10. Dose-activity and dose-lethality relationships of the ferricenium complexes XLIX, LI, LIII and LV against Ehrlich ascites tumor in mice. ▭ Surviving animals

mononuclear complexes XLIX–LV listed in Tables 2 and 6 marked cytostatic properties were found against fluid Ehrlich ascites tumor[16] whereby the picrate and trichloroacetate derivatives LIII and LV exhibited superior antitumor activity and induced a cure rate of 100% at optimum doses (Fig. 10). On the other hand, the poorly water-soluble ferricenium heptamolybdate $[(C_5H_5)_2Fe]^+[H_5Mo_7O_{24}]^- \cdot 2\ H_2O$ did not effect tumor inhibition[16]. As representative of polymeric ferricenium complexes, LVI was tested against Ehrlich ascites tumor. It caused an only low sporadic cure rate of 15%, the toxic threshold values lying higher than with all mononuclear ferricenium salts XLIX–LV.

The third group of antitumor cyclopentadienyl metal complexes are some main group metallocenes of the type of *decasubstituted germanocene and stannocene*. The preliminary results obtained recently indicate tumor-inhibiting properties against Ehrlich ascites tumor[101]. There were broad therapeutic ranges and optimum cure rates between 60 and 100%. Further studies with Ehrlich ascites tumor and other experimental tumor systems are ongoing to confirm and establish the antitumor activity of neutral main group IV metallocenes.

2. *Fluid Sarcoma 180*

Against sarcoma 180 growing as fluid tumor in the peritoneal cavity, the titanocene complexes I, VIII, XIII and XIV were tested[102]. Best results were obtained for I which prolonged the mean survival period of mice, treated with a single injection of optimum doses, by 160–185% and caused cures of 40–50% of the animals. The other titanocene derivatives effected increases in life span of 90–140% and maximum cure rates of 20%. On the other hand, pilot studies with the ferricenium salts XLIX, LI, LIII and LV did not reveal tumor-inhibiting activity of these compounds against ascitic sarcoma 180.

3. *Leukemias L1210 and P388*

In spite of the pronounced antitumor efficacy of titanocene complexes against fluid Ehrlich ascites tumor and fluid sarcoma 180, the metallocene dichlorides I and II showed only marginal activity against the lymphoid leukemia L1210 and the lymphocytic leukemia P388 in vivo[103]. After application of single doses, the survival of animals bearing these tumors was prolonged by only 20–30% (Fig. 11). No further increases in life span were induceable, e.g. by multiple substance administration. Treatment with the ferricenium salts XLIX, LI and LIII–LV did not effect any elongation of the mean survival period of animals bearing L1210 or P388.

At this point it is worth mentioning that a comparable discrepancy in the antitumor activity against the leukemias L1210 and P388 on the one hand and against other experimental tumor systems, e.g. Ehrlich ascites tumor and numerous solid tumors, on the other hand, is not confined to metallocene and ferricenium complexes, but was similarly found in the case of other inorganic and organometallic metal complexes of titanium[104], copper[105], gold[20] or germanium[106]. Therefore, it may be suggested that neither the L1210 nor the P388 system, which belong to the standard test systems of the National Cancer Institute, are appropriate tumor systems for initial screening trials in the course of antitumor testing of metal complexes.

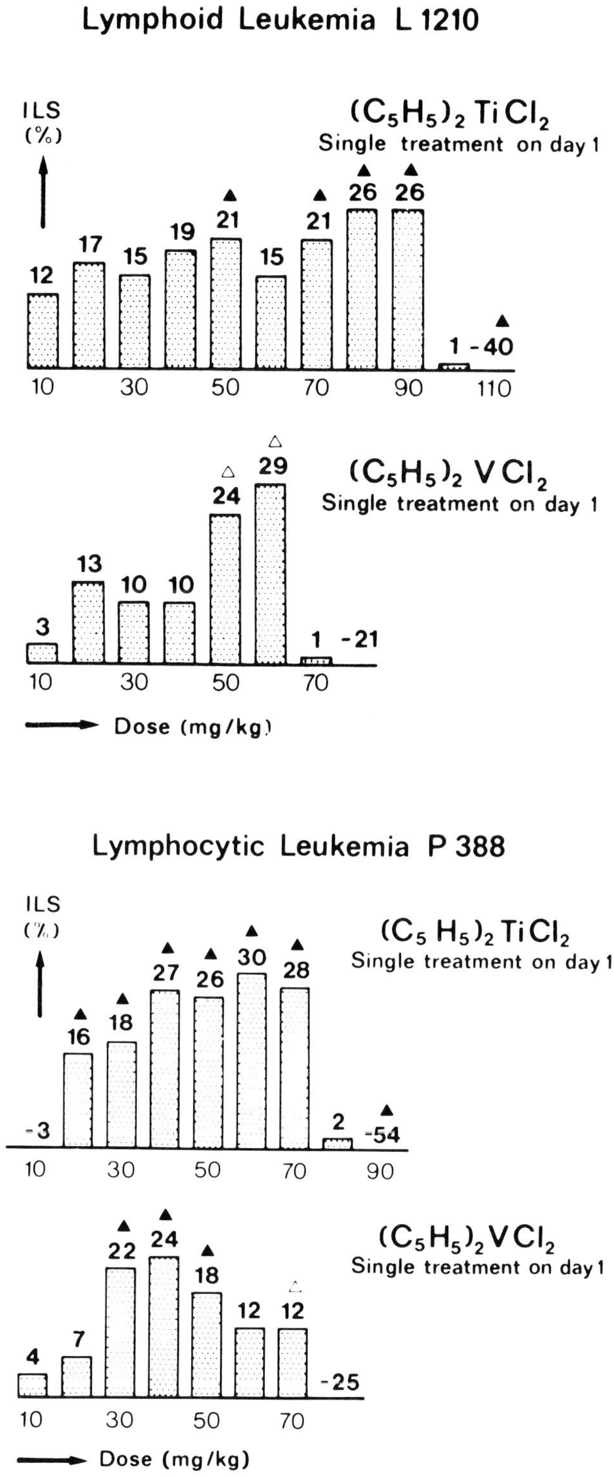

Fig. 11. Dose-dependent increase in life span *(ILS)* after treatment of L1210 *(above)* and P388 *(below)* with $(C_5H_5)_2TiCl_2$ (I) or $(C_5H_5)_2VCl_2$ (II) on day 1 after tumor transplantation. △ Significant ($2P < 0.05$), ▲ highly significant ($2P < 0.01$) difference between survival time of experimental and control animals

4. Solid Ehrlich Ascites Tumor

The investigations with solid experimental tumors were undertaken to study the systemic activity of metallocene and ferricenium compounds.

When animals bearing solid Ehrlich ascites tumor, which grew subcutaneously in the nuchal region, were treated with intraperitoneal injections of I or II[103], the growth of the tumors was inhibited in a clearly dose-dependent manner by 40–86% (Table 8) to T/C ratios ranging between 60 and 14% (Fig. 12). Best results, i.e. tumor inhibition by

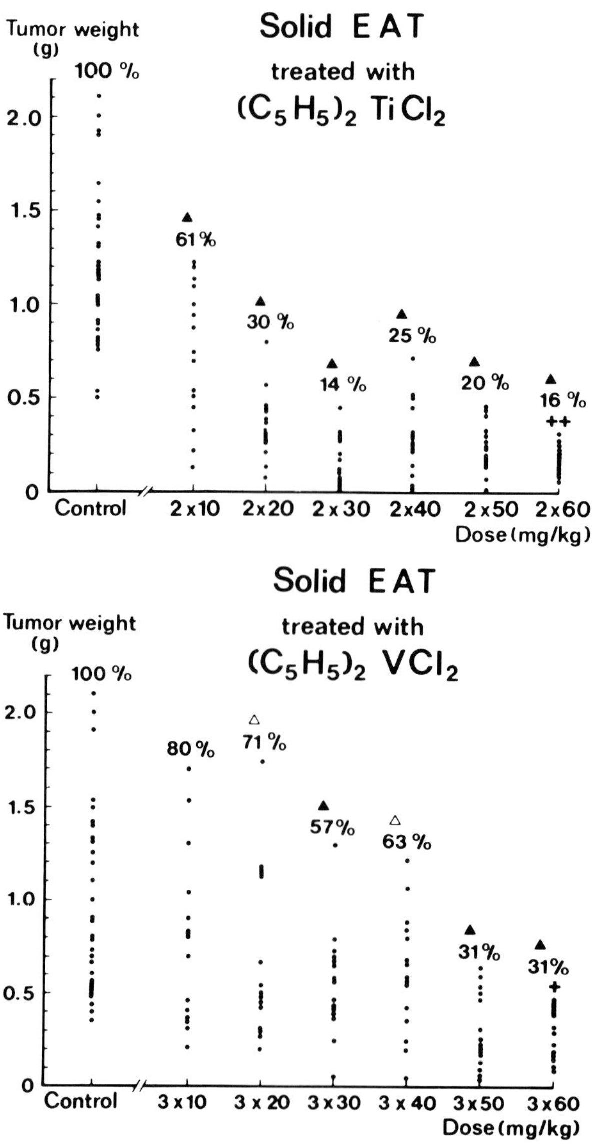

Fig. 12. Ranged weights of solid Ehrlich ascites tumors on day 8 after tumor transplantation after treatment with I *(above)* or II *(below)*. The numbers on the top represent T/C values. △ Significant ($2 P < 0.05$), ▲ highly significant ($2 P < 0.01$) difference of the tumor weights compared to the controls. + Toxic deaths before key-date

80–86%, were obtained by administration of twofold doses of I. Treatment with cisplatin (2 × 10 mg/kg) evoked a comparable T/C value of 19%.
Within the group of ferricenium salts, antitumor activity against solid Ehrlich ascites tumor was tested for the compounds listed in Table 9. They induced less pronounced growth inhibitions against this tumor system than I and II, whereby tumor suppression by about 50% was effected by the tetrachloroferrate, μ-oxo-bis(trichloroferrate) and the trichloroacetate derivatives XLIX, LI and LV[81)].

Table 8. Growth inhibitions[a] effected by some η^5-cyclopentadienyl early transition metal complexes in diverse solid experimental animal tumors

Compound	Applied dose (mg/kg)	*Solid animal tumor systems* Ehrlich ascites tumor	Sarcoma 180	Colon 38 adeno-carcinoma	B16 melanoma	Lewis lung carcinoma
I	1 × 40	–	0	28	28	22
	1 × 50	–	34	59	59	31
	1 × 60	–	–	61	61	65
	2 × 30	86	–	–	–	–
	2 × 40	75	–	–	–	–
	2 × 50	80	–	–	–	–
	2 × 60	84	–	–	–	–
	3 × 30	–	–	63	15	53
	3 × 40	–	42	72	31	71
	3 × 50	–	77	–	63	–
	5 × 20	–	–	72	31	34
	5 × 30	–	–	81	80	66
II	3 × 30	43	–	–	–	–
	3 × 40	37	–	–	–	–
	3 × 50	69	–	–	–	–
	3 × 60	69	–	–	–	–
VIII	1 × 40		0		31	
	1 × 50	–	6	52	51	13
	1 × 60	–	–	64	58	25
	1 × 70	–	–	61	55	52
	3 × 40	–	–	45	28	45
	3 × 50	–	47	63	52	56
	3 × 60	–	63	–	56	–
	5 × 30	–	–	51	68	40
	5 × 40	–	–	58	74	50
XIII	1 × 200	–	4	47	0	–
	1 × 220	–	–	–	15	–
	1 × 320	–	33	–	–	–
	3 × 200	–	0	–	46	–
	3 × 300	–	63	–	59	–

Table 8 (continued)

Compound	Applied dose (mg/kg)	*Solid animal tumor systems* Ehrlich ascites tumor	Sarcoma 180	Colon 38 adeno-carcinoma	B16 melanoma	Lewis lung carcinoma
XIV	1 × 80	–	5	18	0	16
	1 × 100	–	–	–	–	34
	1 × 120	–	27	28	11	–
	3 × 40	–	–	35	6	56
	3 × 60	–	24	43	37	–
	3 × 80	–	–	69	–	–
	3 × 100	–	47	–	50	–
	5 × 40	–	–	63	12	43
	5 × 50	–	–	–	–	55
XLIV	1 × 70	–	–	2	–	31
	1 × 80	–	–	10	10	55
	3 × 40	–	–	70	–	46
	3 × 50	–	–	66	–	66
	3 × 60	–	–	–	54	–
	5 × 30	–	–	62	47	41
	5 × 40	–	–	76	56	75

a Given are values of tumor growth inhibition in % of control tumor size, calculated by 100% – T/C (T/C = mean tumor weight of a dose group × 100/mean tumor weight of the control group)

5. Solid Sarcoma 180

Against solid sarcoma 180 growing subcutaneously in the flanks of mice, the effects of the titanocene complexes[102] I, VIII, XIII and XIV and the ferricenium compounds[86] XLIX, LI and LIII–LV were analyzed after intraperitoneal substance administration. Whereas the ferricenium compounds did only slightly inhibit tumor growth and reduce tumor weights by 35–50% (Table 9), the titanocene compounds suppressed tumor development in a clearly dose-dependent and significant manner (Table 8). After application of sublethal, triple doses of the dihalo complexes I and VIII and the carboxylato derivatives XIII and XIV, tumor growth was reduced by 50–80% thus that the mean weights of treated tumors were diminished to 50–20% of the weights of control tumors corresponding to 100%. Among all cyclopentadienyl metal complexes which were investigated, I effected best activity against solid sarcoma 180.

6. Solid B16 Melanoma

Treatment of animals bearing subcutaneously growing B16 melanoma with diverse bis(cyclopentadienyl) titanium or bis(cyclopentadienyl) iron compounds significantly influenced tumor development[107] and reduced the growth of the tumors by 50–80% (Tables 8, 9). For T/C ratios between 50 and 20%, there was a clear dependence on the doses applied. This was observed after application of sublethal doses of the neutral

Table 9. Growth inhibitions[a] effected by some ferricenium salts in diverse solid experimental animal tumors

Compound	Applied dose (mg/kg)	*Solid animal tumor systems* Ehrlich ascites tumor	Sarcoma 180	Colon 38 adeno-carcinoma	B16 melanoma	Lewis lung carcinoma
XLIX	1 × 120	–	–	30	0	0
	1 × 140	–	–	0	0	30
	3 × 80	13	–	–	–	–
	3 × 100	–	5	38	–	14
	3 × 120	39	43	57	45	13
	3 × 160	56	32	–	–	–
	5 × 80	–	–	46	38	–
	5 × 100	–	–	39	40	–
LI	1 × 120	–	–	18	0	20
	1 × 150	–	–	52	23	58
	1 × 170	–	–	–	–	57
	3 × 100	52	22	42	11	32
	3 × 120	56	34	63	55	56
	3 × 140	52	44	–	–	53
	5 × 80	–	–	37	43	48
	5 × 100	–	–	44	54	–
	5 × 120	–	–	64	56	59
LIII	1 × 120	–	–	6	4	34
	1 × 150	–	–	0	26	50
	1 × 180	–	–	–	–	66
	3 × 100	0	–	21	0	24
	3 × 120	–	0	31	8	54
	3 × 150	4	48	52	0	54
	3 × 200	17	35	–	–	–
	5 × 80	–	–	13	8	13
	5 × 100	–	–	41	14	5
	5 × 120	–	–	53	31	45
LIV	1 × 100	–	–	–	15	20
	1 × 120	–	–	–	0	9
	3 × 40	0	0	–	–	–
	3 × 80	0	35	–	13	–
	5 × 60	–	–	–	0	–
LV	1 × 220	–	–	0	0	39
	1 × 250	–	–	37	22	36
	3 × 100	44	–	–	–	–
	3 × 150	48	–	–	–	–
	3 × 180	–	25	43	0	46
	3 × 200	58	50	73	60	71
	5 × 150	–	–	23	27	60
	5 × 180	–	–	53	29	67

[a] The parameter evaluated is tumor growth inhibition in % of control tumor size, calculated by 100% – T/C (T/C = mean tumor weight of a dose group × 100/mean tumor weight of the control group)

titanocene complexes I, VIII and XIV, the ionic titanocene compound XLIV and, to a less extent, after treatment with the ferricenium salts XLIX, LI and LIII–LV. Again, I exhibited superior antitumor potency among all titanocene and ferricenium complexes administered.

7. Colon 38 Adenocarcinoma

The murine colon 38 carcinoma represents an experimental animal tumor which is inhibited by only a few cytostatic drugs such as 5-fluorouracil and cyclophosphamide[108]. It was found that the titanocene complexes[107] I, VIII, XIV and XLIV as well as the ferricenium salts[86] XLIX, LI, LIII and LV were also able to suppress the proliferation of this tumor in vivo. Maximum reductions of tumor weights by 50–80% (Tables 8, 9) were effected corresponding to T/C ratios of 50–20%, whereby I (Fig. 13), followed by XLIV, obviously induced the best and most pronounced tumor inhibition.

8. Lewis Lung Carcinosarcoma

Lewis Lung carcinoma is another experimental tumor which is sensitive to titanocene complexes. At first, this was found by Chang[109] and confirmed later-on by other groups[86]. When animals bearing Lewis lung carcinoma were treated with titanocene (Table 8) or ferricenium (Table 9) complexes[86], again significant and dose-dependent suppressions of tumor growth were induced by administration of sublethal doses. The

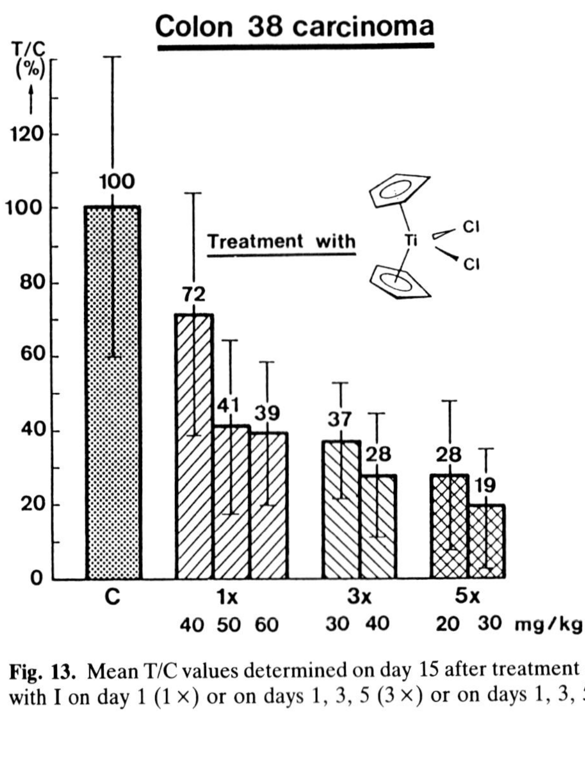

Fig. 13. Mean T/C values determined on day 15 after treatment of mice bearing colon 38 carcinoma with I on day 1 (1 ×) or on days 1, 3, 5 (3 ×) or on days 1, 3, 5, 7, 9 (5 ×)

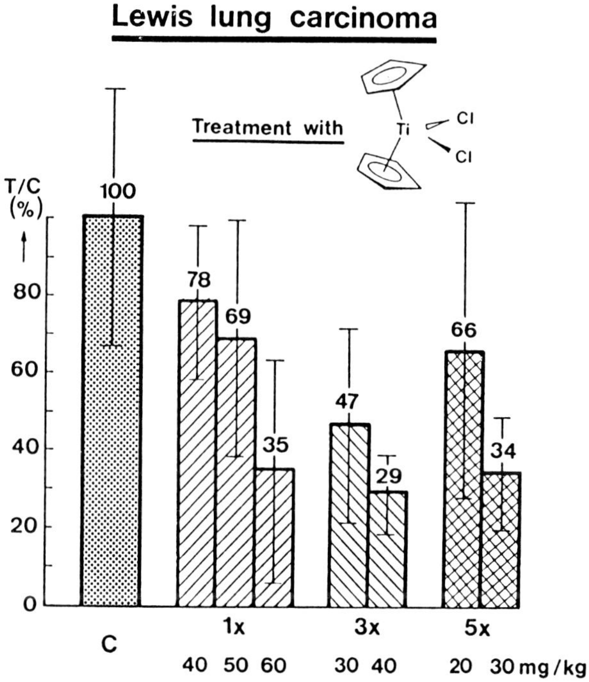

Fig. 14. Mean T/C values determined on day 9 after treatment of mice bearing subcutaneously growing Lewis lung carcinoma with I on day 1 (1 ×) or on days 1, 3, 5 (3 ×) or on days 1, 3, 5, 7, 9 (5 ×)

values of growth inhibition ranged between 50 and 75% in the case of titanocenes, whereby superior activity was detected for the neutral titanocene derivative I (Fig. 14) and the ionic titanocene complex salt XLIV. The ferricenium compounds LI, LIII and LV effected significant tumor suppressions by 50–70%, the trichloroacetate derivative LV exhibiting most pronounced tumor-inhibiting potency.

9. Mouse Mammary Tumor TA3Ha

Vanadocene dichloride (II) was shown to be characterized by marked antiproliferative effects against fluid and solid Ehrlich ascites tumor[92, 103] as well as against other experimental tumor systems such as intraperitoneally growing mouse mammary tumor TA3Ha[87, 88]. The growth of the latter tumor was inhibited remarkably to a similar extent as observed after application of optimum doses of cisplatin. In both cases, the life span of the host animals was prolongated by about 180% in relation to untreated controls[87, 88].

Summary – Animal Antitumor Data

Summarizing the results effected by metallocene diacido complexes and ferricenium salts in experimental animal tumors, it becomes evident

(i) that titanocene, vanadocene and ferricenium complexes exhibit systemic activity against numerous experimental tumors and

(ii) that titanocene and ferricenium compounds are obviously characterized by a similar spectrum of effectivity against experimental animal tumors.

C. Antitumor Properties Against Xenografted Human Tumors

Selected representatives of titanocene, vanadocene and ferricenium complexes were tested against different types of human malignant tumors heterotransplanted to athymic nude mice. During the past years, experiments with nude mice have attained increasing importance for the preclinical screening of cytostatic drugs as it seems to be possible to predict and to encircle the clinical spectrum of activity of newly developed antitumor agents from preclinical testing results with human tumors growing in nude mice[110–112)]. Numerous studies have shown that human tumors xenografted to athymic mice preserve drug susceptibility and histologic reactivity[113–116)] and that, indeed, there exists a high positive correlation between responses obtained with individual human tumors growing in nude mice and the clinical results observed under treatment with the same drugs[110–112, 117–120)].

1. Human Colorectal Carcinomas

Human adenocarcinomas derived from the colon including the colon sigmoideum and the colon rectum are generally rather insensitive to common cytostatic agents. The only drugs exhibiting limited clinical value against these malignancies are, e.g., 5-fluorouracil and mitomycin C.

The influence of titanocene and ferricenium complexes on the development of human colorectal carcinomas was investigated in the case of two rectum carcinomas, four sigma carcinomas and five tumors derived from the upper parts of the colon. All tumors had been xenografted into athymic nude mice, and the compounds were administered according to a Q2D × 5[2] or Q3D × 5[2] schedule. Antitumor activity, i.e. growth suppressions by more than 50%, was observed in one out of two rectum tumors, in all sigma carcinomas and in four of five carcinomas of the upper colon. In Table 10, the results obtained with the nine tumors responding to titanocene complexes are summarized in terms of growth inhibition values.

It becomes obvious that the growth of all colorectal tumors listed was suppressed by I in a clearly dose-dependent manner by 50–94%, corresponding to T/C ratios ranging between 50 and 6%. In the case of CX1 tumor, one of the human standard tumors at the National Cancer Institute[121, 122)], growth suppressions by 50–70% were effected, which persisted several weeks beyond the end of the treatment period (Fig. 15). The colon adenocarcinoma C-Stg2 represented a tumor, the growth of which was inhibited nearly totally during the period of treatment with I, applied in doses of 5 × 15 mg/kg (Fig. 16). Against the rectum carcinoma R85 and the sigma carcinoma S90, some other titanocene derivatives (Table 10) as well as five ferricenium complexes (Table 12) were tested additionally. Whereas the titanocene diacido complexes VIII and XIV caused similar effects against S90 as I, but were less active against R85. The ferricenium salts XLIX, LI and LIII–LV were characterized by an inverse behavior effecting more pronounced growth suppression in R85 (Fig. 17) that in S90. For comparison purposes, cisplatin as well as 5-fluorouraci', one of the clinically approved cytostatic drugs against colorectal car-

2 Q2D × 5, fivefold injection of substance every two days; Q3D × 5, fivefold injection of substance every three days.

Table 10. Growth inhibitions[a] effected by some titanocene diacido complexes in xenografted human colorectal adenocarcinomas, derived from the rectum (R85), the colon sigmoideum (S90, S-Sb1, S-Sb2, S-Stg) and upper parts of the colon (C-Hbg1, C-Stg2, C-Stg3, CX1)

Compound	Schedule	Dose (mg/kg)	*R85* 3 d[b]	*R85* 17 d[b]	*S90* 3 d[b]	*S90* 17 d[b]	*S-Sb1* 3 d[b]	*S-Sb1* 17 d[b]	*S-Sb2* 3 d[b]	*S-Sb2* 17 d[b]	*S-Stg* 3 d[b]	*S-Stg* 17 d[b]	*C-Hbg1* 3 d[b]	*C-Hbg1* 17 d[b]	*C-Stg2* 3 d[b]	*C-Stg2* 17 d[b]	*C-Stg3* 3 d[b]	*C-Stg3* 17 d[b]	*CX1* 3 d[b]	*CX1* 17 d[b]
I	Q3Dx5	10	33	25	–	–	–	–	–	–	57	61	–	–	–	–	18	0	49	42
		15	55	53	64	60	–	–	–	–	48	69	–	–	–	–	71	66	57	70
		20	–	–	–	–	–	–	–	–	65	74	–	–	–	–	–	–	68	65
	Q2Dx5	10	43	36	11	23	0	19	47	52	24	49	16	32	71	71	–	–	15	24
		15	51	55	67	64	42	52	52	48	71	83	49	41	91	94	–	–	41	47
		20	–	–	75	71	58	55	49	47	–	–	44	58	87	92	–	–	68	61
VIII	Q3Dx5	10	0	0	–	–	–	–	–	–	–	–	–	–	–	–	–	–	–	–
		20	35	0	57	58	–	–	–	–	–	–	–	–	–	–	–	–	–	–
	Q2Dx5	10	0	0	2	18	19	27	–	–	–	–	–	–	–	–	–	–	–	–
		20	12	0	68	65	22	42	–	–	–	–	–	–	–	–	–	–	–	–
		40	–	–	65	69	–	–	–	–	–	–	–	–	–	–	–	–	–	–
XIV	Q3Dx5	30	51	50	28	35	46	59	–	–	–	–	–	–	–	–	–	–	–	–
		40	–	–	59	42	–	–	–	–	–	–	–	–	–	–	–	–	–	–
	Q2Dx5	20	–	–	11	30	–	–	–	–	–	–	–	–	–	–	–	–	–	–
		30	–	–	76	59	–	–	–	–	–	–	–	–	–	–	–	–	–	–

[a] The parameter evaluated is tumor growth inhibition in % of control tumor size, calculated by 100% – T/C (T/C = mean relative volume of treated tumors × 100/mean relative volume of control tumors)

[b] The values of tumor growth inhibition listed were determined 3 and 17 days after last substance injection

Table 11. Growth inhibitions[a] effected by some titanocene diacido complexes in xenografted human lung malignancies (L182, L261), a breast carcinoma (M3), a stomach adenocarcinoma (M-Stg4) and a melanoma (Str)

Compound	Schedule	Dose (mg/kg)	*L182* 3 d[b]	*L182* 17 d[b]	*L261* 3 d[b]	*L261* 17 d[b]	*M3* 3 d[b]	*M3* 17 d[b]	*M-Stg4* 3 d[b]	*M-Stg4* 17 d[b]	*Str* 3 d[b]	*Str* 17 d[b]
I	Q3Dx5	10	0	0	9	1	78	80	48	63	41	89
		15	0	0	65	74	51	60	33	23	49	72
	Q2Dx5	10	0	0	48	58	–	–	47	49	50	73
		15	39	68	64	70	–	–	69	76	45	81
		20	61	73	65	72	–	–	58	72	52	82
VIII	Q3Dx5	10	0	0	29	18	69	75	–	–	–	–
		20	4	3	47	41	56	51	–	–	–	–
	Q2Dx5	10	20	11	0	0	–	–	–	–	–	–
		20	36	42	51	52	–	–	–	–	–	–
		40	57	48	46	40	–	–	–	–	–	–
XIV	Q3Dx5	30	15	8	51	66	–	–	–	–	–	–
		40	17	37	65	82	60	48	–	–	–	–
	Q2Dx5	20	48	58	30	29	–	–	–	–	–	–
		30	75	89	50	65	–	–	–	–	–	–

[a, b] See explanations to Table 10

Table 12. Growth inhibitions[a] effected by some ferricenium salts in xenografted human colorectal carcinomas (R85, S90), lung malignancies (L182, L261) and a breast carcinoma (M3)

Compound	Schedule	Dose (mg/kg)	*R85* 3 d[b]	*R85* 17 d[b]	*S90* 3 d[b]	*S90* 17 d[b]	*L182* 3 d[b]	*L182* 17 d[b]	*L261* 3 d[b]	*L261* 17 d[b]	*M3* 3 d[b]	*M3* 17 d[b]
XLIX	Q3Dx5	70	77	44	0	0	0	0	49	75	30	59
		80	–	–	0	0	0	0	71	81	35	35
	Q2Dx5	70	–	–	54	58	0	0	64	70	–	–
LI	Q3Dx5	80	75	71	0	4	21	67	7	0	56	51
		100	–	–	0	25	51	71	2	23	54	79
	Q2Dx5	80	–	–	19	50	0	4	30	30	–	–
		100	–	–	64	56	–	–	–	–	–	–
LIII	Q3Dx5	80	83	83	0	0	49	43	26	49	21	3
		100	–	–	0	0	–	–	49	65	58	73
	Q2Dx5	80	–	–	58	69	–	–	57	72	–	–
LIV	Q3Dx5	50	67	67	0	0	40	40	0	0	0	60
		70	–	–	0	0	–	–	49	63	–	–
	Q2Dx5	50	–	–	37	46	–	–	52	69	–	–
LV	Q3Dx5	120	63	54	0	0	25	20	4	0	0	16
		140	–	–	0	0	–	–	50	55	70	76
	Q2Dx5	120	–	–	39	55	–	–	0	34	–	–

[a, b] See explanations to Table 10

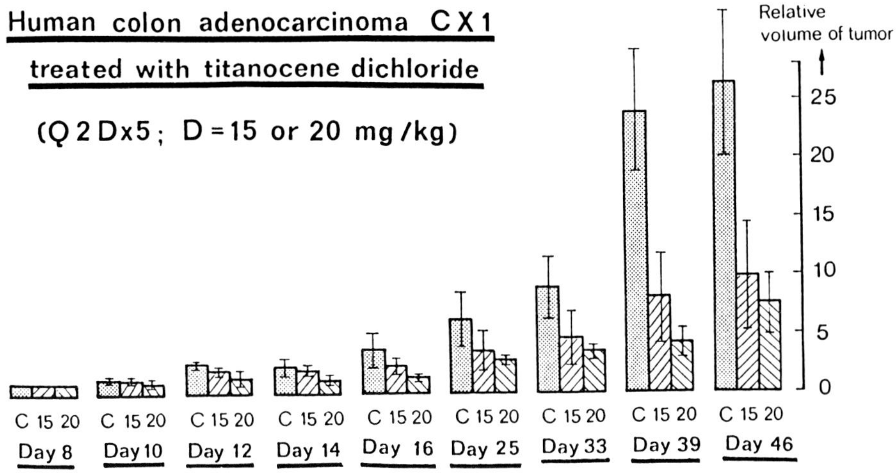

Fig. 15. Influence of I, applied according to the sublethal regimen Q2Dx5 (D = 15 or 20 mg/kg) on days 8, 10, 12, 14 and 16 after tumor transplantation, upon the growth of the human colon adenocarcinoma CX1 heterotransplanted to athymic mice. The schedule 5 × 20 mg/kg corresponds to a LD_{10} regimen. The parameter evaluated is the relative increase of tumor volume, related to the starting tumor size, on day 8. Given are mean values of relative volume of tumor, determined in 4–5 animals, and standard deviations ± s

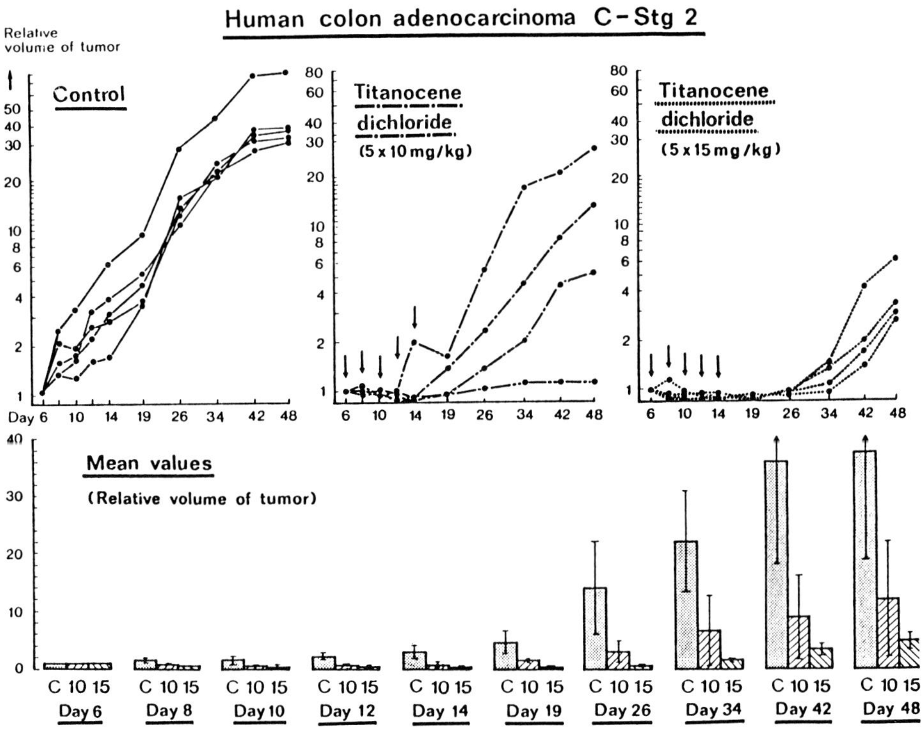

Fig. 16. Growth behavior of a xenografted human colon adenocarcinoma (C-Stg2) under treatment with I, applied at sublethal doses according to Q2Dx5 (D = 10 or 15 mg/kg) on days 6, 8, 10, 12 and 14 after tumor transplantation. *Upper part:* growth curves of individual tumors; on abscissa, days after tumor implant on day 0; arrows indicate substance injections. *Lower part:* mean values of relative volume and standard deviations within control and treatment groups shown in the *upper part*

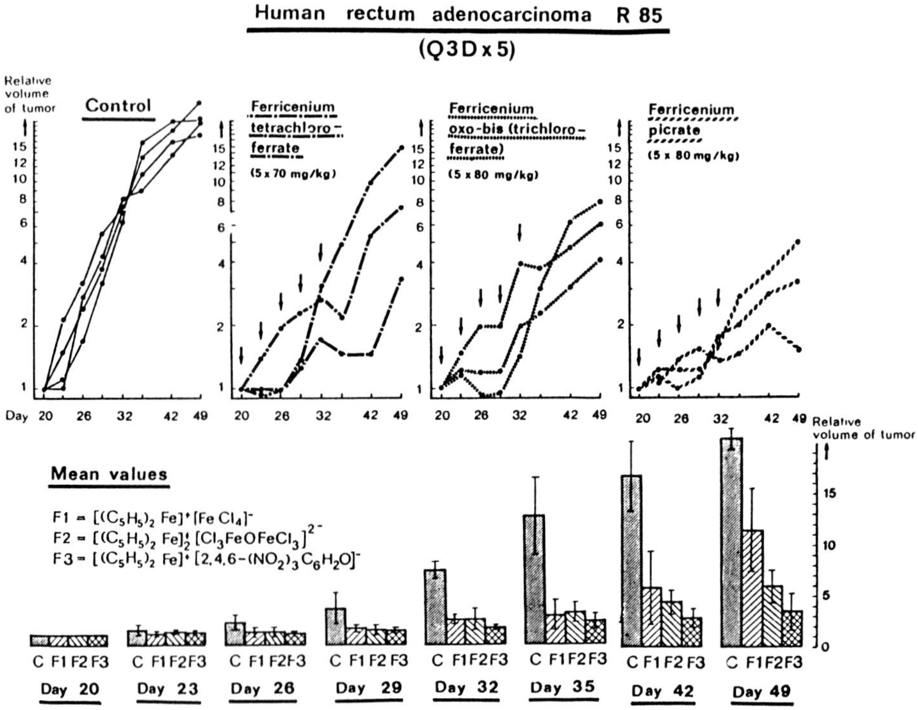

Fig. 17. Growth behavior of a human rectum adenocarcinoma (R85) heterotransplanted to athymic mice under treatment with XLIX (5 × 70 mg/kg), LI (5 × 80 mg/kg) and LIII (5 × 80 mg/kg) applied according to Q3Dx5 on days 20, 23, 26, 29 and 32. For further details, cf. legend to Fig. 16

cinomas, were administered in equitoxic doses to animals bearing the R85 or S90 tumor[123, 124]. None of both compounds were able to induce more pronounced effects than titanocenes. Both compounds caused marginal activity and slowed down tumor proliferation at LD_{10} doses by 20–35% to T/C ratios ranging between 80 and 65%.

2. Human Lung Malignancies

Three human lung malignancies were tested, being represented by the small cell lung carcinoma L182, the lung adenocarcinoma L261 and the tumor LX1, one of the human tumor standard systems of the National Cancer Institute.

In the case of L261[125], significant growth inhibitions by more than 50% to T/C ratios between 50 and 20% were effected by the titanocenes I, VIII and XIV applied at sublethal dose levels (Table 11). These suppressions were stable and persisted beyond the end of the treatment period (Fig. 18). Regarding the ferricenium complexes XLIX, LI and LIII–LV, most of them were less active and induced growth inhibitions by 30–50% on day 3 after the last substance injection[124]. Only XLIV was similarly active as I (Table 12, Fig. 18). For comparison purposes, cisplatin and cyclophosphamide were applied. They caused growth suppression of similar strength as I and XLIX, i.e. of 60–70% of control tumor size, resulting in T/C ratios of 40 and 30%, respectively.

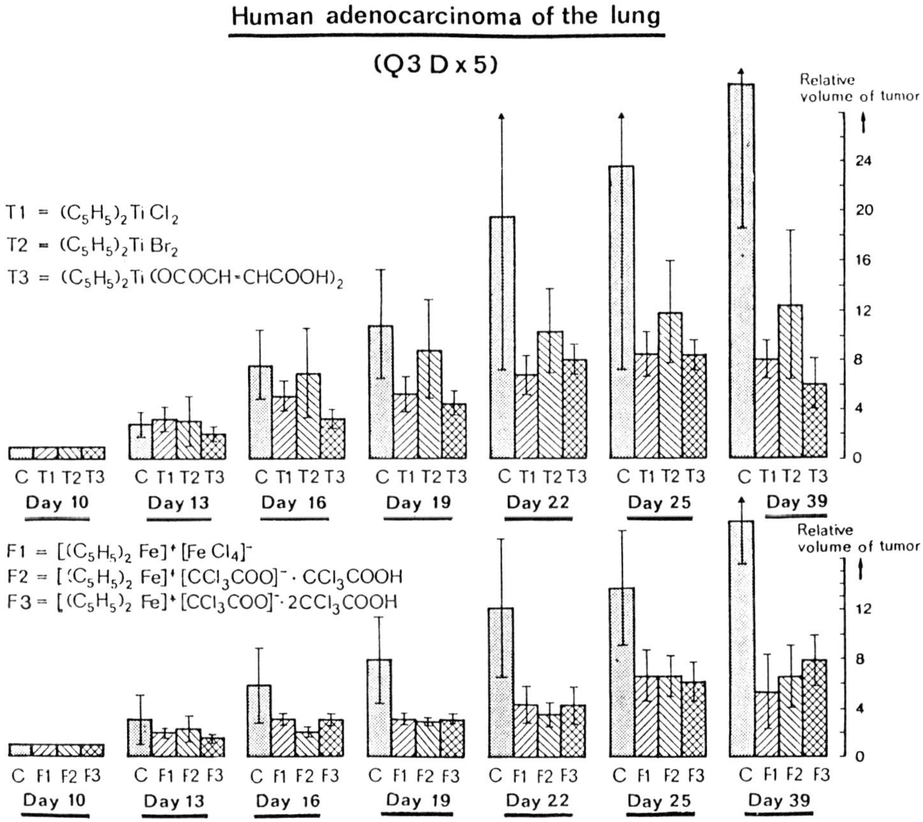

Fig. 18. Influence of I (5 × 15 mg/kg), VIII (5 × 20 mg/kg) and XIV (5 × 40 mg/kg) *(upper part)* and XLIV (5 × 70 mg/kg), LIV (5 × 50 mg/kg) and LV (5 × 120 mg/kg) *(lower part)*, applied according to Q3Dx5 regimen, upon the growth of a human lung adenocarcinoma (L261) heterotransplanted to athymic mice. For further details, cf. legend to Fig. 15

In the case of L182, which proliferated more slowly than L261, sublethal doses of I, VIII and XIV reduced tumor proliferation by about only 50%. LD_{10} doses were necessary to effect growth inhibitions by 70–80% to T/C ratios ranging between 31 and 17%[125] (Table 11). The ferricenium salts XLIX, LI and LIII–LV again caused less pronounced tumor reductions in the case of L182 (Table 12). The control compound cisplatin suppressed tumor development by 70% to a T/C ratio of 31% at LD_{10} doses level[122].

The third lung carcinoma investigated was the LX1 tumor system[121, 122]. No growth retardation became evident after treatment with I even after application of LD_{10} or LD_{20} doses.

3. *Diverse Heterotransplanted Human Tumors*

Selected individual tumors representing different types of human malignancies were xenografted to nude mice and investigated there under the influence of I and some other cyclopentadienyl metal complexes.

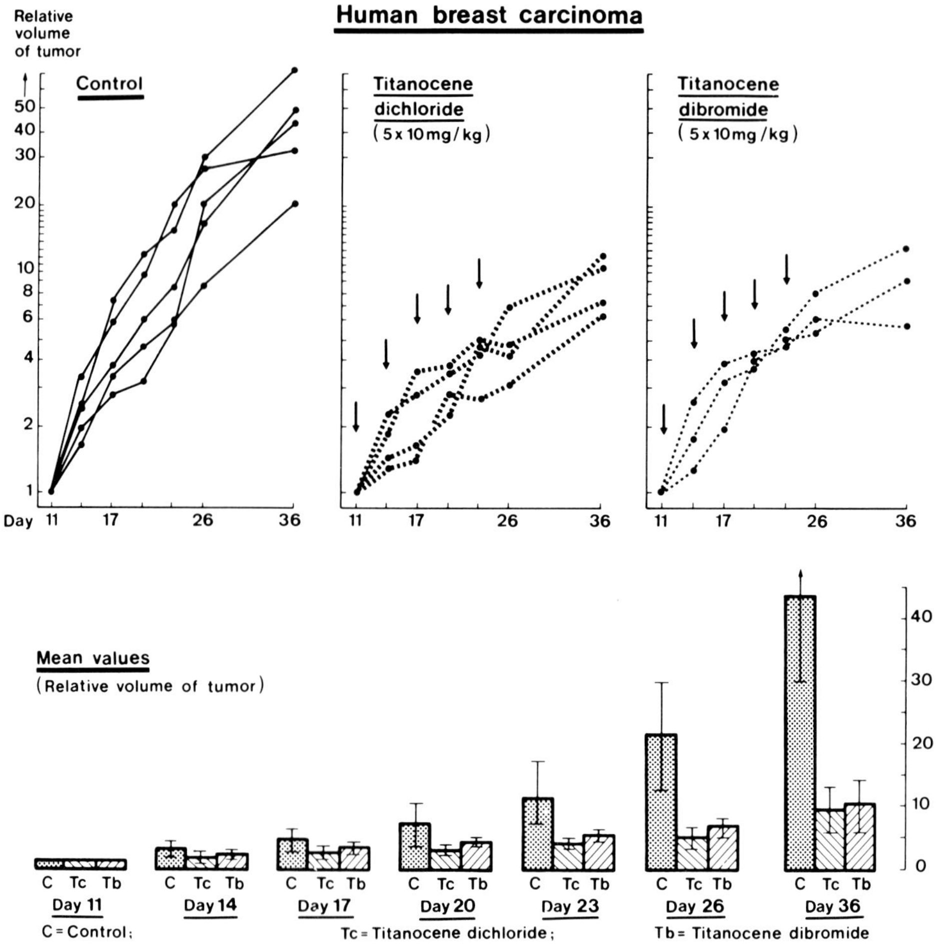

Fig. 19. Growth behavior of a human breast carcinoma (M3) heterotransplanted to athymic mice under treatment with I (5 × 10 mg/kg) and VIII (5 × 10 mg/kg), applied according to Q3Dx5 on days 11, 14, 17, 20 and 23. For further details, cf. legend to Fig. 16

The proliferation of a human *breast carcinoma* (M3) growing rapidly in nude mice was clearly suppressed by the titanocene diacido complexes I and VIII by 70–80% (Table 11) leading to T/C values of 30–20% (Fig. 19). On the other hand, the ferricenium salts XLIX, LI, LIII and LV applied at equitoxic doses suppressed the increase of tumor volume by only 30–60% to T/C ratios of 70–40%.

Another tumor, the growth of which was markedly inhibited by I, was a *stomach adenocarcinoma* (M-Stg4) (Table 11, Fig. 20). Absolute decreases of tumor volume were observed during the treatment period with 5 × 10 mg/kg and 5 × 15 mg/kg. In the case of two animals, treated with 5 × 15 mg I/kg, the tumors even disappeared within one week after the beginning of the application of the substance.

Subsequently, the growth development of a chemosensitive human *melanoma* (Str)[126] was studied under the influence of I. There were again growth reductions which

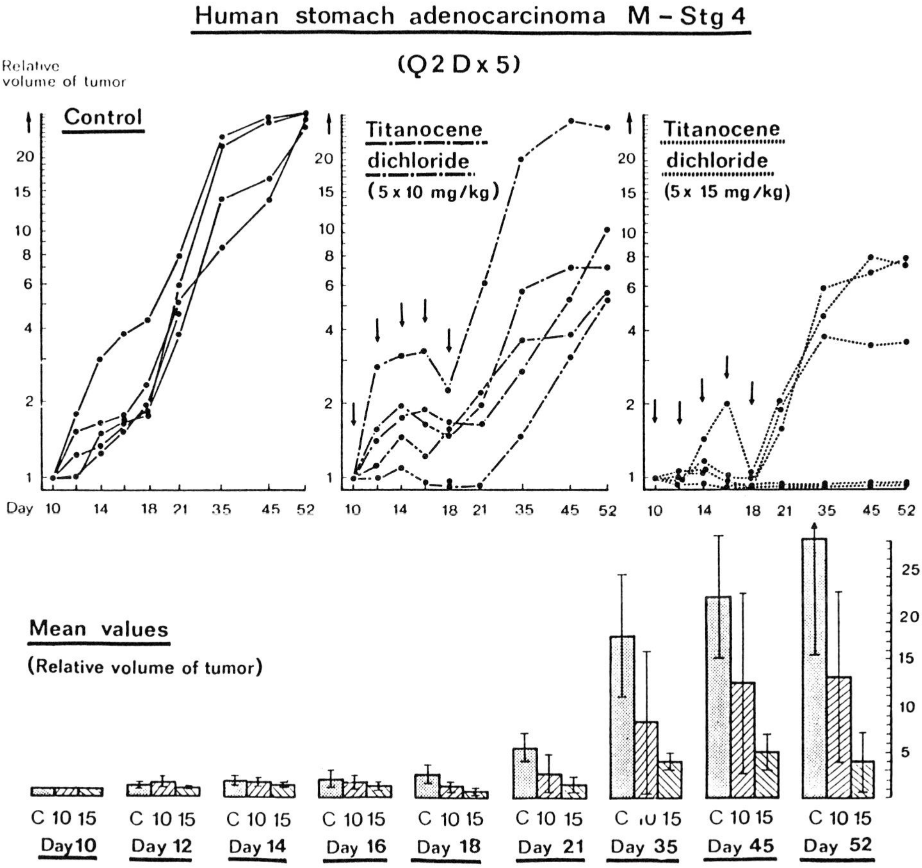

Fig. 20. Growth behavior of a human stomach adenocarcinoma (M-Stg4) heterotransplanted to athymic mice under treatment with I, applied according to the sublethal regimen Q2Dx5 (D = 5 × 10 or 5 × 15 mg/kg) on days 10, 12, 14, 16 and 18. For further details, cf. legend to Fig. 16

ranged between 50 and 89% and were stable for three weeks after the last administration of the substance. The comparison with other cytostatic drugs, e.g. cisplatin, ifosfamide and mitomycin C, which are known to be cytostatically active against this tumor[126)], revealed stronger tumor inhibition and growth suppressions by 91, 97 and 79%, respectively.

Summary – Xenografted Human Tumors

These results obtained with titanocene complexes and ferricenium salts against xenografted human tumors, especially against colorectal carcinomas, underline the antiproliferative effectivity of cyclopentadienyl metal complexes. Because of the noted positive correlation between the findings of preclinical antitumor studies with individual tumors in athymic mice and the clinical results obtained with the same drugs[110–112, 117–120], the described experimental findings are remarkable in suggesting activity of cyclopentadienyl metal compounds against certain human tumors under clinical conditions, the spectra of antitumor activity being apparently similar in the case of titanocene and ferricenium complexes.

IV. Antiviral, Insecticidal and Antiinflammatory Properties

For some cyclopentadienyl metal complexes, recent investigations have uncovered additional biological activities other than antitumor properties.

Significant *antiviral effectivity* was shown in vitro for titanocene dichloride (I) against numerous DNA and RNA viruses in the extracellular phase[127]. Typical examples of viruses, which were inhibited by direct contact with I and lost infectivity up to 100%, were orthopoxvirus (vaccinia) and herpes virus (pseudorabies) as DNA viruses, and rhabdovirus (vesicular stomatitis), paramyxovirus (Newcastle disease) and diverse orthomyxoviruses (e.g. influenza A and B) as RNA viruses. A comparable antiviral effect against herpes viruses was detectable after application of the ferricenium salt LIII[128], whereas, on the other hand, vanadocene dichloride (II)[128] and molybdenocene dichloride (V)[127] failed to show antiviral activity under the same experimental conditions.

Insecticidal properties were ascribed to some modified cyclopentadienyl titanium complexes[129]. They exhibited insecticidal activity against Kapra beetles, causing the deaths of 50% of the animals in concentrations of 0.7–1.4%. In the same experimental trial, the parent compound I was biologically inactive[129].

A recently published study drew attention to *antiinflammatory properties* of some cyclopentadienyl titanium(IV) complexes[130]. The titanocene dihalides I and VII were shown to be potent antioedemic drugs and to exhibit significant acute antiinflammatory properties in rats. The effects observed were equivalent and even more pronounced than those occurring after application of aspirin. Other cyclopentadienyl titanium(IV) complexes which were investigated for acute antiinflammatory activity were the less water-soluble titanocene diisocyanate $(C_5H_5)_2Ti(NCO)_2$ and the cationic species $[(C_5H_5)_2Ti(OH_2)_2]^{2+}$ obtained from I and $AgNO_3$ in water. Both compounds were clearly less active than I.

Though non-irritant doses of I were unable to prevent polyarthritis development induced by injection of Freund's adjuvant, I, VII and $(C_5H_5)_2Ti(NCO)_2$ were effective suppressants of arthritis development when administered just before or after arthritic symptoms appeared[130]. This temporary effectivity of I in suppressing arthritic signs was similar to the action of phenylbutazone, whereas cisplatin and cyclophosphamide showed to be effective prophylactic drugs. Interestingly, cyclopentadiene given under the same conditions had no antiarthritic effect at all. Thus, titanocene complexes are obviously antiinflammatory drugs like phenylbutazone, rather than powerful immunosuppressant agents like cisplatin or cyclophosphamide. This was confirmed in the same study by the observation that graft-versus-host reactions were not influenced and suppressed by titanocene complexes[130].

V. Structure-Activity Relationships

The previously described results of antitumor testing of cyclopentadienyl early transition metal complexes, ferricenium salts and main group IV metallocenes throw some light on the structure-activity relation of η^5-cyclopentadienyl metal complexes. At the first look, the fact is striking that three quite different types of cyclopentadienyl metal complexes (cf. Chap. II) are characterized by antiproliferative properties and that in the case both of titanocene and ferricenium complexes, which were investigated more extensively, a similar spectrum of antitumor activity against experimental and human tumors was found in vivo. Though there exist profound differences concerning the structural characteristics of neutral metallocene diacido complexes $(C_5H_5)_2MX_2$, ionic metallicenium salts $[(C_5H_5)_2M]^+X^-$ and uncharged decasubstituted metallocenes $(C_5R_5)_2M$ (cf. Chap. II), all three types of compounds also exhibit some common structural features:

- They are organometallic compounds containing direct carbon-to-metal bonds.
- They include as organic ligands η^5-bound five-membered cyclopentadienyl rings which can be regarded as aromatic species within the metallocene moieties.

When the metallocene frameworks were varied systematically, the antitumor properties were influenced in a manner being analogous in the case of titanocene and ferricenium compounds (cf. Chap. III).

A. *Influence of the Central Metal Atoms*

On the one hand, different metals are present in the three types of antitumor cyclopentadienyl metal compounds and are able to constitute cytostatically active complexes:

- the early transition metals, e.g. titanium(IV) and vanadium(IV), in the case of metallocene diacido complexes,
- the medium transition metal iron(III) in antitumor metallicenium salts, and
- the main group metals germanium(II) and tin(II) in main group IV metallocenes.

On the other hand, the tumor inhibition effected by $(C_5H_5)_2MX_2$ compounds was distinctly dependent on the central metal atom M present within the molecules. Whereas the metallocene dichloro complexes I–III and V containing the first-row and second-row transition metals titanium, vanadium, niobium or molybdenum effected pronounced antitumor activity with distinct dose-activity relationships and with cure rates of 100% at optimum doses against Ehrlich ascites tumor[13, 14, 91–94], the compounds IV and VI with the third-row homologs tantalum or tungsten only sporadically increased survival of treated tumor-bearing mice[93]. In the case of the analogous zirconium and hafnium complexes, no antitumor activity was detectable against Ehrlich ascites tumor[91].

Other cytostatically active titanium compounds are the six-coordinate bis(benzoylacetonato)titanium(IV) dihalides and bis(alkoxides) LX[131].

In contrast to the tetrahedral titanocene complexes (Fig. 4), the complexes LX are octahedrally configurated, lack direct carbon-to-metal bonds and are, therefore, not organometallic compounds. Their growth-inhibiting properties have been demonstrated against several experimental tumor systems[15, 104, 132]. When the heavier homologs of Ti, i.e. Zr and Hf, were introduced as central metal atoms into the dichloro complex LX, the antitumor activity was preserved against several experimental transplantable tumor sys-

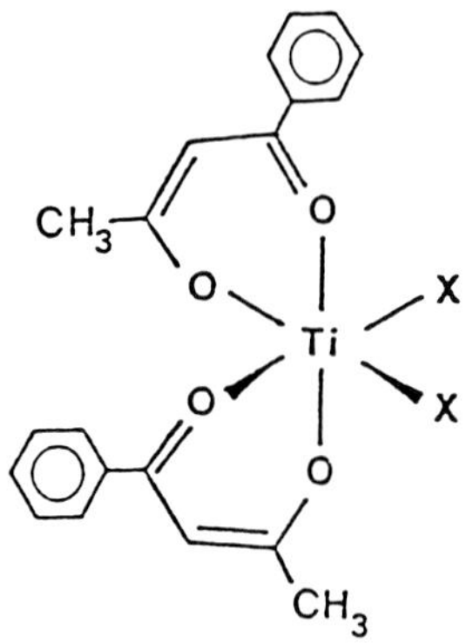

X = F, Cl, Br, OC_2H_5

LX

tems, e.g. against sarcoma 180[133], but reduced in the case of autochthonous tumors, e.g. a chemically induced colorectal tumor of rats[134]. Against the latter tumor, the antiproliferative activity was clearly dependent upon the metal atom present within the bis(benzoylacetonato)metal(IV) dichloride complex and decreased according to the order Ti > Zr > Hf[134].

B. *Influence of the Acido Groups*

Within titanocene complexes $(C_5H_5)_2TiX_2$ and ferricenium compounds $[(C_5H_5)_2Fe]^+X^-$, it is possible to widely vary the acido groups X which are either covalently coordinated as ligands to the central metal titanium or bound electrostatically as anions to the ferricenium cation.

Regarding titanocene complexes, halide and pseudohalide (I, VII–XI)[95], carboxylato (XII–XIV)[37] and certain phenolate, thiophenolate and selenophenolate (XVI, XVII, XXI, XXII)[86] derivatives effected antitumor activity of comparable strength. Other titanocene chalcogeno derivatives, e.g. $(C_5H_5)_2Ti(SR)_2$[96] or $(C_5H_5)_2TiS_5$[97], which were virtually insoluble in aqueous systems, however, were obviously inactive against Ehrlich ascites tumor in vivo[86].A similar result of antitumor inactivity against fluid Ehrlich ascites tumor was found[86] for the pentacoordinate complex titanocene pyridine-2,6-dicarboxylate $(C_5H_5)_2Ti[2,6\text{-}(OCO)_2C_5H_3N]$ containing two carboxyl oxygen atoms and the pyridine nitrogen atom as ligating atoms[135].

Besides the neutral complexes I–XXXVII and LX which all comprise two acido ligands in adjacent positions, the complex cations XLIV–XLVII in which one or both of the acido ligands are replaced by neutral donors, have also proved to be active against experimental tumors (cf. Chap. III.B.). Moreover, the tris(benzoylacetonato)metal(IV) monohalides $(C_6H_5COCHCOCH_3)_3MX$ (M = Zr, Hf; X = Cl, Br), regarded as seven-coordinate and as lacking a *cis*-like diacidometal moiety, have been revealed to be active against ascitic sarcoma 180[133].

In the case of ferricenium complexes which are characterized by the presence of only one ionic acido group, tumor growth inhibition was observed with the water-soluble derivatives XLIX–LV containing the quite different anions tetrachloroferrate(III), μ-oxo-bis[trichloroferrate(III)], picrate and trichloroacetate. When, however, anions such as heptamolybdate $[H_5M_7O_{24}]^-$ are introduced which impair the water solubility of ferricenium complexes, antitumor activity is again markedly diminished and, in extreme cases, totally lost[16]. As representatives of the tetrachloroferrate(III) acido anion lacking the metallicenium counterion, the compounds $[(CH_3)_4N]^+[FeCl_4]^-$ and $[C_6H_5CH_2(C_2H_5)_3N]^+[FeCl_4]^-$ were cytostatically inactive[16].

It is interesting to note that, in continuation of the above mentioned findings, the decasubstituted germanocenes and stannocenes LVII–LIX represent cytostatically active metallocene species which do not possess any acido group functioning as ligand or counterion.

C. Influence of the Cyclopentadienyl Ligands

In principle, the cyclopentadienyl rings in metallocene or metallicenium complexes can be modified by replacing one or more hydrogen atoms of one or both cyclopentadienyl rings by organic or elemento-organic residues or by introduction of a bifunctional bridging group. Examples are monosubstituted (XXIII–XXV), 1,1′-disubstituted (XXVI–XXIX), and 1,1′-bridged (XXX–XXXIV) titanocene dichlorides (cf. Chap. II). At maximum, five substituents can be introduced into each of both cyclopentadienyl rings, resulting in the formation of the decasubstituted complexes LXIa and LXIb.

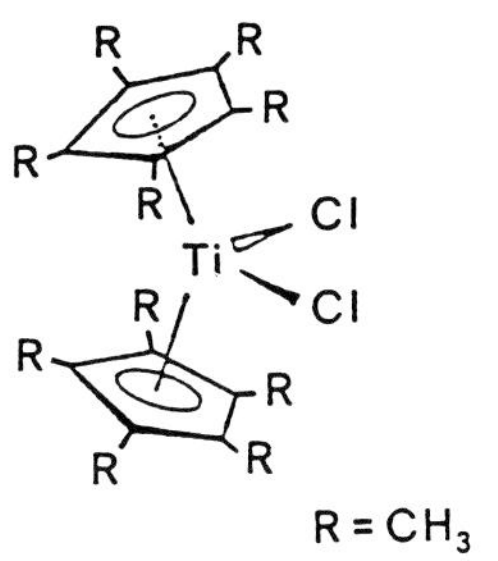

LXIa

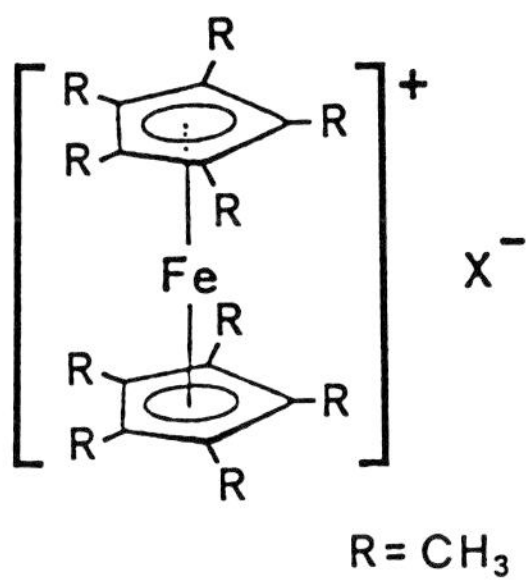

LXIb

One or both of the η^5-bound cyclopentadienyl ring ligands can be wholly exchanged for the related, equally η^5-bonded indenyl (XXXV) or tetrahydroindenyl (XXXVI, XXXVII) or even heterocyclic ring systems such as 2-methyl-1-(trimethylsilyl)-η^5-1,2-azaborolinyl (LXII).

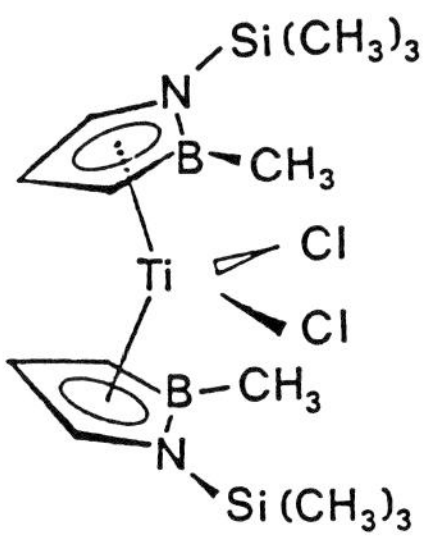

LXII

Moreover, one of the cyclopentadienyl ligands in $(C_5H_5)_2TiX_5$ can be replaced by an additional acido ligand X to give the mono(cyclopentadienyl)triacidotitanium(IV) complexes XXXVIII–XL (cf. Chap. II).

When the antitumor activity of the titanocene complexes XXIII–XL modified at the cyclopentadienyl ligands were investigated against Ehrlich ascites tumor, the general finding was a reduction of antitumor activity, whereby the degree of reduction was clearly dependent on the degree of modification[48, 103].

- The monosubstituted compounds XXIII and XXIV as well as the mono(tetrahydroindenyl) complex XXXVI, still containing one unsubstituted "intact" cyclopentadienyl ring, effected reduced cure rates of 60–80%[48].
- The 1,1′-disubstituted and 1,1′-bridged titanocenes XXVI–XXXIV induced at the best sporadic cures in 10–30% of the animals treated[48, 86]. Comparable values of cure rates of 10–30% were attained with complexes containing either two indenyl (XXXV) or tetrahydroindenyl (XXXVII) rings or with the mono(cyclopentadienyl) complexes XXXVIII–XL containing an additional acido ligand at the position of the replaced cyclopentadienyl ring[48, 103].
- Tetramethylation or pentamethylation of both cyclopentadienyl rings in titanocene as well as in vanadocene diacido complexes led to a total abolition of antitumor potencies against Ehrlich ascites tumor. Neither sporadic cures nor any prolongation of the survival span were observed in the case of $[(CH_3)_5C_5]_2TiCl_2$ (LXIa), $[(CH_3)_5C_5]_2V(NCS)_2$ and $[(CH_3)_4C_5H]_2V(NCS)_2$[98, 99]. Analogous phenomena were obviously induced in ferricenium complexes by pentamethylation of their cyclopentadienyl ring ligands leading to complexes of type LXIb. Applying, e.g., $\{[(CH_3)_5C_5]_2Fe\}^+Br_3^-$ to animals which bore Ehrlich ascites tumor, cancerostatic inactivity was noted[136].
- When both cyclopentadienyl rings in $(C_5H_5)_2TiCl_2$ were exchanged for an η^5-bound five-membered B,N-heterocycle, i.e. the 2-methyl-1-(trimethylsilyl)-η^5-1,2-azaborolinyl ring ligand[137], no tumor-inhibiting effect was detectable against Ehrlich ascites tumor with $\{B(CH_3)N[Si(CH_3)_3]C_3H_3\}_2TiCl_2$ (LXII)[138].

These results indicate a significant role of the unsubstituted, unbulky, disk-shaped cyclopentadienyl ring ligands for the achievement of the antitumor action of η^5-cyclopentadienyl early transition metal and metallicenium complexes. Possibly, the cyclopentadienyl groups may act as carrier ligands[48] which enable the transport of the active complex centers to the intracellular targets, e.g. by facilitating passage through membranes. This function could be impaired by steric or electronic effects of the substituents at the cyclopentadienyl rings.

On the other hand, the main group metallocenes LVII–LIX, which also exhibited antitumor activity against Ehrlich ascites tumor, contained two pentasubstituted cyclopentadienyl rings (Fig. 6), the substituents introduced into the cyclopentadienyl rings being aryl or aralkyl groups such as phenyl (C_6H_5) or benzyl ($C_6H_4CH_2$) residues. Regarding other antitumor non-platinum-group metal complexes, numerous of them actually contain as a common structural feature aromatic hydrocarbons, i.e. six-membered benzene or five-membered, formally anionic cyclopentadienyl rings as constituents of the ligands or as ligands itself (cf. Fig. 3). This situation may be indicative of these hydrocarbon moieties playing a crucial role in the molecular mode of action of some non-platinum-group metal cytostatic agents. In the case of bis[1,2-bis(diphenylphosphino)ethane]gold(I) chloride, antitumor activity was actually demonstrated for the free ligand itself, though the strength of tumor-inhibiting potency was reduced in comparison to the complex[20].

Therefore, the influence of sublethal doses of the fluid hydrocarbons cyclopentadiene (C_5H_6), which is the preparative source of the cyclopentadienyl (C_5H_5) ligand, and of dicyclopentadiene ($C_{10}H_{12}$), which is the Diels-Alder dimer and room temperature-stable form of cyclopentadiene, on the survival of mice bearing fluid Ehrlich ascites tumor was investigated recently[86]. It became obvious, that, indeed, cyclopentadiene as well as dicyclopentadiene effected tumor inhibition by their own and caused the survival of 70 and 100%, respectively, of the animals treated. Thus, both hydrocarbons, especially dicyclopentadiene, effected cure rates comparable to η^5-cyclopentadienyl metal complexes, e.g. I, when they were applied locally into the tumor-containing peritoneal cavity. On injection, both hydrocarbons provoked severe local pain reactions even at low doses. On the other hand, monocyclopentadiene ($LD_{100} > 320$ mg/kg, corresponding to 4.85 mmol/kg) and dicyclopentadiene ($LD_{100} > 391$ mg/kg, corresponding to 2.96 mmol/kg) were clearly less toxic than I ($LD_{100} = 140$ mg/kg, corresponding to 0.56 mmol/kg). A similar discrepancy becomes evident regarding the levels of sublethal doses effecting optimum tumor inhibition:

- In the case of C_5H_6, a dose of 320 mg/kg (corresponding to 4.84 mmol/kg) was necessary to effect 70% cures.
- Applying $C_{10}H_{12}$, a dose of 196 mg/kg (corresponding to 1.48 mmol/kg) induced 100% cures.
- In the case of $(C_5H_5)_2TiCl_2$, a dose of only 60 mg/kg (corresponding to 0.24 mmol/kg) was needed to cause 100% cures.

In the next step, monocyclopentadiene and dicyclopentadiene were administered intraperitoneally to animals bearing solid Ehrlich ascites tumor growing subcutaneously in the nuchal region[86]. No significant growth inhibition was effected by any of both hydrocarbons applied up to lethal doses. The results obtained are summarized in Figs. 21 and 22. These findings confirm that monocyclopentadiene and dicyclopentadiene certainly cause nonspecific, local tumor-inhibiting effects at comparably high molecular doses, but that they do not exhibit systemic antitumor activity which is characteristic for η^5-cyclopentadienyl metal complexes.

Interestingly, the antiinflammatory and antiarthritic properties[130] of I and cyclopentadiene itself followed a similar pattern. Whereas titanocene complexes were potent antiinflammatory agents, cyclopentadiene, given at a dose equivalent to ten times the effective doses of titanocene complexes, had no effect[130].

These results clearly underline that the antitumor activity of cyclopentadienyl metal complexes cannot simply be ascribed to the isolated action of free cyclopentadiene molecules released from a decomposing polymer $[(C_5H_5)TiO]_4O_2$, as was postulated recently[135]. The hypothesis that the cyclopentadienyl ring ligands in metallocene diacido and ferricenium complexes may play an important, perhaps crucial, role for the cytostatic action of these compounds would certainly explain why structurally quite different cyclopentadienyl metal complexes like tetrahedral metallocene (Fig. 4) and linear metallicenium (Fig. 5) complexes are equally effective and characterized by a similar spectrum of activity, especially as cyclopentadiene can be released from the complexes by hydrolytic cleavage of the cyclopentadienyl-metal bond[36, 43, 57, 84, 139]. On the other hand, the complexed metal atoms also seem to play an important role whether as a kind of "carrier" of one or both cyclopentadienyl rings to the site of action, or an "anchor" capable of

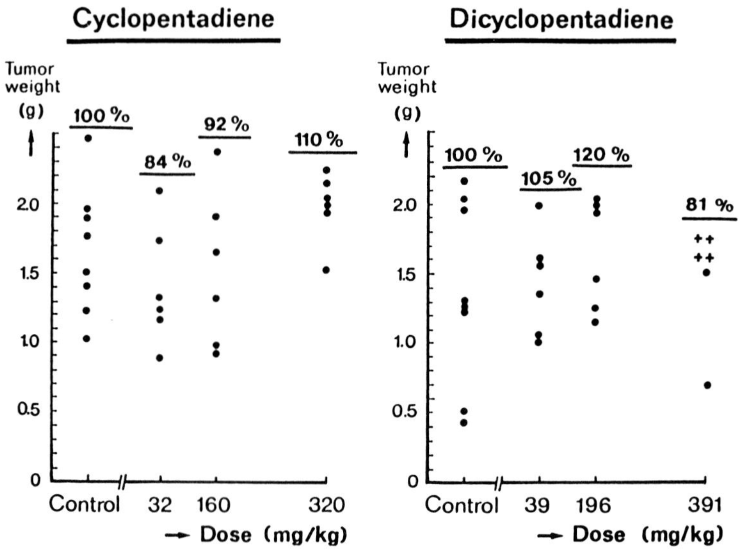

Figs. 21 and **22.** Ranged weights of solid Ehrlich ascites tumors on day 8 after tumor transplantation. Treatment with equimolar doses of monocyclopentadiene (C_5H_6) on day 1 (*left,* **Fig. 21**) or dicyclopentadiene ($C_{10}H_{12}$) on day 1 (*right,* **Fig. 22**). The numbers on top of the columns represent T/C values. + Toxic deaths before day 8

additional, coordinative interaction not accessible to the isolated cyclopentadienyl ligand or free cyclopentadiene molecule itself. This would, moreover, explain the cancerostatic inactivity of zirconocene dichloride[91)], as the enhanced hydrolytic lability of the cyclopentadienyl-metal bond found in this complex[84)] could prevent the transport of cyclopentadiene to the site of action.

Finally, the lack of a systemic tumor-inhibiting effect of cyclopentadiene and dicyclopentadiene stresses that the action of the cyclopentadienyl metal complexes may not only be explained by a mere carrier or anchor function of the metal for the cyclopentadienyl moiety, but may as well be due to a direct interaction of the metal atoms with relevant cellular molecules, or to a potentiating cooperation of both effects.

D. Influence of Charge Effects

Electric neutrality of the compounds, postulated as an essential prerequisite for the achievement of cytostatic properties in the case of platinum complexes[140, 141)], is obviously not a crucial condition for cyclopentadienyl metal complexes to exhibit antitumor properties. There exist, on the one hand, numerous neutral compounds such as the metallocene diacido complexes I–XLIII and the decasubstituted main group IV metallocenes LVII–LIX exhibiting antiproliferative properties; on the other hand, marked tumor-inhibiting potencies were also found with the charged ferricenium salts XLIX–LVI as well as with the ionic cyclopentadienyl titanium complexes XLIV–XLVIII wherein the titanium-containing unit may either form the cationic or the anionic moiety. Because of the generally improved water solubility of the salt-like, ionic ferricenium and titanocene complexes, they are promising compounds for further new developments in the field of antitumor cyclopentadienyl metal complexes.

VI. Cellular Mode of Action

Cytobiological experiments were performed to gain insights into the intracellular and molecular mode of action probably leading to the antiproliferative properties of metallocene diacido complexes.

A. Incorporation Studies

Incorporation studies with tritium-labelled, specific precursors of the DNA, RNA and protein syntheses revealed pronounced and persistent inhibitions of nucleic acid synthesis activities following in vivo and in vitro application of titanocene dichloride (I) or vanadocene dichloride (II). Especially DNA synthesis was suppressed in a significant and long-lasting manner[142, 143], II effecting irreversible inhibition of DNA synthesis by 20% in vitro at doses as low as 5×10^{-6} mol/l[143] (Fig. 23). Interestingly, an actual interaction between metallocenes and nucleic acids leading to an alteration of the secondary structure of the nucleic acids could be demonstrated in vitro by incubation of DNA or RNA with I or II and UV-spectroscopic investigations. The latter revealed an increase of the absorbance maximum of the nucleic acids and its shifting to lower wave-lengths[14].

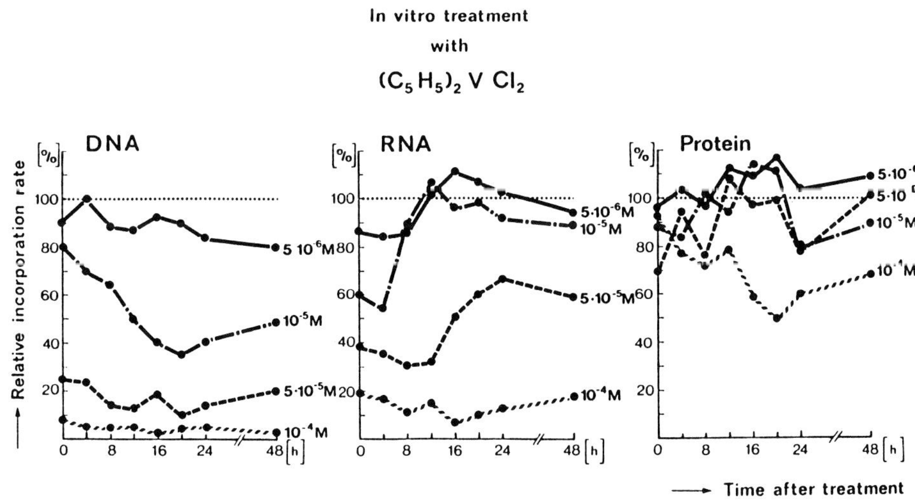

Fig. 23. Incorporation rates related as percentage to the controls (......) of [methyl-^{3}H]thymidine, [5-^{3}H]uridine, and L-[4,5-^{3}H]leucine as a measure of the DNA, RNA, and protein synthesis activities after a 90-min treatment of Ehrlich ascites tumor cells with II in vitro

B. Subcellular Distribution of Central Metal Atoms

The subcellular distributions of titanium and vanadium were analyzed after in vivo and in vitro treatment of Ehrlich ascites tumor cells with I or II by use of the electron energy loss spectroscopy in specimens prepared for electron microscopic purposes. The studies revealed the central metal atoms Ti and V to be mainly accumulated in those cellular regions which are rich in nucleic acids[144–146]. Highest concentrations were found in the nuclear heterochromatin, followed by the euchromatin, the nucleolus and cytoplasmic ribosomes (Fig. 24). In this connection it must be underlined that electron energy loss spectroscopy does not deliver information upon the chemical composition of the metal-containing species accumulated in those cellular regions which are rich in nucleic acids and that, on the other hand, the quantitative enrichment of the central metal atoms within certain cellular areas need not necessarily be indicative of the intracellular site of the crucial molecular action.

C. Cytokinetic Investigations

In the course of cytokinetic studies, performed with Ehrlich ascites tumor cells in vivo[147] and in vitro[148] after treatment with I or II, distinct alterations in the cell transit through the cell cycle were recognized after application of metallocene dichloro complexes.

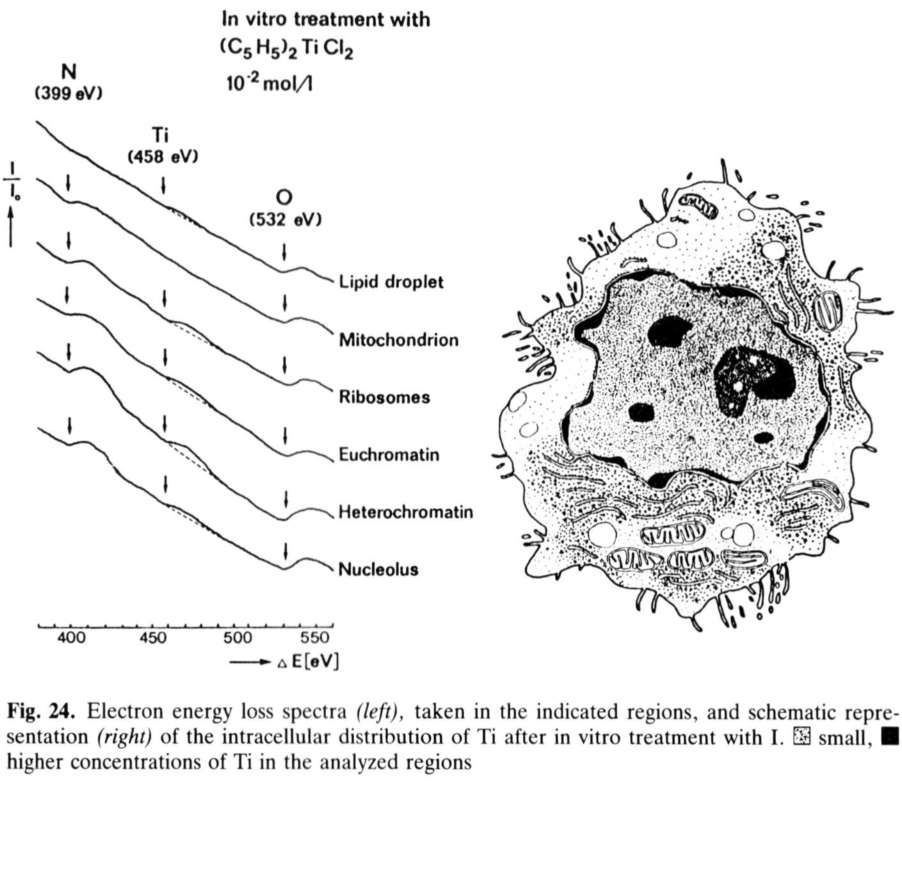

Fig. 24. Electron energy loss spectra *(left),* taken in the indicated regions, and schematic representation *(right)* of the intracellular distribution of Ti after in vitro treatment with I. ▩ small, ■ higher concentrations of Ti in the analyzed regions

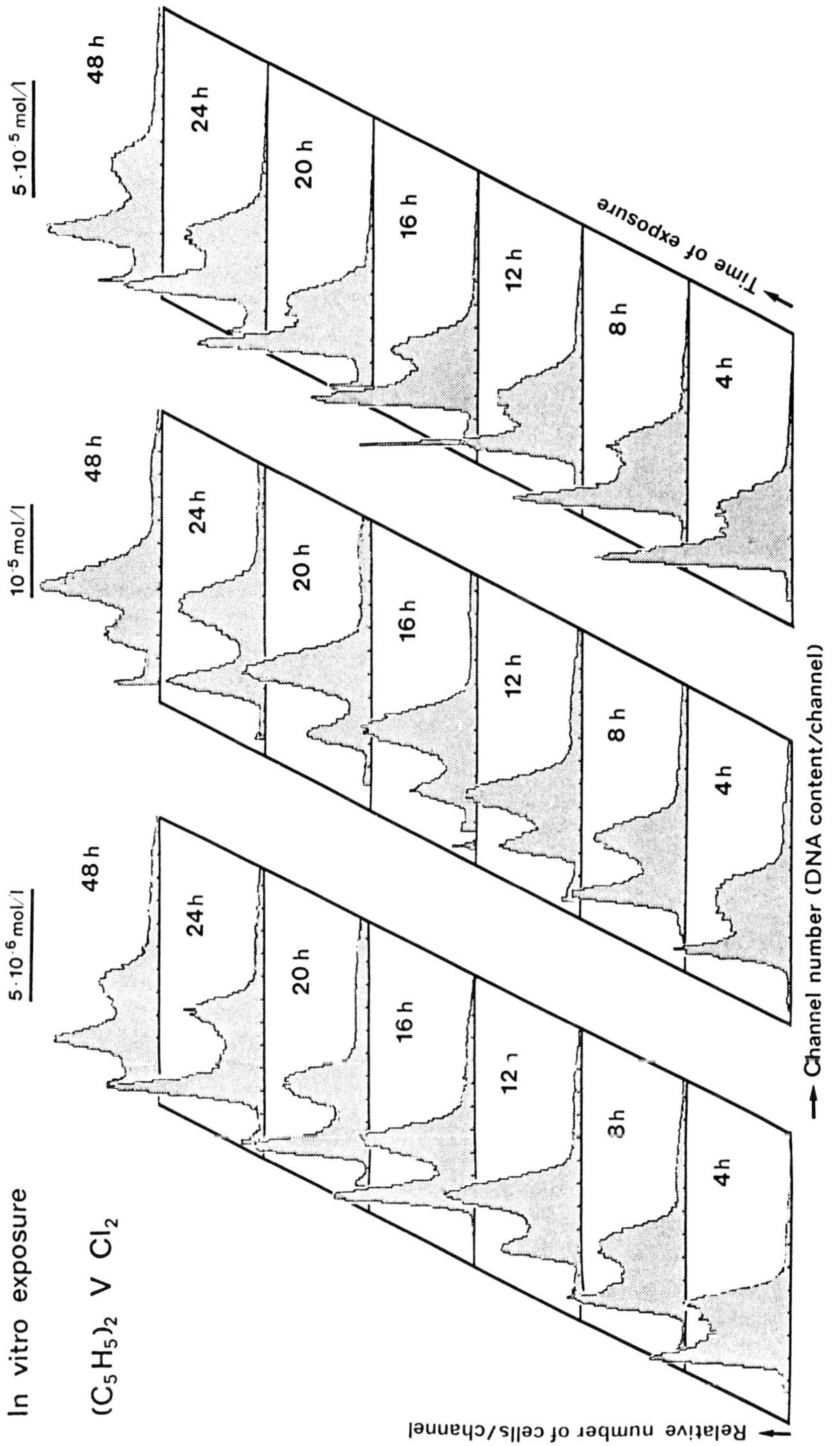

Fig. 25. Pulse-cytophotometric distribution curves of the DNA content of in vitro cultured Ehrlich ascites tumor cells after increasing periods of exposure to II in concentrations of 5×10^{-6} mol/l *(left)*, 10^{-5} mol/l *(middle)*, or 5×10^{-5} mol/l *(right)*

In general, the investigations revealed the appearance of a premitotic G_2 block in vivo and in vitro (Fig. 25) combined with marked mitotic depressions from 3.3% (controls) to 0–1.0% after application of therapeutic doses of I (Fig. 32) or II. During in vitro exposure, an additional accumulation of cells at the G_1/S boundary occurred. Following short exposure periods, these cells escaped the arrest at G_1/S and continued their transit through the cell cycle as a synchronized cell population. The comparison with the cytokinetic behavior of cells treated with other cytostatic drugs shows that many substances, for which the molecular attack upon DNA molecules is generally assumed, provoked similar findings: Cisplatin[147–149], the anthracyclines and alkylating agents such as cyclophosphamide and melphalan caused cell arrests in the G_2 phase. Additional or alternative cell accumulations at the G_1/S boundary were noted under the influence of cisplatin and antimetabolites (for Refs. cf.[148]).

D. Cytological Phenomena in Fibroblasts, Experimental and Human Tumors

Morphological investigations were performed to pursue the histological and ultrastructural alterations of cells under the influence of the cytostatically active compounds I and II, and to compare the phenomena which were induced in non-transformed human fibroblasts with those occurring in animal and human tumor cells.

1. Human Embryonal Fibroblasts

The morphological analysis of human fibroblasts[90] treated in vitro with I or II in concentrations lower than ID_{50} doses (cf. Chap. III.A.) revealed an enlargement of the nuclear volume, pronounced nuclear segmentation (Fig. 27) and increases as well as structural alterations of the nucleoli. Within the cytoplasm of treated fibroblasts, there was evidence of conspicuous protein synthesis which was manifested morphologically by an increase in the size and number of mitochondria, of Golgi apparatuses and of cisternae of the rough endoplasmic reticulum (Figs. 28, 29). Giant cells with voluminous cytoplasms containing either one enlarged nucleus or several nuclei of different size arose (Fig. 31). These phenomena can be interpreted as indicating unbalanced cell growth characterized by a selective inhibition of DNA synthesis, coupled with progressive RNA and protein syntheses. Analogous phenomena were observed after treatment of the same cells with corresponding doses of the inorganic cytostatic drug cisplatin[90]. At concentrations higher than ID_{50} doses, degenerative cytologic phenomena predominated under the influence of I, II and cisplatin leading to nuclear disintegration and vacuolation of cytoplasmic organelles (Figs. 29, 31).

2. Ehrlich Ascites Tumor

Following in vivo treatment of Ehrlich ascites tumor, growing as fluid tumor in the peritoneal cavity of mice, with I or II in therapeutic doses, various developments were discernible by use of light and electron microscopy[82, 150].

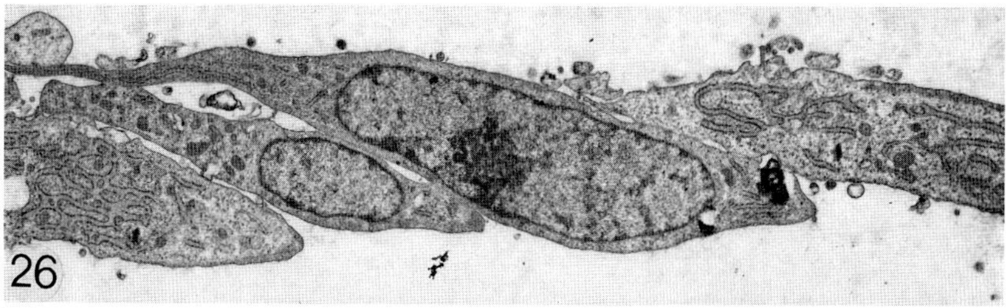

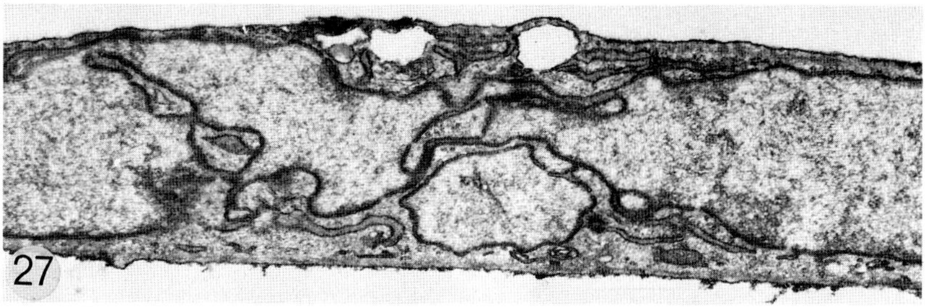

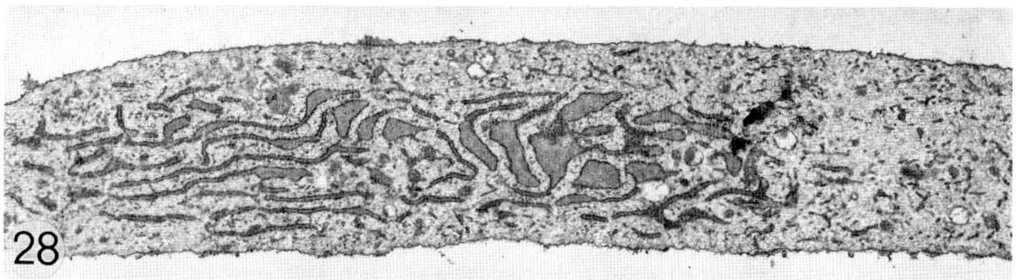

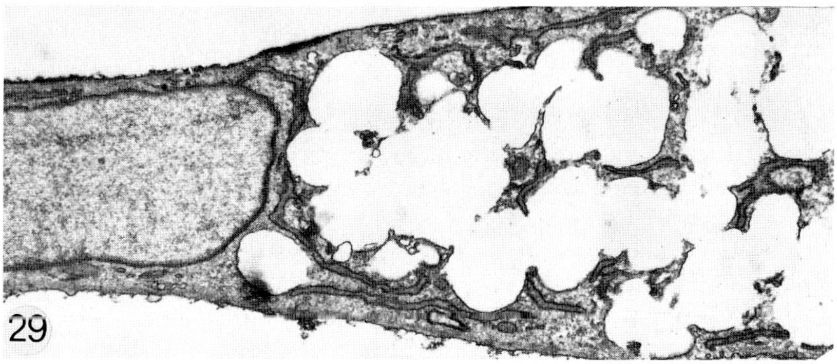

Fig. 26. Untreated human fibroblasts in vitro. × 10,500

Fig. 27. Part of a fibroblast, 24 h after treatment with I (5 × 10^{-4} mol/l). Invaginations of the nuclear envelope and nuclear segmentation, numerous small strands between the nuclear segments. × 16,000

Fig. 28. Part of a fibroblast, 48 h after treatment with I (5 × 10^{-5} mol/l). Numerous distended cisternae of the rough endoplasmic reticulum. × 14,000

Fig. 29. Part of a fibroblast, 24 h after treatment with I (5 × 10^{-5} mol/l). Large numbers of intracytoplasmic vacuoles. × 18,500

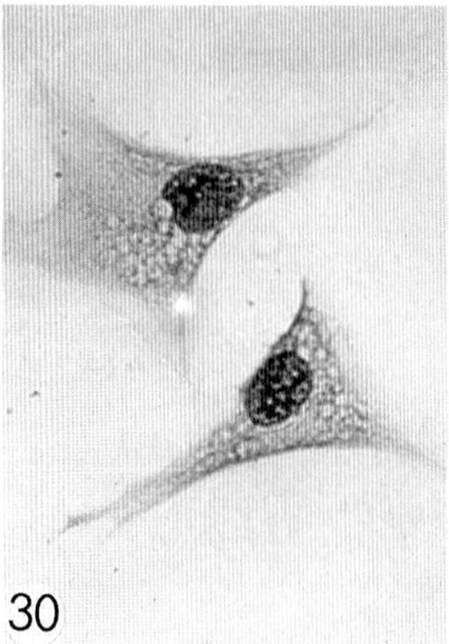

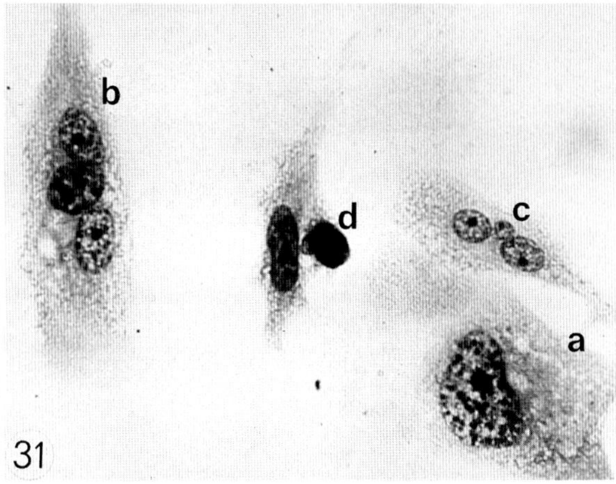

Fig. 30. Untreated human fibroblasts in vitro. × 740

Fig. 31. Fibroblasts, 72 h after treatment with II (10^{-5} mol/l). Mononuclear giant cell with condensed chromatin *(a)*, multinuclear giant cell *(b)*, multinuclear cell of normal size *(c)*, necrotic cell *(d)*. × 740

Within 8 h after application of a single therapeutic dose of I, the mitotic index markedly decreased from 33‰ (control value) to 1‰ (Fig. 32). Mitotic activity remained suppressed, no signs of revival were detectable during the following days.

Regarding the morphologic appearance, similar cytological phenomena were induced by I and II in animal Ehrlich ascites tumor in vivo as they had been observed in non-transformed fibroblasts growing in vitro[90)]. Within two days after substance application, the nuclei increased in size, became segmented (Fig. 34) and were filled with increasingly clumped chromatin (Fig. 34). Unbalanced cellular growth developed and manifested morphologically by the occurrence of nucleoli increased in size and number, enlarged cisternae of the rough endoplasmic reticulum and an augmented number of mitochondria. Numerous giant cells were formed. Most of them were mononucleate (Fig. 35), only a few of them apparently contained several nuclei of different size (Fig. 36). Ultrastructurally, most giant cells exhibited degenerative features in the cytoplasm such as the appearance of large vacuoles or an accumulation of lipid droplets.

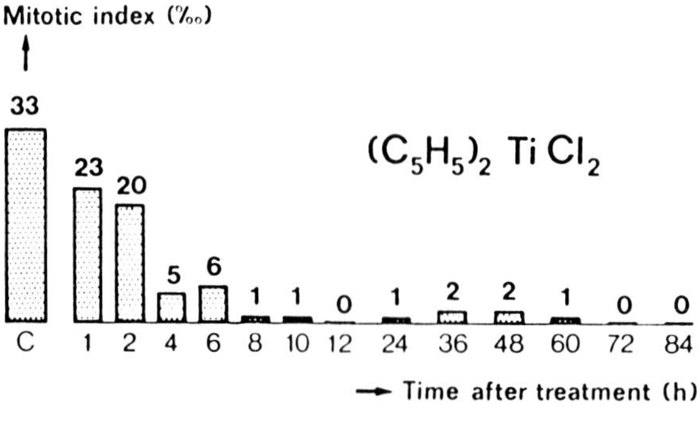

Fig. 32. Mitotic index (‰) in fluid Ehrlich ascites tumor at various intervals after in vivo treatment with I (80 mg/kg) at 0 h. Mitotic index defined as number of mitotic figures per 1000 cells

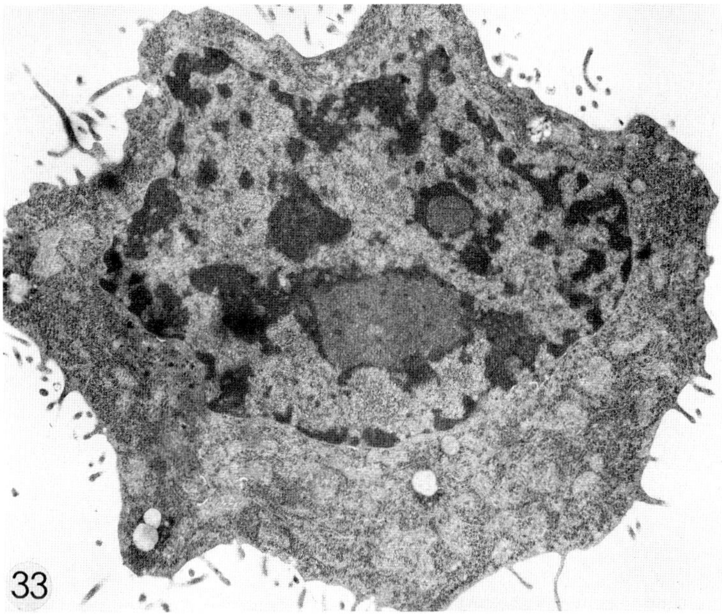

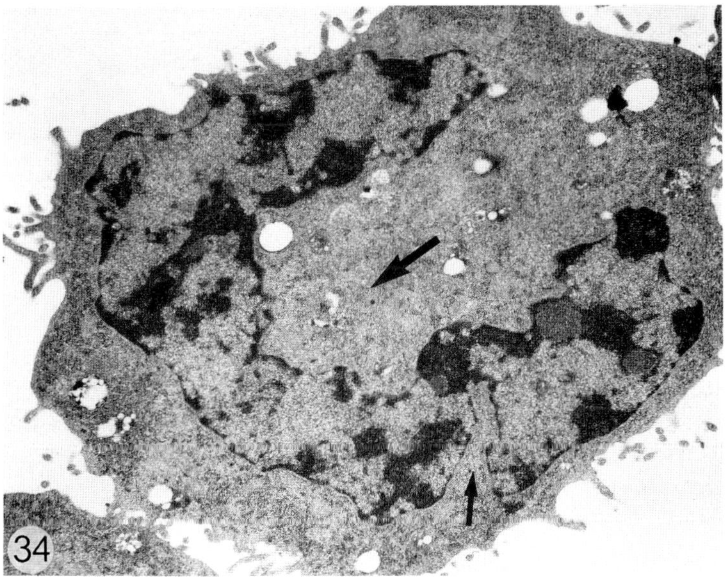

Fig. 33. Electron micrograph of an untreated Ehrlich ascites tumor cell. × 17,000

Fig. 34. Ehrlich ascites tumor cell 8 h after treatment with II (80 mg/kg). Invaginations of the nuclear membrane (→) and slight chromatin clumping within the nucleus. × 15,000

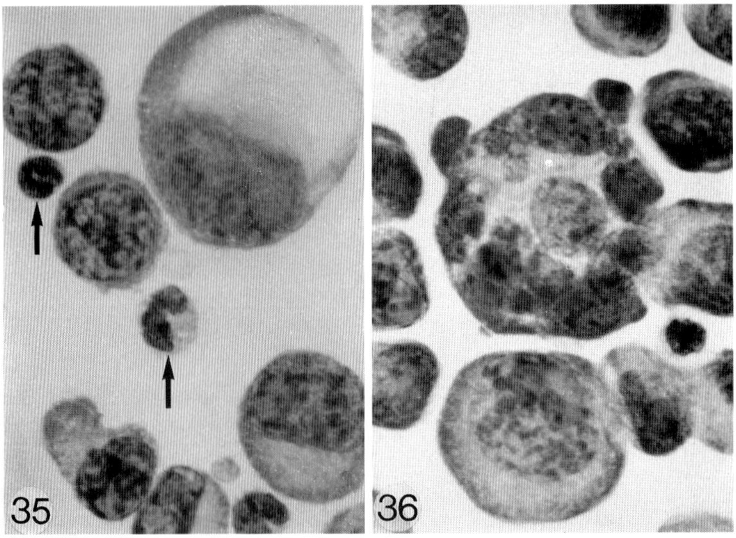

Fig. 35. Ehrlich ascites tumor cells 48 h after treatment with I (80 mg/kg). Mononucleate giant cell surrounded by tumor cells with condensed chromatin and macrophages (→). × 1,000

Fig. 36. Ehrlich ascites tumor cells 72 h after treatment with II (80 mg/kg). Multinucleate giant cell. × 1,000

Additionally to these effects, the morphogenesis of endogenous, previously unexpressed viruses could be observed in Ehrlich ascites tumor cells after in vivo (Figs. 37–39) or in vitro administration of I. Whereas in untreated tumor cells no viruses were detectable by morphological means, complete virus particles of type A were formed within 24 to 28 h after substance application (Figs. 37–39). They crowded the cytoplasm of many tumor cells 2 days after substance application and later.

Two and three days after in vivo treatment with I or II, numerous tumor cells contained bizarre-shaped nuclei with severely condensed chromatin and degenerated cytoplasm. Beginning at 48 h, an increasing number of tumor cells exhibited the features of necrotic cells (Fig. 40). A massive immigration of host cells such as macrophages, leukocytes and lymphocytes belonging to the animal defensive system was observed (Fig. 40). The macrophages attacked giant cells as well as degenerated, necrotic tumor cells and eliminated them by phagocytosis within 72 and 96 h (Fig. 41).

3. Human Colon Adenocarcinoma

The morphological analysis of the human colon adenocarcinoma S90 under the influence of I[151)] again revealed similar cytological phenomena as they had been observed in the case of human fibroblasts treated in vitro and of animal Ehrlich ascites tumor growing in vivo.

Within 24 h after application of a single tumor-inhibiting dose of I, there was a rapid and pronounced decrease of the value of mitotic index of 2.5% (control value) to 0.3%, i.e. to about 10% of the initial value (Fig. 42).

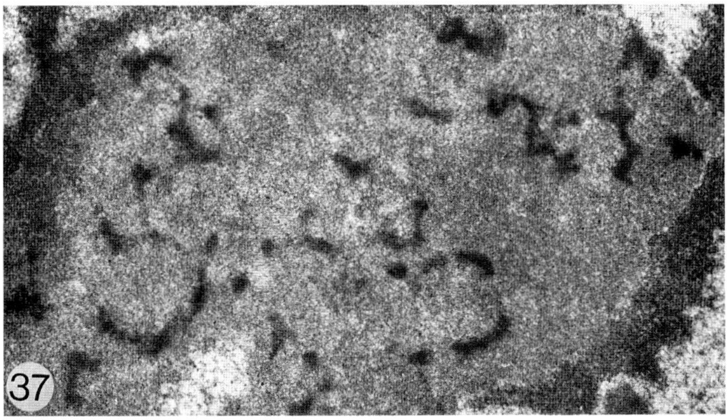

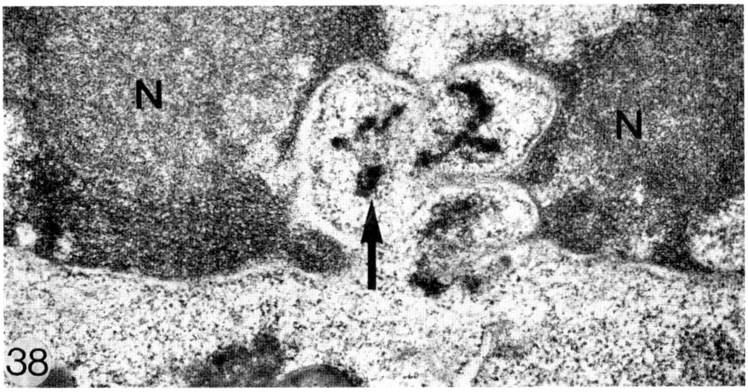

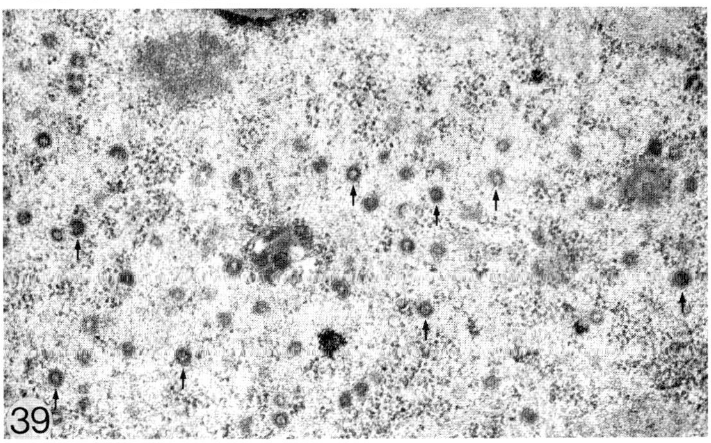

Fig. 37. Nucleolus of an Ehrlich ascites tumor cell 4 h after treatment with I (80 mg/kg). Numerous condensed, worm-shaped figures. × 30,000

Fig. 38. Part of an Ehrlich ascites tumor cell 8 h after treatment with I (80 mg/kg). Accumulated condensed figures (→) within the cytoplasmic invagination. Note the close proximity of nucleoli (N). × 36,000

Fig. 39. Part of the cytoplasm of an Ehrlich ascites tumor cell 36 h after treatment with I (80 mg/kg). Numerous intracytoplasmic spherical particles (→) resembling type A viruses. × 44,000

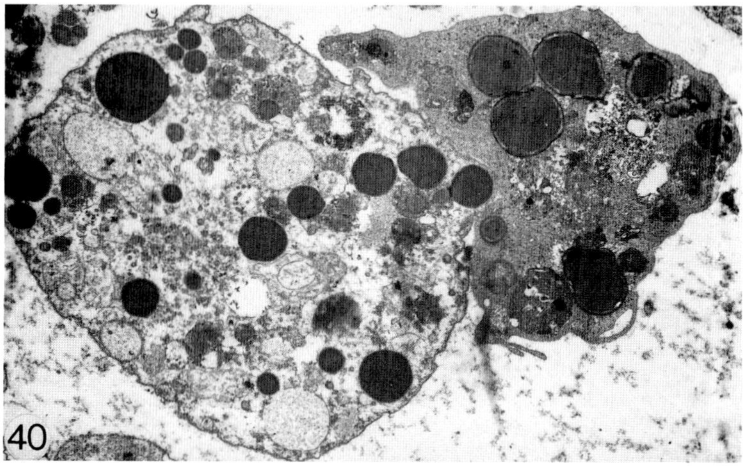

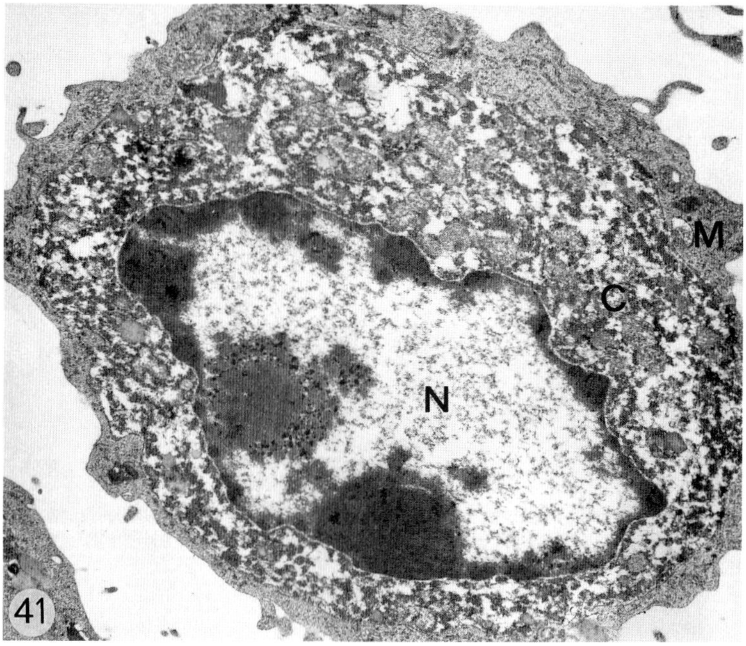

Fig. 40. Ascitic cells 48 h after treatment with I (80 mg/kg). Beginning phagocytosis of a damaged Ehrlich ascites tumor cell *(left)* by a macrophage *(right)*. × 12,500

Fig. 41. Ascitic cells 48 h after treatment with I (80 mg/kg). Performed phagocytosis of an Ehrlich ascites tumor cell (*N* = nucleus, *C* = cytoplasm) by a macrophage *(M)*. × 12,000

Fig. 43. Electron micrograph of untreated human colon adenocarcinoma. × 18,000 ▶

Fig. 44. Tumor cell of human colon adenocarcinoma 24 h after single injection of I (40 mg/kg). Chromatin clumping, enlargment of the nuclear surface, nuclear segmentation. × 16,000 ▶

Fig. 45. Tumor cell 48 h after single injection of I (40 mg/kg). The cytoplasm including a large phagosome and numerous virus particles of type A (→). × 8,600 ▶

Fig. 46. Human colon adenocarcinoma 24 h after threefold injections of I (3 × 30 mg/kg). Degenerated cytoplasm containing lipid droplets and inclusion bodies. × 25,200 ▶

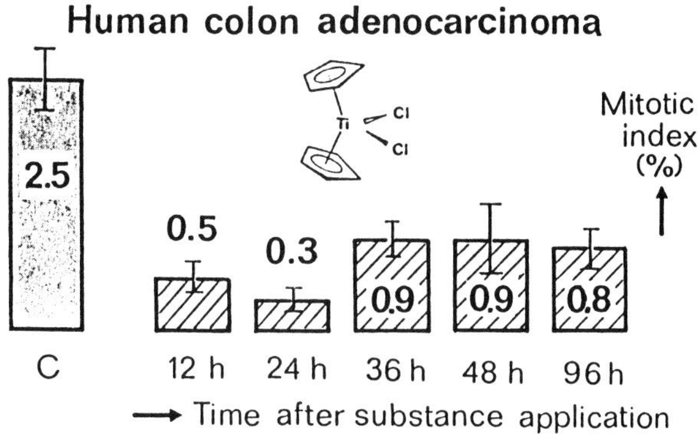

Fig. 42. Mitotic index found in xenografted human colon adenocarcinoma of control animals *(C)* and at different intervals after application of I (40 mg/kg) to nude mice

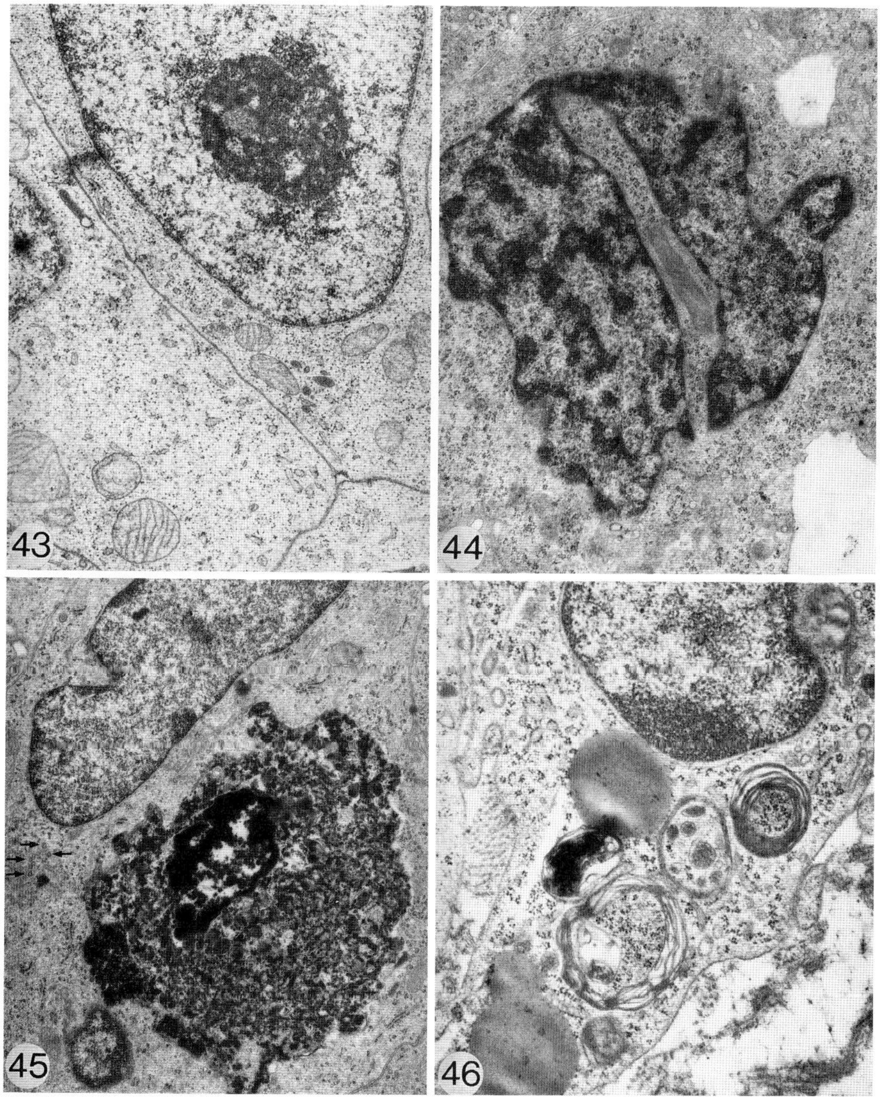

First cytologic alterations of colon adenocarcinoma cells (Fig. 43) manifested 12 h after administration of I and consisted of nuclear changes (Fig. 44) such as chromatin condensation, enlargement of the nuclear envelope, structural aberrations of the nucleoli and formation of segmented nuclei. Cytoplasmic phenomena occurred 12 h later and manifested by the appearance of lipid droplets and inclusion bodies which often contained cellular debris (Figs. 45, 46). These phenomena indicated cytoplasmic degeneration as well as phagocytotic activity of tumor cells.

Intracytoplasmic virus particles of type A were detectable in tumor cells 24 h after substance application and later, forming large assemblages within the cytoplasm (Fig. 47). Because untreated tumor cells were always virus-free, the morphogenesis of type A viruses was apparently stimulated by application of I in a similar way as described in the case of animal Ehrlich ascites tumor cells. Some of the virus particles in human

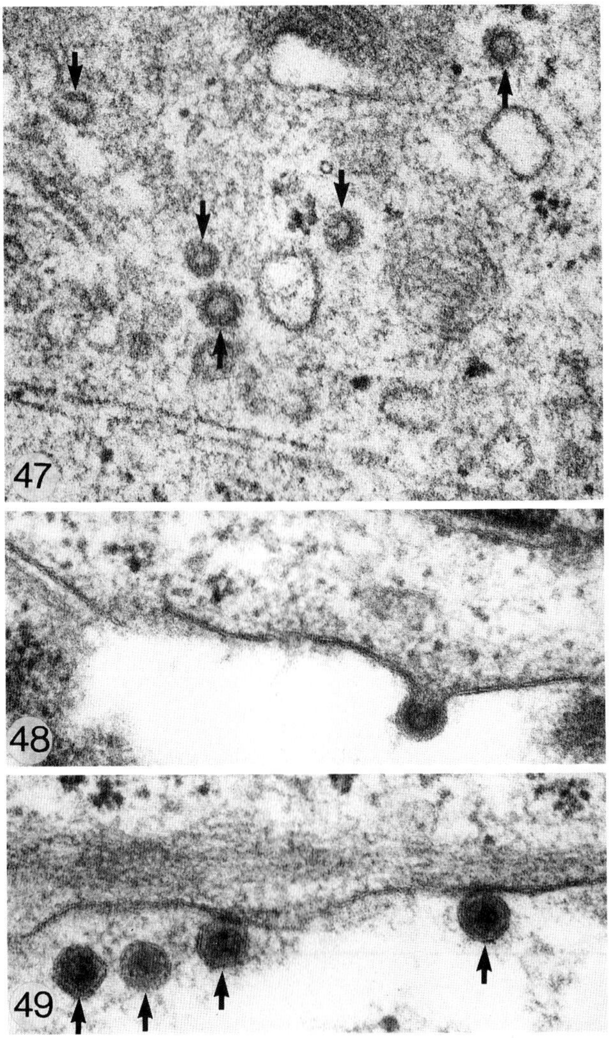

Figs. 47–49. Parts of tumor cells of human colon adenocarcinoma 24 h **(Fig. 47)** and 36 h **(Figs. 48, 49)** after single injection of I (40 mg/kg). Intracytoplasmic virus particles of type A (→) **(Fig. 47)**, budding process **(Fig. 48)**, extracellular viruses of type C (→) **(Fig. 49)**. × 100,000

colon adenocarcinoma cells were extruded by a budding process into the extracellular space (Fig. 48) and became extracellular virus particles of type C (Fig. 49).

Some giant cells were observed 12–48 h after treatment, either containing one enlarged nucleus with a prominent nucleolus or several nuclei of different size.

Beginning 24 h after application of I, numerous inflammatory cells, i.e. granulocytes and macrophages, penetrated the tumor tissue and phagocytosed damaged tumor cells (Fig. 51) thus that, after application of a single dose of I, necrotic cells progressively disappeared within 3 to 5 days. Mitotic activity revived slowly and led to a focal regeneration of tumor tissue 2 to 4 days after administration of a single dose of I.

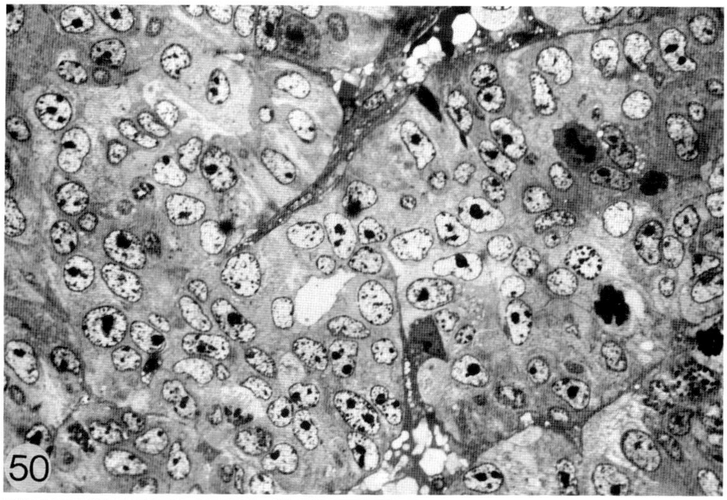

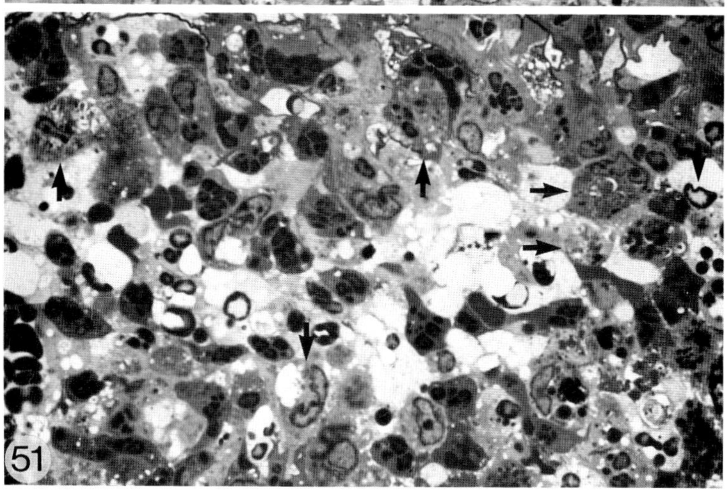

Fig. 50. Untreated human colon adenocarcinoma S90. × 750

Fig. 51. Human colon adenocarcinoma 48 h after single injection of I (40 mg/kg). Profound nuclear and cytoplasmic alterations within tumor cells. Presence of numerous granulocytes and macrophages, some of them surrounding degenerated tumor cells (→). × 1,200

Applying triple doses of I, the morphological alterations provoked were much more pronounced. There was severe destruction and massive degeneration of tumor cells without any signs of regeneration within 5 days after treatment.

Similar to the findings with fibroblasts and Ehrlich ascites tumor, the results of the morphological study with a human colon adenocarcinoma underline the general cytostatic activity of I. Moreover, they clearly confirm the antitumor activity of titanocene complexes against human colorectal carcinomas (cf. Chap. III.C.). Regarding the time sequence of the cytologic phenomena which occurred, the nuclear changes preceding cytoplasmic events hint to a probable primary attack of titanocenes within the nucleus.

VII. Model Complexes

Following the detection of antitumor properties of metallocene complexes, several groups[152–158] have tried to synthesize model compounds of this species with nucleic acid components to elucidate the molecular mode of action possibly involving direct coordination of the metal-containing moiety to DNA or RNA donor sites. Many of these experiments were planned and performed in analogy to those done with cisplatin which apparently induces cytostatic activity by primary attack upon DNA and formation of bifunctional intrastrand cross-links between two adjacent guanine bases[159–163], and were initially stimulated by the observation that a dissociable *cis* or *cis*-like dichlorometal moiety of comparable non-bonding Cl . . . Cl distance ("bite") is present in both cisplatin and the cytostatically active metallocene dichlorides I–III and V[91, 164].

Unfortunately, the synthesis of model complexes of the titanocene system with nucleic acid components as ligands has turned out to be difficult under physiological conditions, i.e. in aqueous solutions, because of hydrolytic side-reactions. Since 1984, Beauchamp and Cozak[152–155] have succeeded in preparing a series of model compounds LXIII–LXIX in organic media using as starting materials not only I, but also the low-valent titanium(III) or titanium(II) species $[(C_5H_5)_2TiCl]_2$ or $(C_5H_5)_2Ti(CO)_2$ (Table 13). They observed as metal-to-nitrogen base linking modes monofunctional bonding of a chlorobis(cyclopentadienyl)titanium(IV) unit to the N-9 atom of the purinato ligand in LXIII (Fig. 52), as well as bifunctional chelation of bis(cyclopentadienyl)titanium(III) units to the N-7 and O-6 atoms of the theophyllinato ligand in LXVIII (Fig. 53), or to both the N-7, O-6 and the N-1, O-2 atoms of the xanthinato dianion in the trinuclear LXIX additionally containing a monodentate dative bond of N-9 to a chlorobis(cyclopentadienyl)titanium(III) centre (Table 13).

Indeed, most of the nitrogen base model complexes listed in Table XIII represent titanium(III) species; but though reduction of titanium(IV) to titanium(III) during biological events does not seem very likely, the models demonstrate that, in principle, the titanocene centre is able to coordinate to nucleobase-related purines and oxopurines by a single monodentate Ti–N bond (Fig. 52), or by bidentate O–Ti–N bonds leading to either five-membered (Fig. 53) or four-membered[155] heterocycles via N-7, O-6 or N-1, O-2 chelation, respectively.

Quite another kind of bonding was pointed out by Marks and coworkers[157] who investigated the interaction of the paramagnetic $(C_5H_5)_2VCl_2$ (II) with mononucleotides in aqueous solution near physiological pH by NMR and ESR methods. They observed

Table 13. Bis(η^5-cyclopentadienyl)metal complexes with nucleobase-related donor molecules

Model complex[a] (M oxidn. state)		Starting compd.[a] (solvent)	Ligand (abbrevn.) reagent	Ligand donor sites[b]	Ref.
LXIII	$Cp_2TiCl(Pur)$ (+4)	Cp_2TiCl_2 (THF)	purine (HPur) HPur + NEt_3 or KPur	N-9	152 153
LXIV	$Cp_2TiCl(HPur)$ (+3)	$(Cp_2TiCl)_2$ (toluene or THF or DME)	HPur		153
LXV	$Cp_2Ti(Pur)(HPur)$ (+3)	$Cp_2Ti(CO)_2$ (THF)	HPur		153
LXVI	$Cp_2TiCl(Ade)$ (+4)	Cp_2TiCl_2 (THF)	adenine (HAde) NaAde		153
LXVII	$(Cp_2Ti)_2(H_{-1}Ade)$ (+3)	$Cp_2Ti(CO)_2$ (THF)	HAde		153
LXVIII	$Cp_2Ti(The)$ (+3)	$Cp_2Ti(CO)_2$ (toluene or THF) or $(Cp_2TiCl)_2$ (THF)	theophylline (HThe) HThe HThe + NaH	N-7, O-6	154
LXIX	$(Cp_2Ti)_3Cl(HXan)$ (+3)	[c]	xanthine (H_3Xan) [c]	N-1, O-2 N-7, O-6 N-9	155
LXX	$(Cp'_2Ti)_2(Ura)$ (+3)	$(Cp'_2TiCl)_2$ (THF)	uracil (H_2Ura) H_2Ura + Na		156
LXXI	$[Cp_2V(OH_2)_2]^{2+}(L^-)_2$ (+4)	Cp_2VCl_2 (H_2O)	$HOP(O)(OC_6H_5)_2$ (HL) HL	$OH_2 \ldots L$ hydrogen bonds	157

[a] Cp = C_5H_5, Cp′ = $CH_3C_5H_4$
[b] By X-ray structure determination
[c] Starting compound, solvent and ligand reagent not published

selective, labile outer-sphere complexation of the vanadocene moiety to the nucleotide phosphate groups, and found that nucleotide-nucleotide Watson-Crick base-pairing was not disrupted by II. Accordingly, the ^1H NMR spectra of alkylated nucleobases such as 9-ethylguanine were not influenced by II in dimethylsulfoxide-d_6. From the X-ray crystal structure of the model complex LXXI (Table 13), obtained from II and the phosphodiester diphenylphosphoric acid simulating the nucleotide phosphate group, the authors concluded the presence of strong hydrogen bonds between the water molecules, coordinated in the vanadocene diaquo cation $[(C_5H_5)_2V(OH_2)_2]^{2+}$, and the oxo oxygen atoms of the diphenylphosphate anions.

In all of the above-mentioned models an intracellular interaction is anticipated between the $[(C_5H_5)_2M]^{+/2+}$ units and the nucleic acid components, which would be analogous to the formation of bonds between the $[(NH_3)_2Pt]^{2+}$ species and nucleobases.

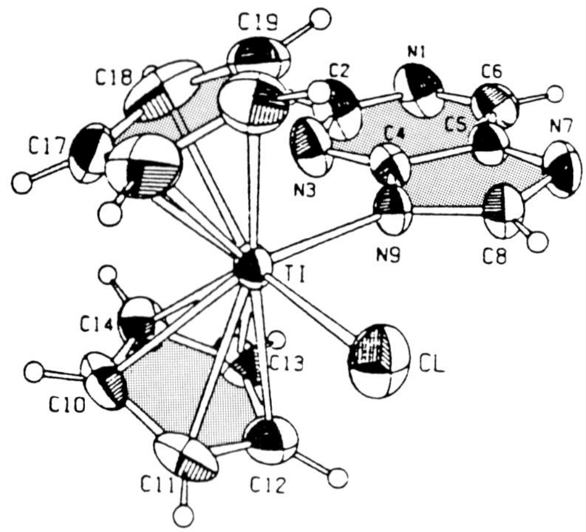

Fig. 52. Molecular structure of chlorobis(cyclopentadienyl)purinato-titanium(IV) $(C_5H_5)_2TiCl(Pur)$ (LXIII). Modified according to[152)]

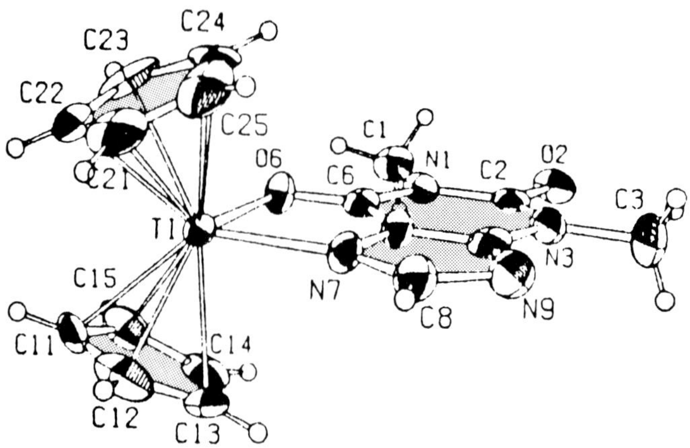

Fig. 53. Molecular structure of bis(cyclopentadienyl)(theophyllinato)titanium(III) $(C_5H_5)_2Ti(The)$ (LXVIII). Modified according to[154)]

On the other hand, our own studies[158)] into the reaction of titanocene dichloride with the nucleobases uracil and guanine in aromatic solvents or in water, respectively, indicated facile loss of one or both cyclopentadienyl ligands of the titanocene moiety (Table 14). The model complexes isolated (LXXII–LXXV, Table 14) contained bridging, dianionic nucleobase ligands bound to mono(cyclopentadienyl)titanium(IV) or even cyclopentadienyl-free titanium(IV). Thus, the possible in vivo interaction not only of intact metallocene, but also of mono(cyclopentadienyl)metal units or even the "naked" metal ions, formed by hydrolytic cleavage of the cyclopentadienyl ligands, with nucleobases or other donor sites at the nucleic acids or even other biological target molecules has also to be envisaged.

The experience that model complexes of metallocenes with nucleic acid components are not easily obtained under physiological conditions, may, however, also be interpreted as a support of the cyclopentadiene hypothesis (cf. Chap. V.C.).

Table 14. Mono(cyclopentadienyl)titanium and titanium complexes with nucleobases[158)]

Model complex[a] (Ti oxidn. state)		Starting compd.[a] (solvent)	Ligand (abbrevn.) reagent	Props.
LXXII	$[CpTi(Ura)]_2(\mu\text{-Ura})$ (+4)	Cp_2TiCl_2 (toluene)	uracil (H_2Ura) $H_2Ura + NaNH_2$	sol., orange
LXXIII	$[CpTiCl(\mu\text{-Ura})_2]_2Ti$ (+4)	Cp_2TiCl_2 (toluene)	$H_2Ura + {}^nBuLi$	sol., orange-brown
LXXIV	$[Ti(\mu\text{-Ura})_2]_n$ (+4)	Cp_2TiCl_2 (benzene)	$H_2Ura + NEt_3$	insol., yellow
LXXV	$[Ti(OH)_2(\mu\text{-Gua})]_n$ (+4)	Cp_2TiCl_2 (H_2O)	Guanine (H_2Gua) K_2Gua	insol., yellow

[a] $Cp = C_5H_5$

VIII. Dissociation and Hydrolysis Reactions

In aqueous solution most of the metallocene complexes are, similar to cisplatin[165, 166)], not stable, but undergo dissociation, aquation and hydrolysis reactions. Thus, titanocene dihalides coexist in reversible equilibria, dependent on pH and concentration, with cationic aquo complexes[36, 83)], according to the following scheme referring to I ($Cp = C_5H_5$):

$$Cp_2TiCl_2 + 2\ H_2O$$

$$\rightleftharpoons$$

$$[Cp_2TiCl(OH_2)]^+ + Cl^- + H_2O$$

$$\rightleftharpoons$$

$$[Cp_2Ti(OH_2)_2]^{2+} + 2\ Cl^-$$

$$\rightleftharpoons$$

$$[Cp_2TiOH(OH_2)]^+ + H^+ + 2\ Cl^-$$

$$\rightleftharpoons$$

$$\tfrac{1}{2}\ [Cp_2(H_2O)Ti\text{–}O\text{–}Ti(OH_2)Cp_2]^{2+} + H^+ + 2\ Cl^- + \tfrac{1}{2}\ H_2O$$

It can be concluded from these equilibria that I, for example, is more stable in acid solutions or in saline than in pure water.

Moreover, neutral oxo-bridged dinuclear and, by hydrolytic cleavage of cyclopentadienyl rings, oligonuclear species, e.g. $(C_5H_5)_2(Cl)Ti\text{–}O\text{–}Ti(Cl)(C_5H_5)_2$ (XLI) and $[(C_5H_5)Ti(Cl)O]_3$, are formed in the course of hydrolysis reactions[57, 83)]. Analogously, by hydrolysis of $(C_5H_5)TiCl_3$ (XXXVIII), the tetranuclear $[(C_5H_5)Ti(Cl)O]_4$ (XLII) was prepared[57]. Therefore, the possibility of hydrolytic cleavage of at least one cyclopentadienyl ring from the titanium centre has to be taken into account (cf. Chap. V.C. and VII.). The diminished antitumor activity of XLI and XLII[43] points to the evidence that these hydrolytically formed products do not represent the intrinsically active species. An analogous result was registered considering the antiinflammatory activity of titanocene dihalides and their hydrolysis products[130]. Whereas titanocene diacido complexes exhibited potent antioedemic and antiinflammatory properties, the products formed by hydrolysis reactions were much less active[130].

On elevating pH values to 7–8 in aqueous injection solutions of the titanocene halides I and VII–IX by buffering before application, decolorizations and the occurrence of white precipitates were observed, and antitumor activity was lost[95]. An insoluble, unstable polymer of the composition $[(C_5H_5)TiO]_4O_2$ has recently been stated to be the only titanium-containing hydrolysis product of bis(cyclopentadienyl) as well as mono(cyclopentadienyl)titanium(IV) complexes in aqueous solution at pH 5.5, and the cancerostatic activity of titanocene complexes has been ascribed to a depot effect of this polymer slowly decomposing and releasing low doses of cyclopentadiene[139] (cf. Chap. V.C.). Accordingly, the same polymeric depot compound should also be hydrolytically formed from the mono(cyclopentadienyl)titanium(IV) complexes XXXVIII–XL and XLII which, however, have been actually found to be only sporadically active against Ehrlich ascites tumor and to exhibit drastically decreased cure rates[43].

In a detailed kinetic study[84] the order of decreasing hydrolytic stability of the M-(C_5H_5) bond of metallocene dichlorides in unbuffered aqueous KNO_3 solution was determined to be $(C_5H_5)_2VCl_2 > (C_5H_5)_2TiCl_2 \gg (C_5H_5)_2ZrCl_2$. Dissociation of the first chloride ion was to be too rapid to be measured. Approximate half-lives for the loss of the second chloride amounted to 50 min for I, 30 min for $(C_5H_5)_2ZrCl_2$ and 24 min for II, the equilibrium constants being $K_2 = 4.2 \times 10^{-2}$ (I) and 2.7×10^{-3} (II).

In the case of cancerostatic (benzoylacetonato)metal(IV) halides of titanium, zirconium and hafnium such as LX, similar halide dissociation and hydrolytic ligand cleavage reactions were found in aqueous systems[131], but no detailed information on this point of interest is yet available.

Finally, it must be emphasized that the experimentally applied aqueous conditions including saline or pH adjustment, do not represent fully physiological conditions. Blood constituents such as proteins or lipids may be able to stabilize transition metal complexes being otherwise sensitive to hydrolysis. In this connection it is worth mentioning that titanocene dichloride is readily soluble in an aqueous lipid emulsion, that this galenical preparation of I is stable against hydrolysis over a long period, and that antitumor activity of I in this preparation is fully preserved[167].

IX. Organ Distribution and Pharmacokinetics

The time-dependent organ distributions of titanium and vanadium were analyzed by flameless atomic absorption spectroscopy in dried organ specimens after single intraperitoneal administrations of therapeutic doses of titanocene dichloride (I, 60 mg/kg)[86] or vanadocene dichloride (II, 80 mg/kg)[168] at time 0.

A. Organ Distribution of Titanium

After treatment with I, initial organ concentration of titanium was highest in the kidneys 1 h after substance application[86] (Fig. 54). The kidney content of titanium fell during the following hours, while the concentrations in the liver and the intestine increased within 24 h after administration of I and exceeded the kidney values at 4 h and later. At 24 and 48 h, about 10% of total titanium injected was accumulated in the liver corresponding to a liver:blood ratio of 8–9. At 96 h, the liver:blood and intestine:blood ratios still amounted to about 5. In the brain, no titanium concentrations higher than control values were measurable at any time during the experimental period, i.e. within four days after substance application.

These findings indicate that the liver and the intestine are obviously the main organs of excretion for titanocene complexes and their metabolites, whereas elimination via the kidneys seems to be less important. On the other side, the results confirm that titanocenes and titanium-containing metabolites are unable to traverse the intact blood-brain barrier.

A phenomenon analogous to the latter result was noted regarding the passage of titanium-containing metabolites across the placental barrier[86]. When pregnant mice were treated with single doses of I at various days of gestation between the phase of organogenesis and late fetal period, there was obviously no transfer of titanium-containing metabolites into the embryonal compartment after treatment on day 10, 12 or 14

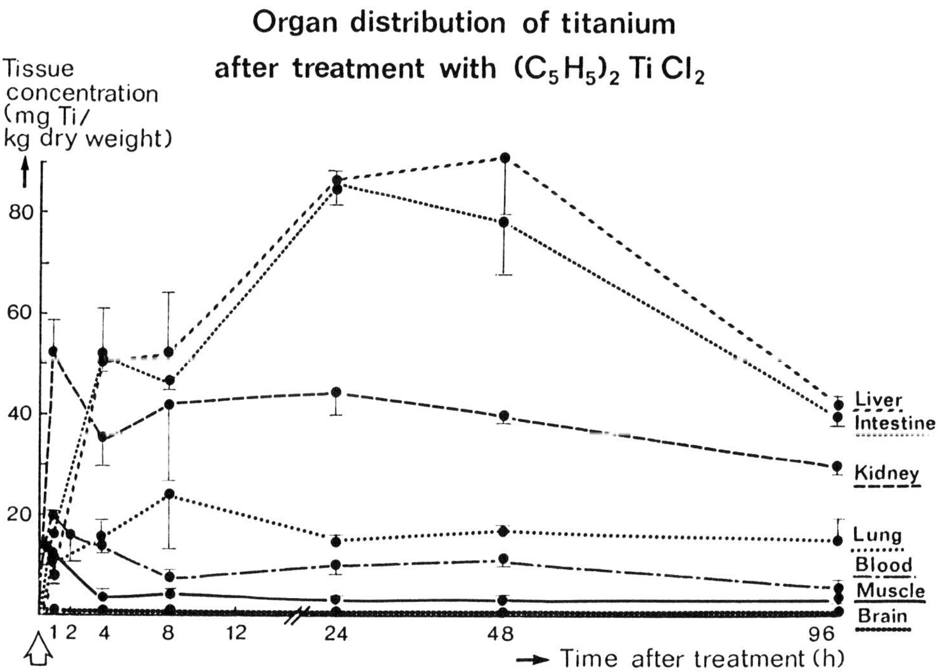

Fig. 54. Time-dependent organ distribution of Ti after single intraperitoneal application of I (60 mg/kg) to NMRI mice at time 0. Control values ranging in all organs between 0.2 ± 0.05 and 0.6 ± 0.35

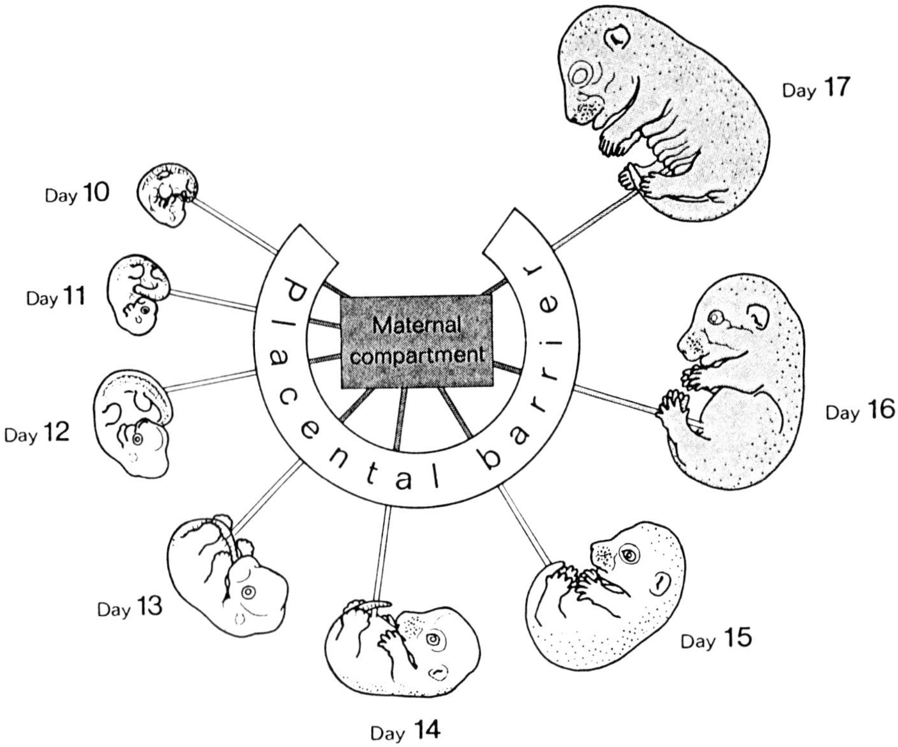

Fig. 55. Schematic representation of the passage of Ti-containing metabolites across the placental barrier in mice in dependence on the days of murine pregnancy. Phase of organogenesis approximately until day 13. Increasing intensity of grey-shaded areas symbolizes increasing concentration of titanium

(Fig. 55). Only when I was applied on day 16, were small amounts of titanium found in the fetuses, the concentrations measured at 8 h after substance application reaching 3.0 mg Ti/kg dry weight, i.e. threefold control values (1.0 mg Ti/kg dry weight). Simultaneously determined titanium concentrations in the maternal blood amounted to 12.3 mg Ti/kg dry weight. In consequence to this apparent inability of titanocene complexes and titanium-containing metabolites to traverse the placental barrier during the sensitive phase of organogenesis, no gross and multiple malformations were inducible in fetuses by application of I to pregnant mice (cf. Chap. X.B.).

Investigating the concentration of titanium in solid, subcutaneously growing tumors, no selective accumulation of titanium within solid tumors was recognizable during two days after substance application[86] (Fig. 56). Thereafter, increasing concentrations of titanium were found in numerous experimental tumors exceeding the values in muscles and, at 96 h, in the blood (Figs. 54, 56). In Fig. 56, the results obtained with solid sarcoma 180 are illustrated. In some human tumors, e.g. in human lung adenocarcinoma L261, even higher concentrations of titanium were found than in sarcoma 180. In all tumors investigated, the concentrations of titanium increased between 8 and 96 h to concentrations of 15–25 mg Ti/kg dry weight at 96 h (Fig. 56) corresponding to 40–60% of the liver concentration registered at this time.

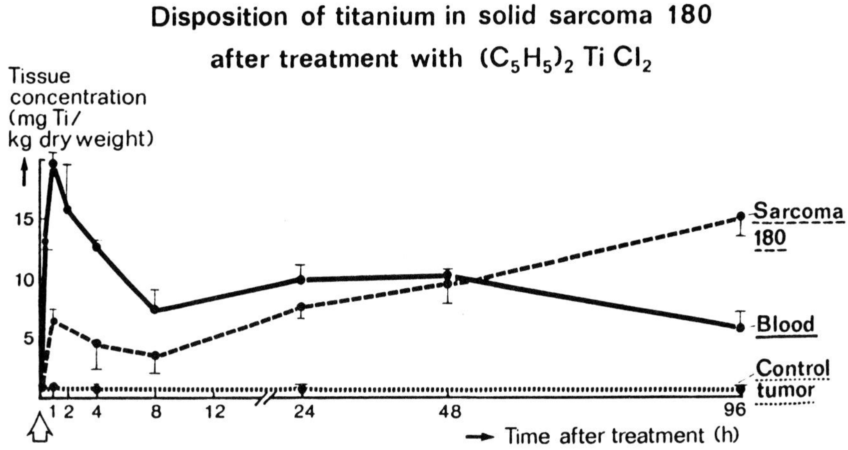

Fig. 56. Time-dependent disposition of Ti in solid sarcoma 180 in comparison to the blood level (for further details cf. legend to Fig. 54)

The clearance of titanium from the blood after application of I was characterized by a clearly biphasic pattern with a rapid-phase half-time of about 5 h and a slow-phase half-time of several days. At 96 h after injection of I, an amount of still 30% of the 1 h-value was recovered in the blood (Fig. 54).

B. Organ Distribution of Vanadium

The analysis of the time-dependent distribution of vanadium in mice after treatment with II[168)] revealed distinct differences to the distribution of titanium following application of I. Main accumulation of vanadium was found in the kidneys, whereas the liver and the intestine contained clearly smaller concentrations. Analogous to the results obtained with I, no transfer of vanadium-containing metabolites across the intact blood-brain barrier occurred thus no vanadium was detectable in the brain over the whole temporal course of the experiment (24 h).

The plasma levels of vanadium declined more rapidly than those of titanium with a half-life of about 2 h. At 24 h, no vanadium was measurable anymore in the plasma[168)]. After application of II to humans, electron spin resonance studies revealed an apparent binding of unaltered II to serum components for more than 12 h[88)].

X. Toxicologic Properties

A. Organ Toxicity

The pattern of organ toxicity induced by titanocene dichloride (I) was analyzed following a single intraperitoneal application of I at ED_{90} (40 mg/kg) and LD_{10} (60 mg/kg) levels.

The toxic pattern differed fundamentally from the toxicologic features provoked by the inorganic cytostatic drug cisplatin at equitoxic dose levels[169–172].

1. Kidneys

It is known that cisplatin is burdened by severe nephrotoxicity which manifests by structural lesions of the proximal and distal tubular cells by functional disturbances, such as increases of blood retention values (Fig. 57) and by proteinuria, glucosuria, and erythrocyturia[171, 173]. In contrast to these effects, I and II did neither disturb renal function nor damage the structure of renal cells. No long-lasting functional impairments of the kidneys, such as long-lasting elevations of blood retention values (BUN, creatinine) (Fig. 57), no changes in the composition of the urine and no histologic and ultrastructural alterations within tubular (Figs. 58, 59) and glomerular cells were detectable after application of metallocene dichloro complexes[88, 170, 171]. This was documented after application of ED_{90}, LD_{10} or even LD_{50} doses of I and II.

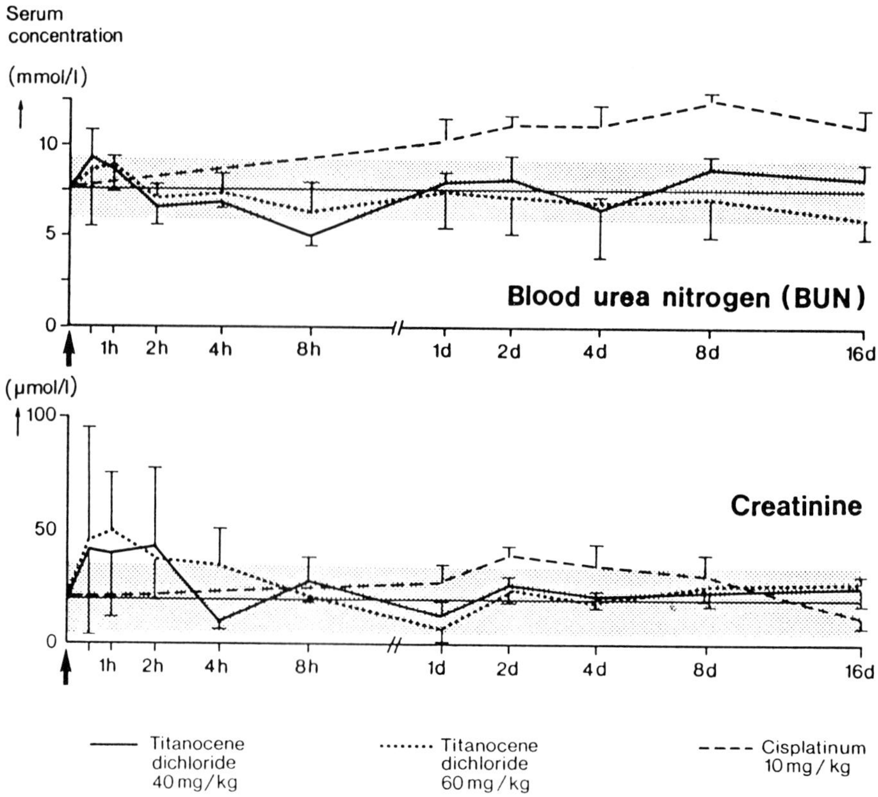

Fig. 57. Serum content of blood urea nitrogen *(BUN)* and creatinine at various intervals (given on *abscissa*) after administration of I, applied at ED_{90} and LD_{10} levels, or cisplatin to mice on time 0. Given are mean values and standard deviations as + s or − s. *Straight continuous lines* and *shaded areas* represent mean values and ranges of standard deviations of control populations

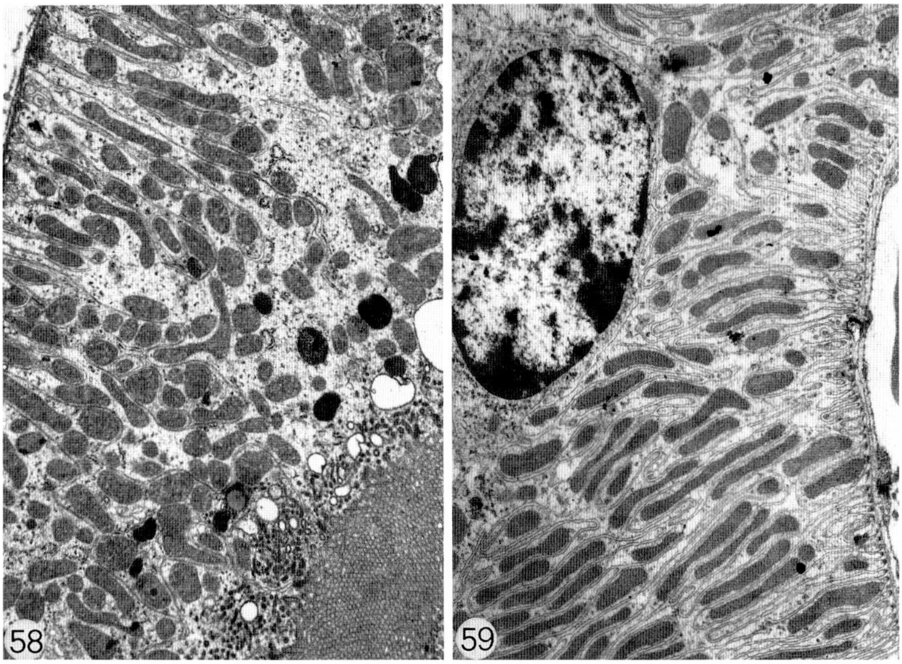

Fig. 58 and **59.** Parts of a proximal **(Fig. 58)** and a distal **(Fig. 59)** convoluted tubule of the kidney 4 days after application of 40 mg I/kg **(Fig. 58)** and 70 mg II/kg **(Fig. 59)**. The ultrastructural features of the renal cells do not differ from those of untreated control cells

2. Liver

Whereas the kidneys were not injured by metallocene dichlorides, the serum levels of some typical enzymes of liver cells, such as glutamate dehydrogenase (GLDH), glutamic-oxaloacetic transaminase (GOT) and glutamic-pyruvic transaminase (GPT), increased within 2 to 4 h to significantly elevated values which exceeded the control values by factors between 3 and 4 (Fig. 60). These findings indicated a short-lasting injury of the integrity of liver cells[169, 170] which was obviously reversible as, 2 to 4 days later, recovery phenomena led to the normalization of the serum levels of liver enzymes. On the other hand, the excretion function of liver cells was apparently not affected. The bilirubin and cholesterol serum levels remained unelevated during the whole experimental period[170] (Fig. 61).

The histologic examination confirmed the findings of functional disturbance of liver parenchyma cells and uncovered the occurrence of small lipid droplets within the cytoplasm of liver cells furnishing evidence for ongoing fatty degeneration of liver parenchyma cells[169] (Fig. 62). At higher doses of I, single cell necroses were observed within liver parenchyma. These structural alterations were also transient and reversible and disappeared within 16 to 32 days after substance application.

In this connection it is worth mentioning that liver toxicity as main side effect was also found after application of therapeutic doses of bis(benzoylacetonato)titanium(IV) dihalides or dialkoxides LX. Increases of the serum values of the enzymes GOT, GPT

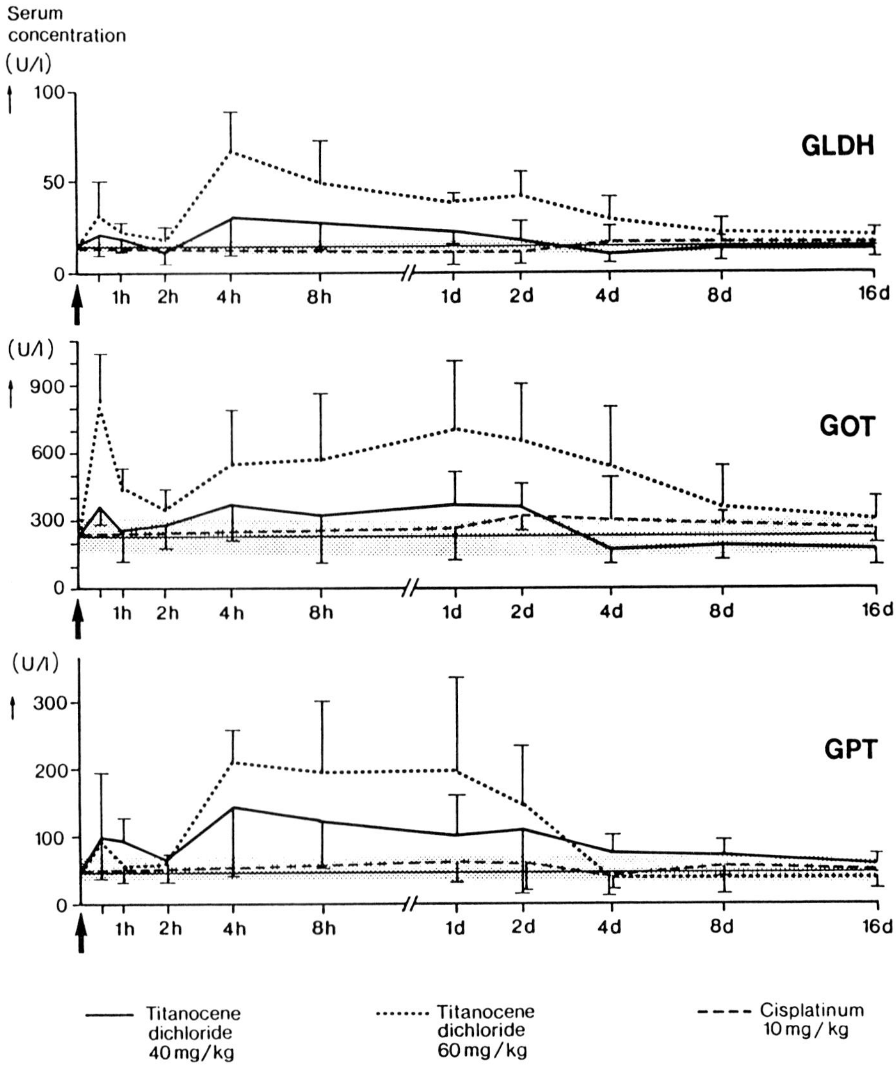

Fig. 60. Serum content of the enzymes GLDH, GOT and GPT. For further explanations, cf. legend to Fig. 57

and LDH and multiple focal necroses in the liver parenchyma were characteristic symptoms[104]. Another kind of organ toxicity induced by higher doses of LX was lung toxicity which manifested macroscopically by hemorrhagic pleural effusions and the appearance of hemorrhagic oedematous areas within the lung[104].

The fact that both kinds of antitumor titanium compounds, i.e. bis(cyclopentadienyl)-titanium and bis(benzoylacetonato)titanium complexes, cause liver toxicity as main side effect may be a hint to titanium or titanium-containing moieties to be responsible for the induction of functional and structural liver damage. This feature is further confirmed by the enrichment of titanium in the liver after injection of I (cf. Chap. IX.A.). On the

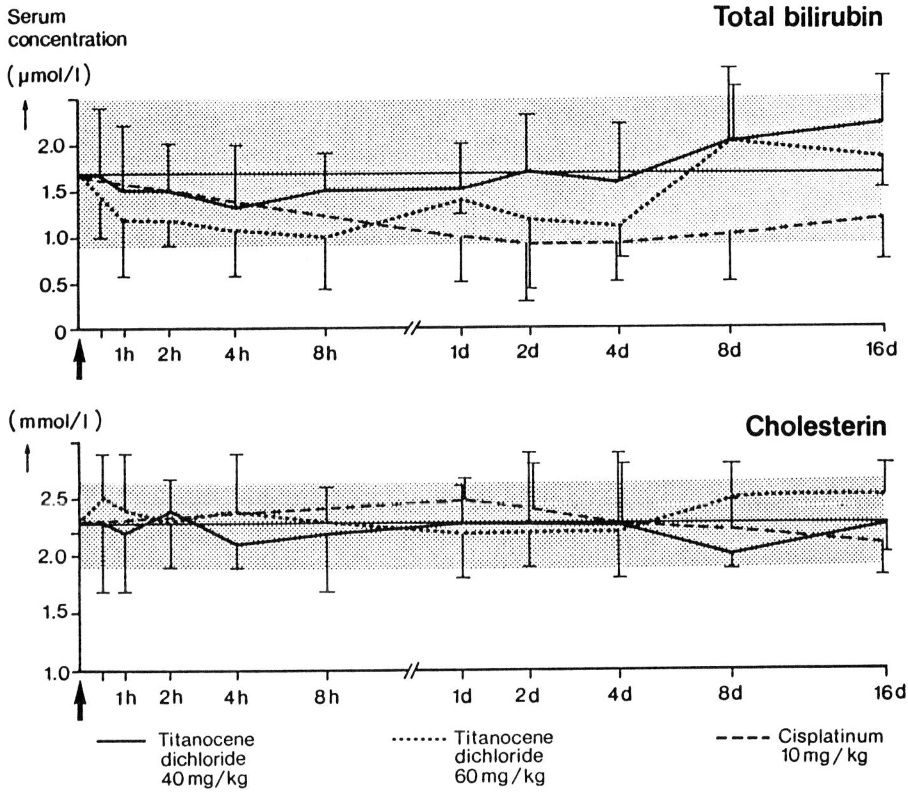

Fig. 61. Serum values of total bilirubin and cholesterol. For further explanations, cf. legend to Fig. 57

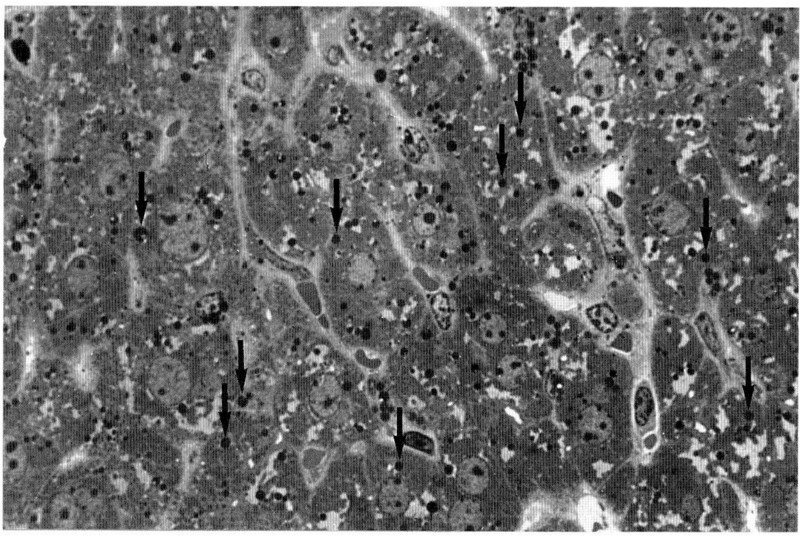

Fig. 62. Histologic appearance of the liver of mice 12 h after application of 60 mg I/kg. Numerous lipid droplets (→) within the cytoplasm of liver parenchyma cells

other hand, it is conceivable that the hydrocarbon moieties present within I and LX are the agents causing liver toxicity. For metallocene complexes of the ferrocene class, $[(C_5H_5)_2Fe]$, metabolization by hydroxylation of the cyclopentadienyl groups inside the liver microsomes via the cytochrome P-450 system and conjugation with glucuronic acid or sulfate molecules were demonstrated, followed by partial degradation of the ferrocene molecules and liberation of Fe^{2+} ions[174, 175].

3. *Endocrine Glands*

Alterations of the hormonal status of animals treated with I were other symptoms of organ toxicity. There were pronounced elevations of the serum concentrations of cortisol (Fig. 63) and glucagon by factors of 3 to 4 within 1 h after substance application, whereas the levels of other hormones such as insulin, aldosterone, catecholamines and progesterone remained unaltered[170, 178]. Possibly, the initial decrease of glucose concentration in the peripheral blood occurring immediately after administration of I was the factor stimulating the regulative output of cortisol and glucagon from suprarenal glands and pancreatic islands.

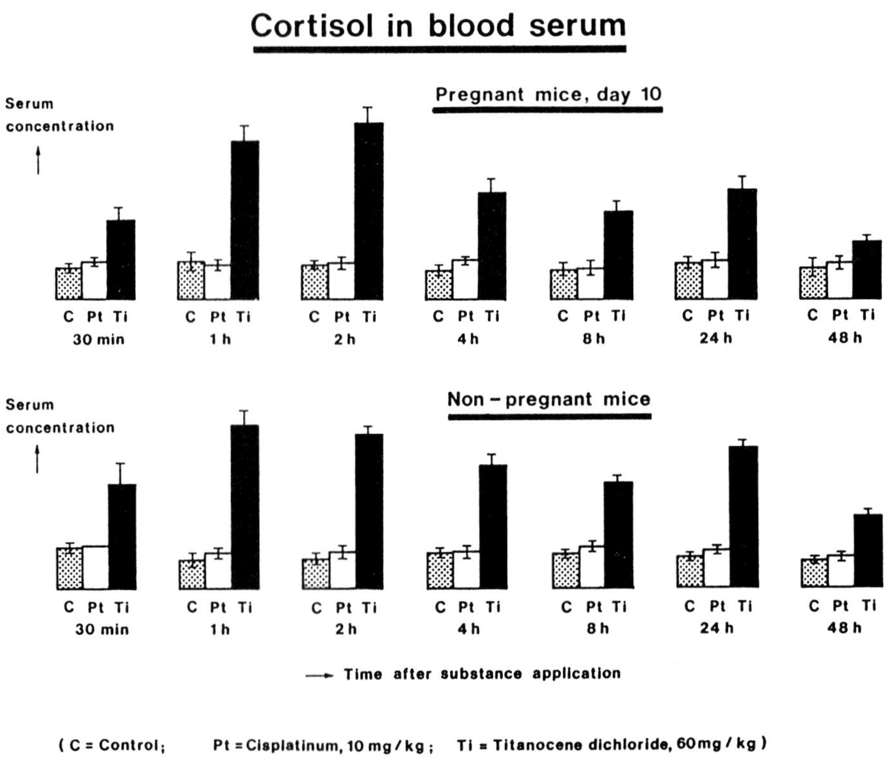

Fig. 63. Mean values and standard deviations of cortisol concentrations in the blood serum of control mice, cisplatin-treated and titanocene dichloride-treated mice at various intervals after substance application at time 0. Treatment of pregnant mice *(upper part)* was performed on day 10 of gestation

In Chap. IX, pharmacokinetic results confirming the inability of I and titanium-containing metabolites to pass the placenta and to enter the embryonal compartment were reported. On the other hand, teratologic experiments (cf. Chap. X.B.) showed the occurrence of cleft palate as single malformation after application of I to pregnant mice[176]. As glucocorticoids are known to induce cleft palate as single malformation in mice[177], the elevation of cortisol concentration observed in a similar manner in the serum of pregnant and non-pregnant mice (Fig. 63) after application of I was probably the "indirect" factor promoting the genesis of cleft palate in mice under the influence of I[178].

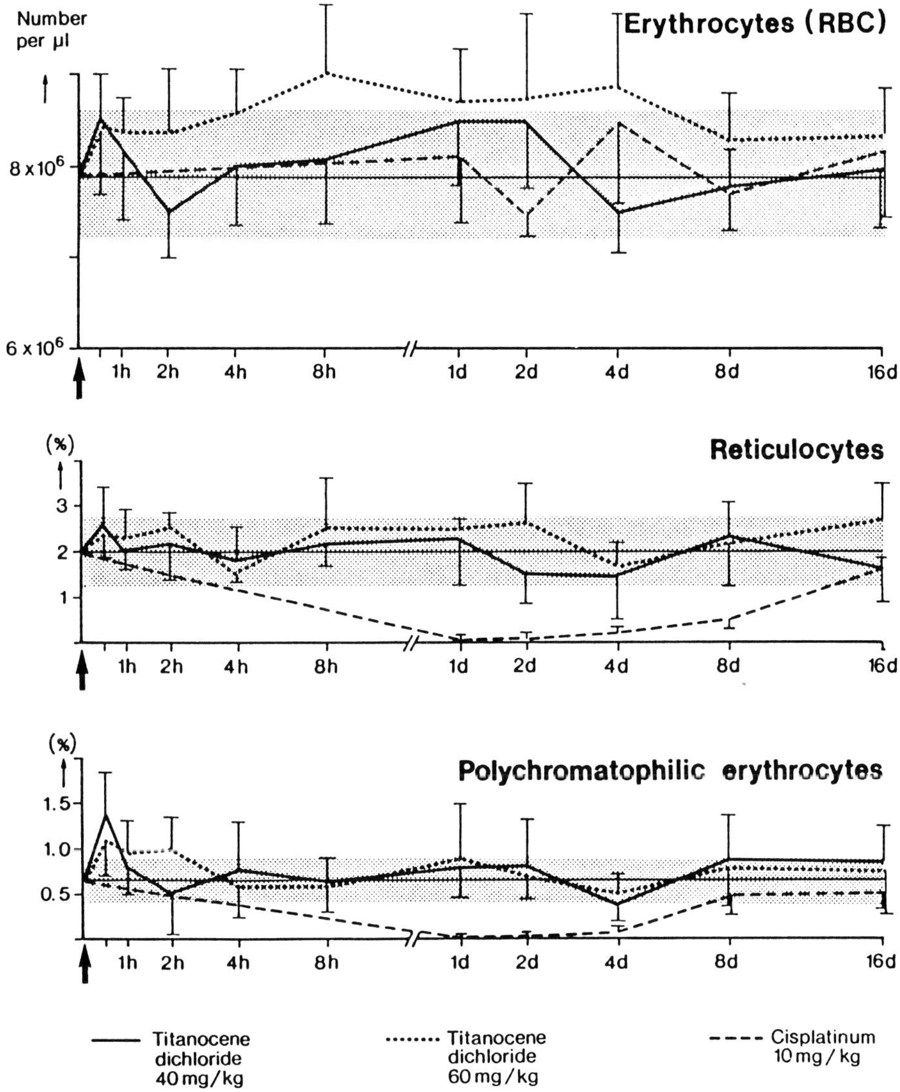

Fig. 64. Counts of mature erythrocytes and young erythrocytes, i.e. of reticulocytes and polychromatophilic erythrocytes, in the peripheral blood. For further explanations, cf. legend to Fig. 57

4. Bone Marrow

Because of its vivacious proliferating activity, the bone marrow generally represents one of the target organs for the toxic action of common cytostatic drugs like alkylating agents, antimetabolites and vinca rosea alkaloids[179]. In contrast to this general feature, myelotoxicity revealed to be of minor significance in the case of the inorganic cytostatic drug cisplatin[180–182]. After application of the organometallic titanium complex I, an even less pronounced depression of bone marrow function was registered than in the case of cisplatin[172]. Neither the numbers of leukocytes and mature erythrocytes in the peripheral blood nor the supply of young erythrocytes from bone marrow were obviously influenced and diminished by the treatment with I (Figs. 64, 65). Only a slight and transient

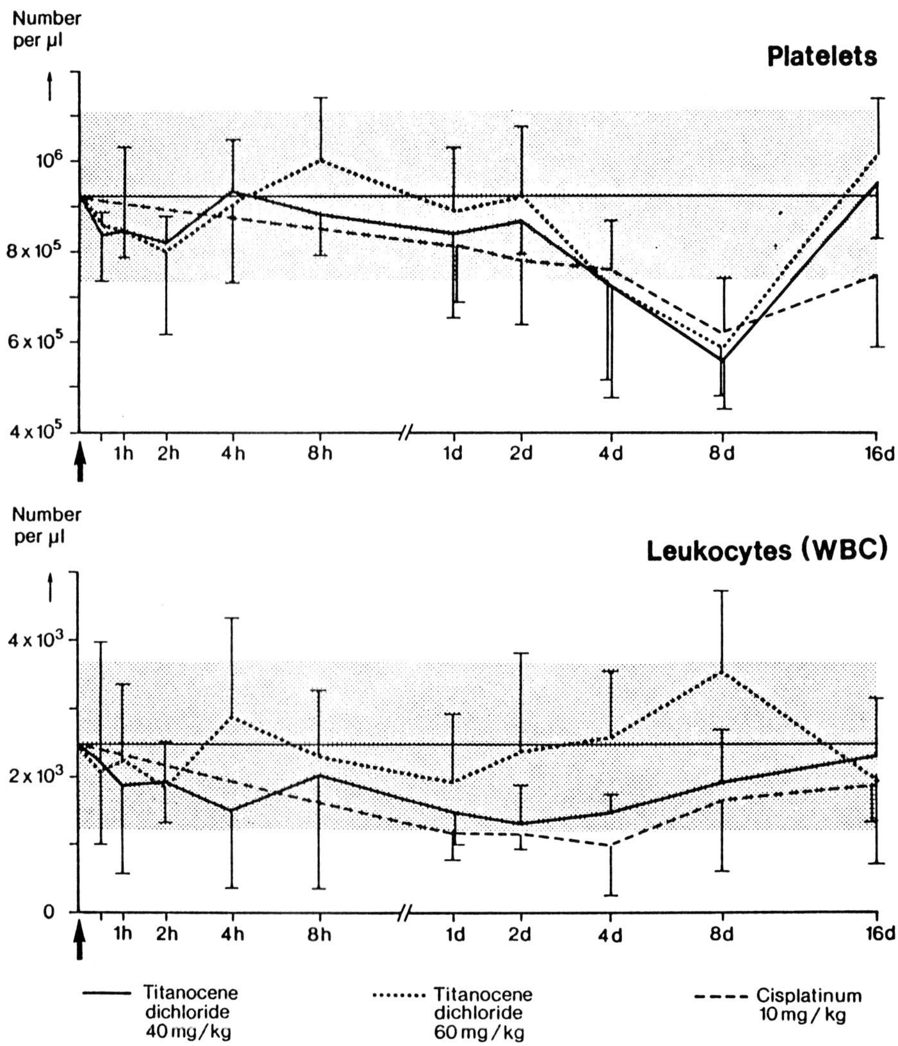

Fig. 65. Counts of platelets and leukocytes in the peripheral blood. For further explanations, cf. legend to Fig. 57

decrease in the count of circulating platelets beneath the control range was discernible 8 days after application of a single dose of I (Fig. 65). This result of an only very mild myelotoxicity by a cytostatic agent is a quite unusual finding. It was analogously observed in the course of preclinical studies with the dichloro and diethoxy derivatives of the titanium complex LX[104].

B. *Embryotoxicity*

Titanocene dichloride effected a general embryotoxic influence when it was applied to pregnant mice. It caused diminution of the number of live fetuses per litter, a marked and dose-dependent reduction of mean fetal body weight after application on day 8 through day 16 of murine pregnancy, a dose-dependent delay of fetal growth and development and a distinct retardation of skeletal ossification[176] (Fig. 66). At higher doses of I, abortions were induced within few hours after substance application.

At no dose level up to LD_{50} doses, multiple and gross malformations of the skeleton and/or the viscera were detectable, as they were usually caused by treatment of pregnant mice with sublethal doses of cytostatic drugs, e.g. alkylating agents, hydroxyurea, 6-mercaptopurine or vinca rosea alkaloids, during the sensitive phase of organogenesis (for Refs. cf.[176, 177]). The only malformation observed after application of I was the dose-dependent occurrence of cleft palate (Fig. 67) in 10–50% of those fetuses, the mothers of which had been treated with I during organogenesis (Fig. 68).

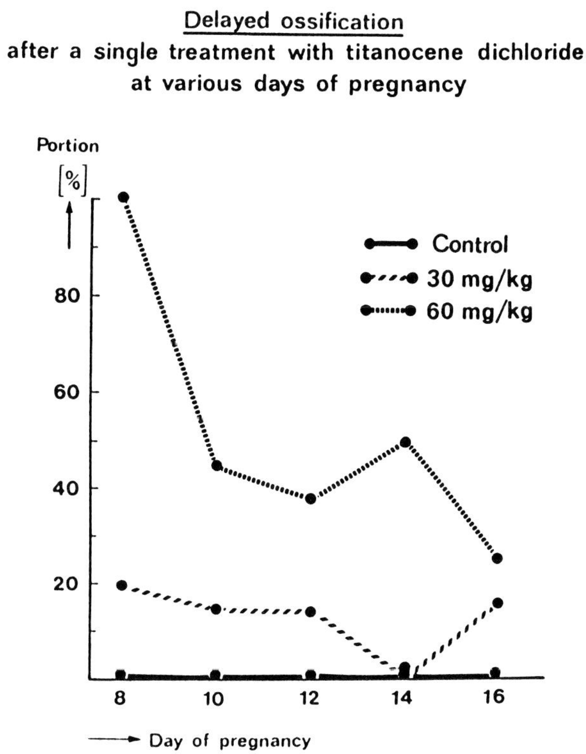

Fig. 66. Portion of fetuses with signs of delayed ossification (no metatarsal ossification centers present) on day 18 of gestation after treatment of pregnant mice with I (30 or 60 mg/kg) on day 8, 10, 12, 14 or 16 of gestation

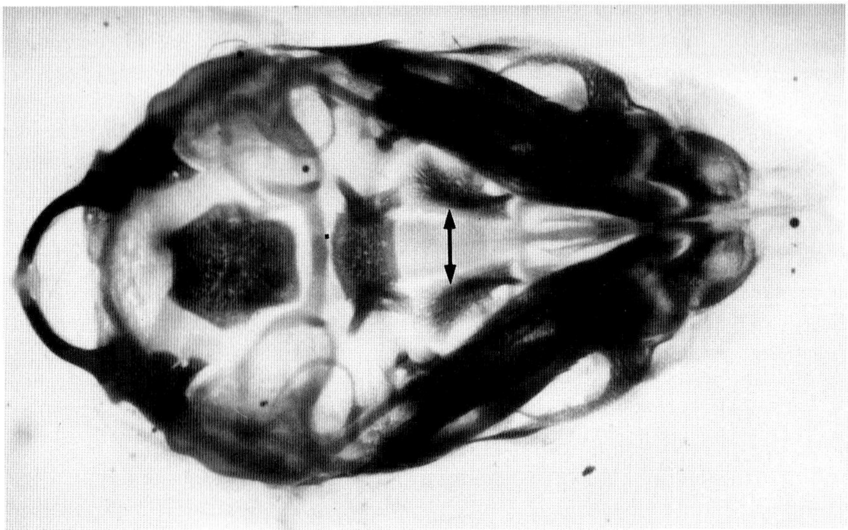

Fig. 67. Double-stained skull of a fetus on day 18 after maternal treatment with I (60 mg/kg on day 12). Base of the skull, showing a broad gap between the palatine bones (↔) which normally contact each other on day 18

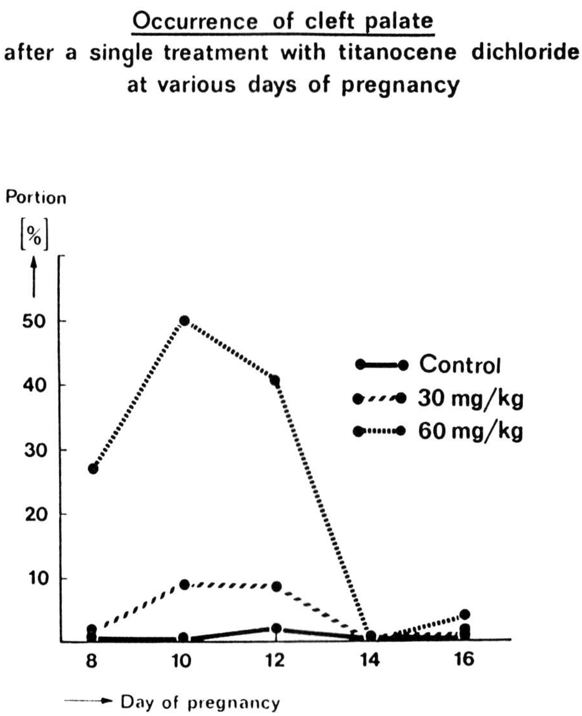

Fig. 68. Portion of fetuses with cleft palate on day 18 of gestation after treatment with I (30 or 60 mg/kg) on day 8, 10, 12, 14 or 16 of gestation

Taking into account that I is able to suppress DNA metabolism in a pronounced and persistent manner[142, 143] and to inhibit cellular proliferation[82, 147, 148, 150, 151], the lack of multiple and variable malformations after treatment of pregnant mice during the sensitive phase of organogenesis is surprising and only explainable by the inability of I and its metabolites to traverse the placental barrier. Recently, this was confirmed experimentally[86] (cf. Chap. IX.A.). Therefore, it must be assumed that the genesis of cleft palate is mediated by indirect mechanisms and that, probably, the increase in the concentration of glucocorticoids in the maternal serum (cf. Chap. X.A.) is the factor responsible for the induction of this single malformation in mice after treatment with I[178].

XI. Summary

Cyclopentadienyl metal complexes represent a group of antitumor agents comprising organometallic complexes of different structural type:

(i) neutral bis(η^5-cyclopentadienyl)metal ("metallocene") diacido complexes $(C_5H_5)_2MX_2$ containing an early transition metal such as titanium(IV) or vanadium(IV) as central metal atom, two cyclopentadienyl rings as organic ligands and two uninegative acido ligands X bound coordinatively to the central atom;

(ii) ionic metallicenium salts $[(C_5H_5)_2M]^+X^-$ consisting of a medium transition metal, e.g. iron(III), as central atom, two η^5-cyclopentadienyl ligands and an anion X^- linked by electrostatic forces in a salt-like crystal lattice;

(iii) uncharged decasubstituted metallocenes $(C_5R_5)_2M$ including a main group element, e.g. tin(II) or germanium(II), as central metal and two cyclopentadienyl ring ligands decasubstituted by the aryl or aralkyl groups C_6H_5 or $C_6H_4CH_2$, but neither containing acido ligands bound coordinatively nor counterions linked by electrostatic forces.

In vitro, cyclopentadienyl metal complexes were able to suppress the proliferation of normal or transformed tumor cells. Best activity was found for vanadocene dichloride in this respect. In vivo, numerous of the cyclopentadienyl metal complexes inhibited the development of diverse experimental animal tumors (e.g., Ehrlich ascites tumor, sarcoma 180, B16 melanoma, colon 38 carcinoma and Lewis lung carcinoma) and the growth of human carcinomas xenografted to nude mice. Especially certain titanocene and ferricenium compounds were cytostatically effective against human colorectal carcinomas. Moreover, titanocene complexes were shown to be antiviral agents and potent antiinflammatory compounds comparable to phenylbutazone.

Concerning the structure-activity relation of cyclopentadienyl metal complexes, the following conclusions can be drawn from the experimental data known:

- Different metals can be present within the three types of cyclopentadienyl metal compounds possessing antitumor activity. The metals range from first-row and second-row early and medium transition metals to main group elements of group IV of the Periodic Table.
- Within titanocene complexes $(C_5H_5)_2TiX_2$ and ferricenium salts $[(C_5H_5)_2Fe]^+X^-$, the acido groups X are widely variable.
- Modification of the cyclopentadienyl rings in metallocene diacido and metallicenium complexes by monosubstitution, 1,1′-disubstitution, or decasubstitution with alkyl

groups, or the exchange of one cyclopentadienyl ring by an additional acido ligand, results in distinct reductions of the antitumor activity.

These results indicate a significant role of the unsubstituted, unbulky cyclopentadienyl ring ligands for the achievement of the antitumor action of cyclopentadienyl early transition metal and metallicenium complexes. Because, on the other hand, cyclopentadiene itself did not effect systemic tumor-inhibiting activity, the complexed metal atoms themselves also appear to be important whether as a kind of "carrier" of one or both cyclopentadienyl rings to the site of action, or as "anchor" capable of additional interactions not accessible to the isolated cyclopentadienyl ligand or the free cyclopentadiene molecule itself, or as the intrinsic molecular centre building up coordinative bonds to relevant cellular molecules.

Biological experiments pointed to the nucleic acids as probable primary target for metallocene diacido complexes.

- After in vivo and in vitro treatment with titanocene or vanadocene dichloride, precursor incorporation studies showed a pronounced and persistent inhibition of nucleic acid synthesis, especially of DNA synthesis.
- Electron energy loss-spectroscopic studies illustrated accumulation of the central metal atoms titanium and vanadium in those cellular regions which are rich in nucleic acids.
- Cytokinetic investigations revealed the appearance of a G_2 block and the immigration of inflammatory cells belonging to the host defensive system after in vivo application of titanocene or vanadocene dichloride. After treatment in vitro, cell arrests at the G_1/S boundary and in G_2 were induced. These results correspond to the findings observed after administration of other cytostatic agents interfering with the nucleic acid metabolism.
- The mitotic activity significantly decreased after treatment of animal or human tumors with titanocene or vanadocene dichloride. Numerous giant cells were formed containing one enlarged nucleus or several nuclei of different size. The nuclear chromatin condensed and cytoplasmic degeneration developed. Virus particles appeared in animal and human tumor cells. Immigrating macrophages and leukocytes phagocytosed degenerated tumor cells.

Diverse model complexes were synthesized mostly containing the titanium(III) centre $[(C_5H_5)_2Ti]^+$ coordinated to nucleobase-related purines and oxopurines by a single monodentate Ti–N bond or by bidentate O–Ti–N bonds. With the vanadocene moiety $[(C_5H_5)_2V]^{2+}$, a labile outer-sphere complexation to nucleotide phosphate groups was observed. Other experiments indicated the facile loss of one or both cyclopentadienyl ligands of the titanocene moiety and led to the isolation of model complexes containing bridging, dianionic nucleobase ligands bound to mono(cyclopentadienyl)titanium(IV) or even cyclopentadienyl-free titanium(IV).

Pharmacokinetic studies uncovered a main accumulation of titanium in the liver and the intestine, whereas lower amounts were found in the kidneys and the lungs. No titanium-containing metabolites obviously entered the brain or passed across the placental barrier during organogenesis and early fetal period. The clearance of titanium from the blood was characterized by a biphasic pattern with a rapid-phase half-time of about 5 h and a slow-phase half-time of several days.

Toxicological studies with titanocene dichloride showed a different pattern of organ toxicity in comparison to organic antitumor agents and platinum cytostatic drugs. Dose-limiting toxicity was due to hepatotoxicity manifested by significant increases of the serum levels of typical liver enzymes and by fatty degeneration and necrotization of liver parenchyma cells. No functional and structural alterations of the kidneys were detectable even after application of toxic doses of titanocene or vanadocene dichloride. Bone marrow function was only slightly impaired, the thrombocytes being the only cells the number of which decreased after treatment with titanocene dichloride.

The described results underline that organometallic cyclopentadienyl metal complexes are characterized by antitumor activity against experimental and human tumors and exhibit an unusual spectrum of organ toxicity. These biological features confirm the metallocene and metallicenium complexes to be an independent group of non-platinum-group metal antitumor agents which clearly differ from known organic and inorganic cytostatics.

Acknowledgements. The authors' work on cyclopentadienyl metal complexes was supported by financial grants of the Deutsche Forschungsgemeinschaft, the Fonds der Chemischen Industrie, the Medac GmbH, Hamburg, and the Trude Goerke Heritage Foundation for the benefit of cancer research at the Freie Universität Berlin. The authors are indebted to Mrs. A. Stiller for typing the manuscript.

XII. Abbreviations

ID_{50}	concentration effecting 50% inhibition of cellular proliferation in vitro
ED_{90}	dose effecting complete tumor regression in 90% of the animals treated
LD_{50}, LD_{100}	doses killing 50 or 100%, resp., of the animals treated
T.I.	therapeutic index, defined as the relation of a lethal dose (LD_{50}) to a therapeutic dose (ED_{90})
T/C	ratio of tumor weights of treated and untreated (control) tumors. Values less than 50% are considered significant.

XIII. References

1. Rosenberg, B., VanCamp, L., Trosko, J. E., Mansour, V. H.: Nature *222,* 385 (1969)
2. Rosenberg, B.: Naturwissenschaften *60,* 399 (1973)
3. Prestayko, A. W., Crooke, S. T., Carter, S. K. (eds.): Cisplatin: Current Status and New Developments. New York: Academic Press 1980
4. Hacker, M. P., Douple, E. B., Krakoff, I. H. (eds.): Platinum Coordination Complexes in Cancer Chemotherapy. Boston: Nijhoff 1984
5. Weiss, G. R., Kisner, D. L., Kuhn, J. G., Melink, T. J., Myers, J. W., Von Hoff, D. D., in: Cancer Chemotherapy Annual *6,* p. 133 (Pinedo, H. M., Chabner, B. A., eds.). New York: Elsevier 1984
6. Nicolini, M., Bandoli, G. (eds.): Fifth International Symposium on Platinum and Other Metal Coordination Compounds in Cancer Chemotherapy. Abstracts. Padua: Cleup 1987
7. Smith, I. E., Harland, S. J., Robinson, B. A., Evans, B. D., Goodhart, L. C., Calvert, A. H., Yarnold, J., Glees, J. P., Baker, J., Ford, H. T.: Cancer Treat. Rep. *69,* 43 (1985)
8. Adamson, R. H., Canellos, G. R., Sieber, S. M.: Cancer Chemother. Rep. (Part 1) *59,* 599 (1975)
9. Mulinos, M. G., Amin, P.: Fed. Proc. *39,* 747 (1980)
10. Kumano, N., Nakai, Y., Ishikawa, T., Koinumaru, S., Suzuki, S., Konno, K.: Sci. Rep. Res. Inst. Tohoku Univ. *25,* 89 (1978)

11. Crowe, A. J., Smith, P. J., Atassi, G.: Chem. Biol. Interact. *32,* 171 (1980)
12. Crowe, A. J., Smith, P. J.: Inorg. Chim. Acta *93,* 179 (1984)
13. Köpf, H., Köpf-Maier, P.: Angew. Chem. *91,* 509 (1979); Angew. Chem. Int. Ed. Engl. *18,* 477 (1979)
14. Köpf, H., Köpf-Maier, P.: ACS Symp. Ser. *209,* 315 (1983)
15. Keller, H. J., Keppler, B., Schmähl, D.: Arzneim.-Forsch./Drug Res. *32,* 806 (1982)
16. Köpf-Maier, P., Köpf, H., Neuse, E. W.: J. Cancer Res. Clin. Oncol. *108,* 336 (1984)
17. Sadler, P. J., Nasr, M., Narayanan, V. L.: Dev. Oncol. *17,* 290 (1984)
18. Elo, H. O., Lumme, P. O.: Cancer Treat. Rep. *69,* 1021 (1984)
19. Shaw, C. F., III., Beery, A., Stocco, G. C.: Inorg. Chim. Acta *123,* 213 (1986)
20. Berner-Price, S. J., Mirabelli, C. K., Johnson, R. K., Mattern, M. R., McCabe, F. L., Faucette, L. F., Sung, C. M., Mong, S. M., Sadler, P. J., Crooke, S. T.: Cancer Res. *46,* 5486 (1986)
21. Köpf-Maier, P., Köpf, H.: Chem. Rev. *87,* 1137 (1987)
22. Köpf-Maier, P.: Naturwissenschaften *74,* 374 (1987)
23. Köpf-Maier, P., Janiak, C., Schumann, H.: Inorg. Chim. Acta *152,* 75 (1988)
24. Wilkinson, G., Birmingham, J. M.: J. Am. Chem. Soc. *76,* 4281 (1954)
25. Clearfield, A., Warner, D. K., Saldarriaga-Molina, C. H., Ropal, R., Bernal, I.: Can. J. Chem. *53,* 1622 (1975)
26. Siegert, F. W., De Liefde Meijer, H. J.: J. Organomet. Chem. *23,* 177 (1970)
27. Green, J. C., Green, M. L. H., Prout, C. K.: J. Chem. Soc., Chem. Commun. 421 (1972)
28. Van Baalen, A., Groenenboom, C. J., De Liefde Meijer, H. J.: J. Organomet. Chem. *74,* 245 (1974)
29. Green, M. L. H., Moreau, J. J. E.: J. Organomet. Chem. *161,* C25 (1978)
30. Cooper, R. L., Green, M. L. H.: J. Chem. Soc. A 1155 (1967)
31. Samuel, E.: Bull. Soc. Chim. 3548 (1966)
32. Chiesi Villa, A. C., Manfredotti, A. G., Guastini, C.: Acta Cryst. B *32,* 909 (1976)
33. Coutts, R. S. P., Wailes, P. C.: Aust. J. Chem. *24,* 1075 (1971)
34. De Gil, E. R., de Burguera, M., Rivera, A. V., Maxfield, P.: Acta Cryst. B *33,* 578 (1977)
35. Drozdov, G. V., Klebanski, A. L., Bartashov, V. A.: Zh. Obsch. Khim. *32,* 2390 (1962)
36. Döppert, K.: Makromol. Chem., Rapid Commun. *1,* 519 (1980)
37. Köpf-Maier, P., Grabowski, S., Liegener, J., Köpf, H.: Inorg. Chim. Acta *108,* 99 (1985)
38. Döppert, K., Sanchez-Delgado, R., Klein, H. P., Thewalt, U.: J. Organomet. Chem. *233,* 205 (1982)
39. Nesmeyanov, A. N., Nogina, O. V., Lazareva, N. A., Dubovitskii, V. A., Lokshin, B. V.: Izv. Akad. Nauk SSSR Ser. Khim. 2482 (1971)
40. Nesmeyanov, A. N., Nogina, O. V., Dubovitskii, V. A., Kvasov, B. A., Petrovskii, P. V., Lazareva, N. A.: Izv. Akad. Nauk SSSR Ser. Khim. 2729 (1971)
41. Klapötke, T., Köpf, H.: Inorg. Chim. Acta *133,* 115 (1987)
42. Besançon, J., Camboli, D., Trimaille, B., Colette, M., Tirouflet, J.: Compt. Rend. C *287,* 573 (1978)
43. Köpf-Maier, P., Grabowski, S., Köpf, H.: Eur. J. Med. Chem. *19,* 347 (1984)
44. Köpf, H., Klapötke, T.: Naturforsch. *41b,* 971 (1986)
45. Köpf, H., Schmidt, M.: J. Organomet. Chem. *4,* 426 (1965)
46. Klapötke, T., Köpf, H.: Monatsh. Chem. *118,* 671 (1987)
47. Marcellus, C. G., Oakley, R. T., Cordes, A. W., Pennington, W. T.: J. Chem. Soc., Chem. Commun. 1451 (1983)
48. Köpf-Maier, P., Kahl, W., Klouras, N., Hermann, G., Köpf, H.: Eur. J. Med. Chem. *16,* 275 (1981)
49. Boche, G., Stahl, K. P.: Personal communication
49a. Stahl, K. P., Boche, G., Massa, W.: J. Organomet. Chem. *277,* 113 (1984)
50. Katz, T. J., Acton, N., Martin, G.: J. Am. Chem. Soc. *95,* 2934 (1973)
51. Köpf, H., Kahl, W.: J. Organomet. Chem. *64,* C37 (1974)
52. Samuel, E., Setton, R.: J. Organomet. Chem. *4,* 156 (1965)
53. Gorsich, R. D.: J. Am. Chem. Soc. *82,* 4211 (1960)
54. Engelhardt, L. M., Papasergio, R. I., Raston, C. L., White, A. H.: Organometallics *3,* 18 (1984)

55. LePage, Y.: J. Organomet. Chem. *193,* 201 (1980)
56. Skapski, A. C., Troughton, P. G. H.: Acta Cryst. B *26,* 716 (1970)
57. Köpf, H., Grabowski, S., Voigtländer, R.: J. Organomet. Chem. *216,* 185 (1981)
58. Köpf, H., Grabowski, S.: Z. Anorg. Allgem. Chem. *496,* 167 (1983)
59. Meirim, M. G., Neuse, E. W.: Trans. Met. Chem. *9,* 337 (1984)
60. Thewalt, U., Berhalter, K., Neuse, E. W.: Trans. Met. Chem. *10,* 393 (1985)
61. Thewalt, U., Berhalter, K.: J. Organomet. Chem. *302,* 193 (1986)
62. Klapötke, T.: Dissertation, Technische Universität Berlin, 1986
63. Köpf, H., Klapötke, T.: J. Organomet. Chem. *307,* 319 (1986)
64. Nesmeyanov, A. N., Perewalova, E. G., Jurjeva, L. P.: Chem. Ber. *93,* 2729 (1960)
65. Paulus, E. F., Schäfer, L.: J. Organomet. Chem. *144,* 205 (1978)
66. Riemschneider, R., Helm, D.: Chem. Ber. *89,* 155 (1956)
67. Boeyens, J. C. A., Lalloo, P. C., Neuse, E. W., Wei, H. H.: S. Afr. J. Chem. *37,* 32 (1984)
68. Neuse, E. W., Meirim, M. G.: Trans. Met. Chem. *9,* 205 (1984)
69. Bullen, G. J., Howlin, B. J., Silver, J., Fitzsimmons, B. W., Sayer, I., Larkworthy, L. F.: J. Chem. Soc. Dalton Trans. 1937 (1986)
70. Cowell, G. W., Ledwith, A., White, A. C., Woods, H. J.: J. Chem. Soc. B 227 (1970)
71. Pavlík, I., Klikorka, J.: Coll. Czech. Chem. Commun. *30,* 664 (1965)
72. Hendrickson, D. N., Sohn, Y. S., Gray, H. B.: Inorg. Chem. *10,* 1559 (1971)
73. Churchill, M. R., Landers, A. G., Rheingold, A. L.: Inorg. Chem. *20,* 849 (1981)
74. Castagnola, M., Floris, B., Illuminati, G., Ortaggi, G.: J. Organomet. Chem. *60,* C17 (1973)
75. Neuse, E. W., Loonat, R., Boeyens, J. C. A.: Trans. Met. Chem. *9,* 12 (1984)
76. Neuse, E. W.: Personal communication
77. Schumann, H., Janiak, C., Zuckerman, J. J.: Personal communication
78. Heeg, M. J., Janiak, C., Zuckerman, J. J.: J. Am. Chem. Soc. *106,* 4259 (1984)
79. Schumann, H., Janiak, C., Hahn, E., Kolax, C., Loebel, J., Rausch, M. D., Zuckerman, J. J., Heeg, M. J.: Chem. Ber. *119,* 2656 (1986)
80. Köpf-Maier, P., Wagner, W., Köpf, H.: Cancer Chemother. Pharmacol. *5,* 237 (1981)
81. Köpf-Maier, P.: Z. Naturforsch. *40c,* 843 (1985)
82. Köpf-Maier, P.: J. Cancer Res. Clin. Oncol. *103,* 145 (1982)
83. Döppert, K.: J. Organomet. Chem. *178,* C3 (1979)
84. Toney, J. H., Marks, T. J.: J. Am. Chem. Soc. *107,* 947 (1985)
85. Nitschke, I.: Inaugural-Dissertation, Freie Universität Berlin, 1984
86. Köpf-Maier, P.: Unpublished results
87. Toney, J. H., Rao, L. N., Murthy, M. S., Marks, T. J.: Breast Cancer Res. Treat. *6,* 185 (1985)
88. Murthy, M. S., Toney, J. H., Rao, L. N., Kuo, L. Y., Marks, T. J.: Proc. Am. Assoc. Cancer Res. *27,* 279 (1986)
89. Hermann, G.: Inaugural-Dissertation, Freie Universität Berlin, 1984
90. Köpf-Maier, P., Hermann, G.: Virchows Arch. [Cell Pathol.] *47,* 107 (1984)
91. Köpf-Maier, P., Hesse, B., Köpf, H.: J. Cancer Res. Clin. Oncol. *96,* 43 (1980)
92. Köpf-Maier, P., Köpf, H.: Z. Naturforsch. *34b,* 805 (1979)
93. Köpf-Maier, P., Leitner, M., Köpf, H.: J. Inorg. Nucl. Chem. *42,* 1789 (1980)
94. Köpf-Maier, P., Leitner, M., Voigtländer, R., Köpf, H.: Z. Naturforsch. *34c,* 1174 (1979)
95. Köpf-Maier, P., Hesse, B., Voigtländer, R., Köpf, H.: J. Cancer Res. Clin. Oncol. *97,* 31 (1980)
96. Köpf, H., Schmidt, M.: Z. Anorg. Allg. Chem. *340,* 139 (1965)
97. Köpf, H., Block, B., Schmidt, M.: Chem. Ber. *101,* 272 (1968)
98. Köpf-Maier, P., Klapötke, T., Köpf, H.: Unpublished results
99. Köpf-Maier, P., Fernández, V.: Unpublished results
100. Köpf-Maier, P., Thewalt, U.: Unpublished results
101. Köpf-Maier, P., Janiak, C., Schumann, H.: J. Cancer Res. Clin. Oncol., in press
102. Köpf-Maier, P., Preiß, F., Marx, T., Klapötke, T., Köpf, H.: Anticancer Res. *6,* 33 (1986)
103. Köpf-Maier, P., Wagner, W., Hesse, B., Köpf, H.: Eur. J. Cancer *17,* 665 (1981)
104. Keppler, B. K., Schmähl, D.: Arzneim.-Forsch./Drug Res. *36,* 1822 (1986)
105. Elo, H. O., Lumme, P. O.: Cancer Treat. Rep. *69,* 1021 (1985)
106. Slavik, M., Blanc, O., Davis, J.: Invest. New Drugs *1,* 225 (1987)
107. Köpf-Maier, P., Köpf, H.: Arzneim.-Forsch./Drug Res. *37,* 532 (1987)

108. Wolpert-De Fillippes, M. K.: Cancer Treat. Rep. *63,* 1453 (1979)
109. Chang, S. I.: Yao Hsueh T'ung Pao *16,* 57 (1981)
110. Bellet, R. E., Danna, V., Mastrangelo, M. J., Berd, D.: J. Natl. Cancer Inst. *63,* 1185 (1979)
111. Shorthouse, A. J., Smyth, J. F., Steel, G. G., Ellison, M., Mills, J., Peckham, M. J.: Brit. J. Surg. *67,* 715 (1980)
112. Fiebig, H. H., Schuchhardt, C., Henss, H., Fiedler, L., Löhr, G. W.: Behr. Inst. Mitt. *74,* 343 (1984)
113. Berenbaum, M. C., Sheard, C. E., Reittie, J. R., Bundick, R. V.: Brit. J. Cancer *30,* 13 (1974)
114. Povlsen, C. O., Jacobsen, G. K.: Cancer Res. *35,* 2790 (1975)
115. Azar, H. A., Fernandez, S. B., Bros., L. M., Sullivan, J. L.: Ann. Clin. Lab. Sci. *12,* 51 (1982)
116. Kyriazis, A. A., Kyriazis, A. P., Kereiakes, J. G., Soloway, M. S., McCombs, W. B.: Exp. Cell Biol. *51,* 83 (1983)
117. Fujita, M., Hayata, S., Taguchi, T.: J. Surg. Oncol. *15,* 211 (1980)
118. Nowak, K., Peckham, M. J., Steel, G. G.: Brit. J. Cancer *37,* 576 (1978)
119. Osieka, R., Houchens, D. P., Goldin, A., Johnson, R. K.: Cancer *40,* 2640 (1977)
120. Steel, G. G., Courtenay, V. D., Peckham, M. J.: Brit. J. Cancer *47,* 1 (1983)
121. Goldin, A., Wolpert-de Fillippes, M. K.: Bull. Cancer (Paris) *66,* 61 (1979)
122. Venditti, J. M.: Sem. Oncol. *8,* 349 (1981)
123. Köpf-Maier, P., Moormann, A., Köpf, H.: Europ. J. Cancer Clin. Oncol. *21,* 853 (1985)
124. Köpf-Maier, P., in: ESO Monograph Human Tumor Xenograft in Anticancer Drug Development (Winograd, B., Peckham, M. J., Pinedo, H. M., eds.), in press
125. Köpf-Maier, P.: Cancer Res. Clin. Oncol. *113,* 342 (1987)
126. Osieka, R.: Cancer Treat. Rev. *11* (Suppl. A), 85 (1984)
127. Tonew, E., Tonew, M., Heyn, B., Schröer, H. P.: Zbl. Bakt. Hyg., I. Abt. Orig. A *250,* 425 (1981)
128. Taylor, C.: Personal communication
129. Saxena, S., Saxena, P. N., Rai, A. K., Saxena, S. C.: Toxicol. *35,* 241 (1985)
130. Fairlie, D. P., Whitehouse, M. W., Broomhead, J. A.: Chem. Biol. Interact. *61,* 277 (1987)
131. Serpone, N., Fay, R. C.: Inorg. Chem. *6,* 1835 (1967)
132. Keppler, B. K., Diez, A., Seifried, V.: Arzneim.-Forsch./Drug Res. *35,* 1832 (1985)
133. Keppler, B. K., Michels, K.: Arzneim.-Forsch./Drug Res. *35,* 1837 (1985)
134. Keppler, B. K.: Personal communication
135. Leik, R., Zsolnai, L., Huttner, G., Neuse, E. W., Brintzinger, H. H.: J. Organomet. Chem. *312,* 177 (1986)
136. Köpf-Maier, P., Mohtachemi, R., Schumann, H.: Unpublished results
137. Schmid, G., Kampmann, D., Meyer, W., Boese, R., Paetzold, P., Delpy, K.: Chem. Ber. *118,* 2418 (1985)
138. Köpf-Maier, P., Schmid, G.: Unpublished results
139. Döppert, K.: J. Organomet. Chem. *319,* 351 (1987)
140. Cleare, M. J.: Coord. Chem. Rev. *12,* 349 (1974)
141. Cleare, M. J., Hydes, P. C., Hepburn, D. R., Malerbi, B. W., in: Cisplatin, Current Status and New Developments, p. 149 (Prestayko, A. W., Crooke, S. T., Carter, S. K., eds.). New York: Academic Press 1980
142. Köpf-Maier, P., Köpf, H.: Naturwiss. *67,* 415 (1980)
143. Köpf-Maier, P., Wagner, W., Köpf, H.: Naturwiss. *68,* 272 (1981)
144. Köpf-Maier, P., Krahl, D., in: Culture Techniques, p. 509 (Neubert, D., Merker, H. J., eds.). Berlin: Walter de Gruyter 1981
145. Köpf-Maier, P., Krahl, D.: Chem. Biol. Interact. *44,* 317 (1983)
146. Köpf-Maier, P., Krahl, D.: Naturwiss. *68,* 273 (1981)
147. Köpf-Maier, P., Wagner, W., Liss, E.: J. Cancer Res. Clin. Oncol. *102,* 21 (1981)
148. Köpf-Maier, P., Wagner, W., Liss, E.: J. Cancer Res. Clin. Oncol. *106,* 44 (1983)
149. Bergerat, J. P., Barlogie, B., Göhde, W., Johnston, D. A., Drewinko, B.: Cancer Res. *39,* 4356 (1979)
150. Köpf-Maier, P., Köpf, H.: Dev. Oncol. *17,* 279 (1984)
151. Köpf-Maier, P.: Verh. Anat. Ges. *82,* in press
152. Beauchamp, A. L., Cozak, D., Mardhy, A.: Inorg. Chim. Acta *92,* 191 (1984)
153. Cozak, D., Mardhy, A., Morneau, A.: Can. J. Chem. *64,* 751 (1986)
154. Cozak, D., Mardhy, A., Olivier, M. J., Beauchamp, A. L.: Inorg. Chem. *25,* 2600 (1986)

155. Beauchamp, A. L., Bélanger-Gariépy, F., Mardhy, A., Cozak, D.: Inorg. Chim. Acta *124,* L23 (1986)
156. Fieselmann, B. F., Hendrickson, D. N., Stucky, G. D.: Inorg. Chem. *17,* 1841 (1978)
157. Toney, J. H., Brock, C. P., Marks, T. J.: J. Am. Chem. Soc. *108,* 7263 (1986)
158. Köpf, H., Tekeş, Z., Grabowski, S.: Unpublished results; cf. Tekeş, Z., Dissertation Techn. Univ. Berlin, 1985
159. Caradonna, J. P., Lippard, S. J.: Dev. Oncol. *17,* 14 (1984)
160. Kistenmacher, T. J., Orbell, J. D., Marzilli, L. G.: ACS Symp. Ser. *209,* 191 (1983)
161. Lippert, B.: ACS Symp. Ser. *209,* 147 (1983)
162. Macquet, J. P., Butour, J. L., Johnson, N. P., Razaka, H., Salles, B., Vieussens, C., Wright, M.: Dev. Oncol. *17,* 27 (1984)
163. Plooy, A. C. M., Fichtinger-Schepman, A. M. J., Schutte, H. H., van Dijk, M., Lohmann, P. H. M.: Carcinogenesis *6,* 561 (1985)
164. Köpf-Maier, P., Köpf, H.: Drugs Fut. *11,* 297 (1986)
165. Lippert, B., Lock, C. J. L., Rosenberg, B.: Inorg. Chem. *16,* 1525 (1977)
166. Rosenberg, B.: Cancer Treat. Rep. *63,* 1433 (1979)
167. Köpf-Maier, P., Mai, W., Saß, G.: Unpublished results
168. Toney, J. H., Murthy, M. S., Marks, T. J.: Chem. Biol. Interact. *56,* 45 (1985)
169. Köpf-Maier, P., Köpf, H.: Anticancer Res. *6,* 227 (1986)
170. Köpf-Maier, P., Gerlach, S.: J. Cancer Res. Clin. Oncol. *111,* 243 (1986)
171. Köpf-Maier, P., Funke-Kaiser, P.: Toxicol. *38,* 81 (1986)
172. Köpf-Maier, P., Gerlach, S.: Anticancer Res. *6,* 227 (1986)
173. Goldstein, R. S., Mayor, G. H.: Life Sci. *32,* 685 (1983)
174. Hanzlik, R. P., Robert, P., Soine, W. H.: J. Am. Chem. Soc. *180,* 1290 (1978)
175. Dombrowski, K. E., Baldwin, W., Sheats, J. E.: J. Organomet. Chem. *302,* 281 (1986)
176. Köpf-Maier, P., Erkenswick, P.: Toxicol. *33,* 171 (1984)
177. Neubert, D., Barrach, H. J., Merker, H. J., in: Drug-Induced Pathology, p. 241 (Grundmann, E., ed.). New York: Springer 1980
178. Köpf-Maier, P.: Toxicol. *37,* 111 (1985)
179. Calabresi, P., Parks, R. E., in: The Pharmacological Basis of Therapeutics, p. 1247 (Goodman, L. S., Gilman, A., eds.). London: Macmillan 1985
180. Schaeppi, U., Heyman, I. A., Fleischman, R. W., Rosenkrantz, H., Ilievski, V., Phelau, R., Cooney, D. A., Davis, R. D.: Toxicol. Appl. Pharmacol. *25,* 230 (1973)
181. Lippman, A. J., Helson, C., Helson, L., Krakoff, I. H.: Cancer Chemother. Rep. (Part 1) *57,* 191 (1973)
182. Von Hoff, D. D., Schilsky, R., Reichert, C. M., Reddick, R. L., Rozencweig, M., Young, R. C., Muggia, F. M.: Cancer Treat. Rep. *63,* 1527 (1979)

Author Index Volumes 1–70

Ahrland, S.: Factors Contributing to (b)-behaviour in Acceptors. Vol. 1, pp. 207–220.

Ahrland, S.: Thermodynamics of Complex Formation between Hard and Soft Acceptors and Donors. Vol. 5, pp. 118–149.

Ahrland, S.: Thermodynamics of the Stepwise Formation of Metal-Ion Complexes in Aqueous Solution. Vol. 15, pp. 167–188.

Allen, G. C., Warren, K. D.: The Electronic Spectra of the Hexafluoro Complexes of the First Transition Series. Vol. 9, pp. 49–138.

Allen, G. C., Warren, K. D.: The Electronic Spectra of the Hexafluoro Complexes of the Second and Third Transition Series. Vol. 19, pp. 105–165.

Alonso, J. A., Balbás, L. C.: Simple Density Functional Theory of the Electronegativity and Other Related Properties of Atoms and Ions. Vol. 66, pp. 41–78.

Ardon, M., Bino, A.: A New Aspect of Hydrolysis of Metal Ions: The Hydrogen-Oxide Bridging Ligand ($H_3O_2^-$). Vol. 65, pp. 1–28.

Augustynski, J.: Aspects of Photo-Electrochemical and Surface Behaviour of Titanium(IV) Oxide. Vol. 69, pp. 1–61.

Averill, B. A.: Fe–S and Mo–Fe–S Clusters as Models for the Active Site of Nitrogenase. Vol. 53, pp. 57–101.

Babel, D.: Structural Chemistry of Octahedral Fluorocomplexes of the Transition Elements. Vol. 3, pp. 1–87.

Bacci, M.: The Role of Vibronic Coupling in the Interpretation of Spectroscopic and Structural Properties of Biomolecules. Vol. 55, pp. 67–99.

Baker, E. C., Halstead, G.W., Raymond, K. N.: The Structure and Bonding of 4*f* and 5*f* Series Organometallic Compounds. Vol. 25, pp. 21–66.

Balsenc, L. R.: Sulfur Interaction with Surfaces and Interfaces Studied by Auger Electron Spectrometry. Vol. 39, pp. 83–114.

Banci, L., Bencini, A., Benelli, C., Gatteschi, D., Zanchini, C.: Spectral-Structural Correlations in High-Spin Cobalt(II) Complexes. Vol. 52, pp. 37–86.

Bartolotti, L. J.: Absolute Electronegativities as Determined from Kohn-Sham Theory. Vol. 66, pp. 27–40.

Baughan, E. C.: Structural Radii, Electron-cloud Radii, Ionic Radii and Solvation. Vol. 15, pp. 53–71.

Bayer, E., Schretzmann, P.: Reversible Oxygenierung von Metallkomplexen. Vol. 2, pp. 181–250.

Bearden, A. J., Dunham, W. R.: Iron Electronic Configurations in Proteins: Studies by Mössbauer Spectroscopy. Vol. 8, pp. 1–52.

Bergmann, D., Hinze, J.: Electronegativity and Charge Distribution. Vol. 66, pp. 145–190.

Berners-Price, S. J., Sadler, P. J.: Phosphines and Metal Phosphine Complexes: Relationship of Chemistry to Anticancer and Other Biological Activity. Vol. 70, pp. 27–102.

Bertini, I., Luchinat, C., Scozzafava, A.: Carbonic Anhydrase: An Insight into the Zinc Binding Site and into the Active Cavity Through Metal Substitution. Vol. 48, pp. 45–91.

Blasse, G.: The Influence of Charge-Transfer and Rydberg States on the Luminescence Properties of Lanthanides and Actinides. Vol. 26, pp. 43–79.

Blasse, G.: The Luminescence of Closed-Shell Transition Metal-Complexes. New Developments. Vol. 42, pp. 1–41.

Blauer, G.: Optical Activity of Conjugated Proteins. Vol. 18, pp. 69–129.

Bleijenberg, K. C.: Luminescence Properties of Uranate Centres in Solids. Vol. 42, pp. 97–128.

Boeyens, J. C. A.: Molecular Mechanics and the Structure Hypothesis. Vol. 63, pp. 65–101.

Bonnelle, C.: Band and Localized States in Metallic Thorium, Uranium and Plutonium, and in Some Compounds, Studied by X-Ray Spectroscopy. Vol. 31, pp. 23–48.

Bradshaw, A. M., Cederbaum, L. S., Domcke, W.: Ultraviolet Photoelectron Spectroscopy of Gases Adsorbed on Metal Surfaces. Vol. 24, pp. 133–170.

Braterman, P. S.: Spectra and Bonding in Metal Carbonyls. Part A: Bonding. Vol. 10, pp. 57–86.

Braterman, P. S.: Spectra and Bonding in Metal Carbonyls. Part B: Spectra and Their Interpretation. Vol. 26, pp. 1–42.

Bray, R. C., Swann, J. C.: Molybdenum-Containing Enzymes. Vol. 11, pp. 107–144.

Brooks, M. S. S.: The Theory of 5f Bonding in Actinide Solids. Vol. 59/60, pp. 263–293.

van Bronswyk, W.: The Application of Nuclear Quadrupole Resonance Spectroscopy to the Study of Transition Metal Compounds. Vol. 7, pp. 87–113.

Buchanan, B. B.: The Chemistry and Function of Ferredoxin. Vol. 1, pp. 109–148.
Buchler, J. W., Kokisch, W., Smith, P. D.: *Cis, Trans,* and *Metal* Effects in Transition Metal Porphyrins. Vol. 34, pp. 79–134.
Bulman, R. A.: Chemistry of Plutonium and the Transuranics in the Biosphere. Vol. 34, pp. 39–77.
Bulman, R. A.: The Chemistry of Chelating Agents in Medical Sciences. Vol. 67, pp. 91–141.
Burdett, J. K.: The Shapes of Main-Group Molecules; A Simple Semi-Quantitative Molecular Orbital Approach. Vol. 31, pp. 67–105.
Burdett, J. K.: Some Structural Problems Examined Using the Method of Moments. Vol. 65, pp. 29–90.
Campagna, M., Wertheim, G. K., Bucher, E.: Spectroscopy of Homogeneous Mixed Valence Rare Earth Compounds. Vol. 30, pp. 99–140.
Chasteen, N. D.: The Biochemistry of Vanadium, Vol. 53, pp. 103–136.
Cheh, A. M., Neilands, J. P.: The δ-Aminolevulinate Dehydratases: Molecular and Environmental Properties. Vol. 29, pp. 123–169.
Ciampolini, M.: Spectra of 3d Five-Coordinate Complexes. Vol. 6, pp. 52–93.
Chimiak, A., Neilands, J. B.: Lysine Analogues of Siderophores. Vol. 58, pp. 89–96.
Clack, D. W., Warren, K. D.: Metal-Ligand Bonding in 3d Sandwich Complexes, Vol. 39, pp. 1–41.
Clark, R. J. H., Stewart, B.: The Resonance Raman Effect. Review of the Theory and of Applications in Inorganic Chemistry. Vol. 36, pp. 1–80.
Clarke, M. J., Fackler, P. H.: The Chemistry of Technetium: Toward Improved Diagnostic Agents. Vol. 50, pp. 57–78.
Cohen, I. A.: Metal-Metal Interactions in Metalloporphyrins, Metalloproteins and Metalloenzymes. Vol. 40, pp. 1–37.
Connett, P. H., Wetterhahn, K. E.: Metabolism of the Carcinogen Chromate by Cellular Constitutents. Vol. 54, pp. 93–124.
Cook, D. B.: The Approximate Calculation of Molecular Electronic Structures as a Theory of Valence. Vol. 35, pp. 37–86.
Cotton, F. A., Walton, R. A.: Metal-Metal Multiple Bonds in Dinuclear Clusters. Vol. 62, pp. 1–49.
Cox, P. A.: Fractional Parentage Methods for Ionisation of Open Shells of *d* and *f* Electrons. Vol. 24, pp. 59–81.
Crichton, R. R.: Ferritin. Vol. 17, pp. 67–134.
Daul, C., Schläpfer, C. W., von Zelewsky, A.: The Electronic Structure of Cobalt(II) Complexes with Schiff Bases and Related Ligands. Vol. 36, pp. 129–171.
Dehnicke, K., Shihada, A.-F.: Structural and Bonding Aspects in Phosphorus Chemistry-Inorganic Derivates of Oxohalogeno Phosphoric Acids. Vol. 28, pp. 51–82.
Dobiáš, B.: Surfactant Adsorption on Minerals Related to Flotation. Vol. 56, pp. 91–147.
Doi, K., Antanaitis, B. C., Aisen, P.: The Binuclear Iron Centers of Uteroferrin and the Purple Acid Phosphatases . Vol. 70, pp. 1–26.
Doughty, M. J., Diehn, B.: Flavins as Photoreceptor Pigments for Behavioral Responses. Vol. 41, pp. 45–70.
Drago, R. S.: Quantitative Evaluation and Prediction of Donor-Acceptor Interactions. Vol. 15, pp. 73–139.
Duffy, J. A.: Optical Electronegativity and Nephelauxetic Effect in Oxide Systems. Vol. 32, pp. 147–166.
Dunn, M. F.: Mechanisms of Zinc Ion Catalysis in Small Molecules and Enzymes. Vol. 23, pp. 61–122.
Emsley, E.: The Composition, Structure and Hydrogen Bonding of the β-Deketones. Vol. 57, pp. 147–191.
Englman, R.: Vibrations in Interaction with Impurities. Vol. 43, pp. 113–158.
Epstein, I. R., Kustin, K.: Design of Inorganic Chemical Oscillators. Vol. 56, pp. 1–33.
Ermer, O.: Calculations of Molecular Properties Using Force Fields. Applications in Organic Chemistry. Vol. 27, pp. 161–211.
Ernst, R. D.: Structure and Bonding in Metal-Pentadienyl and Related Compounds. Vol. 57, pp. 1–53.
Erskine, R. W., Field, B. O.: Reversible Oxygenation. Vol. 28, pp. 1–50.
Fajans, K.: Degrees of Polarity and Mutual Polarization of Ions in the Molecules of Alkali Fluorides, SrO, and BaO. Vol. 3, pp. 88–105.
Fee, J. A.: Copper Proteins – Systems Containing the "Blue" Copper Center. Vol. 23, pp. 1–60.
Feeney, R. E., Komatsu, S. K.: The Transferrins. Vol. 1, pp. 149–206.
Felsche, J.: The Crystal Chemistry of the Rare-Earth Silicates. Vol. 13, pp. 99–197.

Ferreira, R.: Paradoxical Violations of Koopmans' Theorem, with Special Reference to the 3d Transition Elements and the Lanthanides. Vol. 31, pp. 1–21.
Fidelis, I. K., Mioduski, T.: Double-Double Effect in the Inner Transition Elements. Vol. 47, pp. 27–51.
Fournier, J. M.: Magnetic Properties of Actinide Solids. Vol. 59/60, pp. 127–196.
Fournier, J. M., Manes, L.: Actinide Solids. 5f Dependence of Physical Properties. Vol. 59/60, pp. 1–56.
Fraga, S., Valdemoro, C.: Quantum Chemical Studies on the Submolecular Structure of the Nucleic Acids. Vol. 4, pp. 1–62.
Fraústo da Silva, J. J. R., Williams, R. J. P.: The Uptake of Elements by Biological Systems. Vol. 29, pp. 67–121.
Fricke, B.: Superheavy Elements. Vol. 21, pp. 89–144.
Fuhrhop, J.-H.: The Oxidation States and Reversible Redox Reactions of Metalloporphyrins. Vol. 18, pp. 1–67.
Furlani, C., Cauletti, C.: He(I) Photoelectron Spectra of *d*-metal Compounds. Vol. 35, pp. 119–169.
Gázquez, J. L., Vela, A., Galván, M.: Fukui Function, Electronegativity and Hardness in the Kohn-Sham Theory. Vol. 66, pp. 79–98.
Gerloch, M., Harding, J. H., Woolley, R. G.: The Context and Application of Ligand Field Theory. Vol. 46, pp. 1–46.
Gillard, R. D., Mitchell, P. R.: The Absolute Configuration of Transition Metal Complexes. Vol. 7, pp. 46–86.
Gleitzer, C., Goodenough, J. B.: Mixed-Valence Iron Oxides. Vol. 61, pp. 1–76.
Gliemann, G., Yersin, H.: Spectroscopic Properties of the Quasi One-Dimensional Tetracyanoplatinate(II) Compounds. Vol. 62, pp. 87–153.
Golovina, A. P., Zorov, N. B., Runov, V. K.: Chemical Luminescence Analysis of Inorganic Substances. Vol. 47, pp. 53–119.
Green, J. C.: Gas Phase Photoelectron Spectra of *d*- and *f*-Block Organometallic Compounds. Vol. 43, pp. 37–112.
Grenier, J. C., Pouchard, M., Hagenmuller, P.: Vacancy Ordering in Oxygen-Deficient Perovskite-Related Ferrities. Vol. 47, pp. 1–25.
Griffith, J. S.: On the General Theory of Magnetic Susceptibilities of Polynuclear Transitionmetal Compounds. Vol. 10, pp. 87–126.
Gubelmann, M. H., Williams, A. F.: The Structure and Reactivity of Dioxygen Complexes of the Transition Metals. Vol. 55, pp. 1–65.
Guilard, R., Lecomte, C., Kadish, K. M.: Synthesis, Electrochemistry, and Structural Properties of Porphyrins with Metal-Carbon Single Bonds and Metal-Metal Bonds. Vol. 64, pp. 205–268.
Gütlich, P.: Spin Crossover in Iron(II)-Complexes. Vol. 44, pp. 83–195.
Gutmann, V., Mayer, U.: Thermochemistry of the Chemical Bond. Vol. 10, pp. 127–151.
Gutmann, V., Mayer, U.: Redox Properties: Changes Effected by Coordination. Vol. 15, pp. 141–166.
Gutmann, V., Mayer, H.: Application of the Functional Approach to Bond Variations under Pressure. Vol. 31, pp. 49–66.
Hall, D. I., Ling, J. H., Nyholm, R. S.: Metal Complexes of Chelating Olefin-Group V Ligands. Vol. 15, pp. 3–51.
Harnung, S. E., Schäffer, C. E.: Phase-fixed 3-Γ Symbols and Coupling Coefficients for the Point Groups. Vol. 12, pp. 201–255.
Harnung, S. E., Schäffer, C. E.: Real Irreducible Tensorial Sets and their Application to the Ligand-Field Theory. Vol. 12, pp. 257–295.
Hathaway, B. J.: The Evidence for "Out-of-the-Plane" Bonding in Axial Complexes of the Copper(II) Ion. Vol. 14, pp. 49–67.
Hathaway, B. J.: A New Look at the Stereochemistry and Electronic Properties of Complexes of the Copper(II) Ion. Vol. 57, pp. 55–118.
Hellner, E. E.: The Frameworks (Bauverbände) of the Cubic Structure Types. Vol. 37, pp. 61–140.
von Herigonte, P.: Electron Correlation in the Seventies. Vol. 12, pp. 1–47.
Hemmerich, P., Michel, H., Schug, C., Massey, V.: Scope and Limitation of Single Electron Transfer in Biology. Vol. 48, pp. 93–124.
Hider, R. C.: Siderophores Mediated Absorption of Iron. Vol. 58, pp. 25–88.
Hill, H. A. O., Röder, A., Williams, R. J. P.: The Chemical Nature and Reactivity of Cytochrome P-450. Vol. 8, pp. 123–151.

Hogenkamp, H. P. C., Sando, G. N.: The Enzymatic Reduction of Ribonucleotides. Vol. 20, pp. 23–58.

Hoffmann, D. K., Ruedenberg, K., Verkade, J. G.: Molecular Orbital Bonding Concepts in Polyatomic Molecules – A Novel Pictorial Approach. Vol. 33, pp. 57–96.

Hubert, S., Hussonnois, M., Guillaumont, R.: Measurement of Complexing Constants by Radiochemical Methods. Vol. 34, pp. 1–18.

Hudson, R. F.: Displacement Reactions and the Concept of Soft and Hard Acids and Bases. Vol. 1, pp. 221–223.

Hulliger, F.: Crystal Chemistry of Chalcogenides and Pnictides of the Transition Elements. Vol. 4, pp. 83–229.

Ibers, J. A., Pace, L. J., Martinsen, J., Hoffman, B. M.: Stacked Metal Complexes: Structures and Properties. Vol. 50, pp. 1–55.

Iqbal, Z.: Intra- und Inter-Molecular Bonding and Structure of Inorganic Pseudohalides with Triatomic Groupings. Vol. 10, pp. 25–55.

Izatt, R. M., Eatough, D. J., Christensen, J. J.: Thermodynamics of Cation-Macrocyclic Compound Interaction. Vol. 16, pp. 161–189.

Jain, V. K., Bohra, R., Mehrotra, R. C.: Structure and Bonding in Organic Derivatives of Antimony(V). Vol. 52, pp. 147–196.

Jerome-Lerutte, S.: Vibrational Spectra and Structural Properties of Complex Tetracyanides of Platinum, Palladium and Nickel. Vol. 10, pp. 153–166.

Jørgensen, C. K.: Electric Polarizability, Innocent Ligands and Spectroscopic Oxidation States. Vol. 1, pp. 234–248.

Jørgensen, C. K.: Recent Progress in Ligand Field Theory. Vol. 1, pp. 3–31.

Jørgensen, C. K.: Relations between Softness, Covalent Bonding, Ionicity and Electric Polarizability. Vol. 3, pp. 106–115.

Jørgensen, C. K.: Valence-Shell Expansion Studied by Ultra-violet Spectroscopy. Vol. 6, pp. 94–115.

Jørgensen, C. K.: The Inner Mechanism of Rare Earths Elucidated by Photo-Electron Spectra. Vol. 13, pp. 199–253.

Jørgensen, C. K.: Partly Filled Shells Constituting Anti-bonding Orbitals with Higher Ionization Energy than their Bonding Counterparts. Vol. 22, pp. 49–81.

Jørgensen, C. K.: Photo-electron Spectra of Non-metallic Solids and Consequences for Quantum Chemistry. Vol. 24, pp. 1–58.

Jørgensen, C. K.: Narrow Band Thermoluminescence (Candoluminescence) of Rare Earths in Auer Mantles. Vol. 25, pp. 1–20.

Jørgensen, C. K.: Deep-lying Valence Orbitals and Problems of Degeneracy and Intensities in Photoelectron Spectra. Vol. 30, pp. 141–192.

Jørgensen, C. K.: Predictable Quarkonium Chemistry. Vol. 34, pp. 19–38.

Jørgensen, C. K.: The Conditions for Total Symmetry Stabilizing Molecules, Atoms, Nuclei and Hadrons. Vol. 43, pp. 1–36.

Jørgensen, C. K., Reisfeld, R.: Uranyl Photophysics. Vol. 50, pp. 121–171.

O'Keeffe, M., Hyde, B. G.: An Alternative Approach to Non-Molecular Crystal Structures with Emphasis on the Arrangements of Cations. Vol. 61, pp. 77–144.

Kahn, O.: Magnetism of the Heteropolymetallic Systems. Vol. 68, pp. 89–167.

Kimura, T.: Biochemical Aspects of Iron Sulfur Linkage in None-Heme Iron Protein, with Special Reference to "Adrenodoxin". Vol. 5, pp. 1–40.

Kitagawa, T., Ozaki, Y.: Infrared and Raman Spectra of Metalloporphyrins. Vol. 64, pp. 71–114.

Kiwi, J., Kalyanasundaram, K., Grätzel, M.: Visible Light Induced Cleavage of Water into Hydrogen and Oxygen in Colloidal and Microheterogeneous Systems. Vol. 49, pp. 37–125.

Kjekshus, A., Rakke, T.: Considerations on the Valence Concept. Vol. 19, pp. 45–83.

Kjekshus, A., Rakke, T.: Geometrical Considerations on the Marcasite Type Structure. Vol. 19, pp. 85–104.

König, E.: The Nephelauxetic Effect. Calculation and Accuracy of the Interelectronic Repulsion Parameters I. Cubic High-Spin d^2, d^3, d^7 and d^8 Systems. Vol. 9, pp. 175–212.

Köpf-Maier, P., Köpf, H.: Transition and Main-Group Metal Cyclopentadienyl Complexes: Preclinical Studies on a Series of Antitumor Agents of Different Structural Type. Vol. 70, pp. 103–185.

Koppikar, D. K., Sivapullaiah, P. V., Ramakrishnan, L., Soundararajan, S.: Complexes of the Lanthanides with Neutral Oxygen Donor Ligands. Vol. 34, pp. 135–213.

Krause, R.: Synthesis of Ruthenium(II) Complexes of Aromatic Chelating Heterocycles: Towards the Design of Luminescent Compounds. Vol. 67, pp. 1–52.

Krumholz, P.: Iron(II) Diimine and Related Complexes. Vol. 9, pp. 139–174.
Kustin, K., McLeod, G. C., Gilbert, T. R., Briggs, LeB. R., 4th.: Vanadium and Other Metal Ions in the Physiological Ecology of Marine Organisms. Vol. 53, pp. 137–158.
Labarre, J. F.: Conformational Analysis in Inorganic Chemistry: Semi-Empirical Quantum Calculation vs. Experiment. Vol. 35, pp. 1–35.
Lammers, M., Follmann, H.: The Ribonucleotide Reductases: A Unique Group of Metalloenzymes Essential for Cell Proliferation. Vol. 54, pp. 27–91.
Lehn, J.-M.: Design of Organic Complexing Agents. Strategies towards Properties. Vol. 16, pp. 1–69.
Linarès, C., Louat, A., Blanchard, M.: Rare-Earth Oxygen Bonding in the $LnMO_4$Xenotime Structure. Vol. 33, pp. 179–207.
Lindskog, S.: Cobalt(II) in Metalloenzymes. A Reporter of Structure-Function Relations. Vol. 8, pp. 153–196.
Liu, A., Neilands, J. B.: Mutational Analysis of Rhodotorulic Acid Synthesis in *Rhodotorula pilimanae*. Vol. 58, pp. 97–106.
Livorness, J., Smith, T.: The Role of Manganese in Photosynthesis. Vol. 48, pp. 1–44.
Llinás, M.: Metal-Polypeptide Interactions: The Conformational State of Iron Proteins. Vol. 17, pp. 135–220.
Lucken, E. A. C.: Valence-Shell Expansion Studied by Radio-Frequency Spectroscopy. Vol. 6, pp. 1–29.
Ludi, A., Güdel, H. U.: Structural Chemistry of Polynuclear Transition Metal Cyanides. Vol. 14, pp. 1–21.
Lutz, H. D.: Bonding and Structure of Water Molecules in Solid Hydrates. Correlation of Spectroscopic and Structural Data. Vol. 69, pp. 125.
Maggiora, G. M., Ingraham, L. L.: Chlorophyll Triplet States. Vol. 2, pp. 126–159.
Magyar, B.: Salzebullioskopie III. Vol. 14, pp. 111–140.
Makovicky, E., Hyde, B. G.: Non-Commensurate (Misfit) Layer Structures. Vol. 46, pp. 101–170.
Manes, L., Benedict, U.: Structural and Thermodynamic Properties of Actinide Solids and Their Relation to Bonding. Vol. 59/60, pp. 75–125.
Mann, S.: Mineralization in Biological Systems. Vol. 54, pp. 125–174.
Mason, S. F.: The Ligand Polarization Model for the Spectra of Metal Complexes: The Dynamic Coupling Transition Probabilities. Vol. 39, pp. 43–81.
Mathey, F., Fischer, J., Nelson, J. H.: Complexing Modes of the Phosphole Moiety. Vol. 55, pp. 153–201.
Mayer, U., Gutmann, V.: Phenomenological Approach to Cation-Solvent Interactions. Vol. 12, pp. 113–140.
Mildvan, A. S., Grisham, C. M.: The Role of Divalent Cations in the Mechanism of Enzyme Catalyzed Phosphoryl and Nucleotidyl. Vol. 20, pp. 1–21.
Mingos, D. M. P., Hawes, J. C.: Complementary Spherical Electron Density Model. Vol. 63, pp. 1–63.
Mingos, D. M. P., Johnston, R. L.: Theoretical Models of Cluster Bonding. Vol. 68, pp. 29–87.
Moreau-Colin, M. L.: Electronic Spectra and Structural Properties of Complex Tetracyanides of Platinum, Palladium and Nickel. Vol. 10, pp. 167–190.
Morgan, B., Dolphin, D.: Synthesis and Structure of Biometic Porphyrins. Vol. 64, pp. 115–204.
Morris, D. F. C.: Ionic Radii and Enthalpies of Hydration of Ions. Vol. 4, pp. 63–82.
Morris, D. F. C.: An Appendix to Structure and Bonding. Vol. 4 (1968). Vol. 6, pp. 157–159.
Mortensen, O. S.: A Noncommuting-Generator Approach to Molecular Symmetry. Vol. 68, pp. 1–28.
Mortier, J. W.: Electronegativity Equalization and its Applications. Vol. 66, pp. 125–143.
Müller, A., Baran, E. J., Carter, R. O.: Vibrational Spectra of Oxo-, Thio-, and Selenometallates of Transition Elements in the Solid State. Vol. 26, pp. 81–139.
Müller, A., Diemann, E., Jørgensen, C. K.: Electronic Spectra of Tetrahedral Oxo, Thio and Seleno Complexes Formed by Elements of the Beginning of the Transition Groups. Vol. 14, pp. 23–47.
Müller, U.: Strukturchemie der Azide. Vol. 14, pp. 141–172.
Müller, W., Spirlet, J.-C.: The Preparation of High Purity Actinide Metals and Compounds. Vol. 59/60, pp. 57–73.
Mullay, J. J.: Estimation of Atomic and Group Electronegativities. Vol. 66, pp. 1–25.
Murrell, J. N.: The Potential Energy Surfaces of Polyatomic Molecules. Vol. 32, pp. 93–146.
Naegele, J. R., Ghijsen, J.: Localization and Hybridization of 5f States in the Metallic and Ionic Bond as Investigated by Photoelectron Spectroscopy. Vol. 59/60, pp. 197–262.

Nag, K., Bose, S. N.: Chemistry of Tetra- and Pentavalent Chromium. Vol. 63, pp. 153–197.
Neilands, J. B.: Naturally Occurring Non-porphyrin Iron Compounds. Vol. 1, pp. 59–108.
Neilands, J. B.: Evolution of Biological Iron Binding Centers. Vol. 11, pp. 145–170.
Neilands, J. B.: Methodology of Siderophores. Vol. 58, pp. 1–24.
Nieboer, E.: The Lanthanide Ions as Structural Probes in Biological and Model Systems. Vol. 22, pp. 1–47.
Novack, A.: Hydrogen Bonding in Solids. Correlation of Spectroscopic and Christallographic Data. Vol. 18, pp. 177–216.
Nultsch, W., Häder, D.-P.: Light Perception and Sensory Transduction in Photosynthetic Prokaryotes. Vol. 41, pp. 111–139.
Odom, J. D.: Selenium Biochemistry. Chemical and Physical Studies. Vol. 54, pp. 1–26.
Oelkrug, D.: Absorption Spectra and Ligand Field Parameters of Tetragonal $3d$-Transition Metal Fluorides. Vol. 9, pp. 1–26.
Oosterhuis, W. T.: The Electronic State of Iron in Some Natural Iron Compounds: Determination by Mössbauer and ESR Spectroscopy. Vol. 20, pp. 59–99.
Orchin, M., Bollinger, D. M.: Hydrogen-Deuterium Exchange in Aromatic Compounds. Vol. 23, pp. 167–193.
Peacock, R. D.: The Intensities of Lanthanide $f \longleftrightarrow f$ Transitions. Vol. 22, pp. 83–122.
Penneman, R. A., Ryan, R. R., Rosenzweig, A.: Structural Systematics in Actinide Fluoride Complexes. Vol. 13, pp. 1–52.
Powell, R. C., Blasse, G.: Energy Transfer in Concentrated Systems. Vol. 42, pp. 43–96.
Que, Jr., L.: Non-Heme Iron Dioxygenases. Structure and Mechanism. Vol. 40, pp. 39–72.
Ramakrishna, V. V., Patil, S. K.: Synergic Extraction of Actinides. Vol. 56, pp. 35–90.
Raymond, K. N., Smith, W. L.: Actinide-Specific Sequestering Agents and Decontamination Applications. Vol. 43, pp. 159–186.
Reedijk, J., Fichtinger-Schepman, A. M. J., Oosterom, A. T. van, Putte, P. van de: Platinum Amine Coordination Compounds as Anti-Tumor Drugs. Molecular Aspects of the Mechanism of Action. Vol. 67, pp. 53–89.
Reinen, D.: Ligand-Field Spectroscopy and Chemical Bonding in Cr^{3+}-Containing Oxidic Solids. Vol. 6, pp. 30–51.
Reinen, D.: Kationenverteilung zweiwertiger $3d^n$-Ionen in oxidischen Spinell-, Granat- und anderen Strukturen. Vol. 7, pp. 114–154.
Reinen, D., Friebel, C.: Local and Cooperative Jahn-Teller Interactions in Model Structures. Spectroscopic and Structural Evidence. Vol. 37, pp. 1–60.
Reisfeld, R.: Spectra and Energy Transfer of Rare Earths in Inorganic Glasses. Vol. 13, pp. 53–98.
Reisfeld, R.: Radiative and Non-Radiative Transitions of Rare Earth Ions in Glasses. Vol. 22, pp. 123–175.
Reisfeld, R.: Excited States and Energy Transfer from Donor Cations to Rare Earths in the Condensed Phase. Vol. 30, pp. 65–97.
Reisfeld, R., Jørgensen, C. K.: Luminescent Solar Concentrators for Energy Conversion. Vol. 49, pp. 1–36.
Reisfeld, R., Jørgensen, C. K.: Excited States of Chromium(III) in Translucent Glass-Ceramics as Prospective Laser Materials. Vol. 69, pp. 63–96.
Russo, V. E. A., Galland, P.: Sensory Physiology of *Phycomyces Blakesleeanus.* Vol. 41, pp. 71–110.
Rüdiger, W.: Phytochrome, a Light Receptor of Plant Photomorphogenesis. Vol. 40, pp. 101–140.
Ryan, R. R., Kubas, G. J., Moody, D. C., Eller, P. G.: Structure and Bonding of Transition Metal-Sulfur Dioxide Complexes. Vol. 46, pp. 47–100.
Sadler, P. J.: The Biological Chemistry of Gold: A Metallo-Drug and Heavy-Atom Label with Variable Valency. Vol. 29, pp. 171–214.
Schäffer, C. E.: A Perturbation Representation of Weak Covalent Bonding. Vol. 5, pp. 68–95.
Schäffer, C. E.: Two Symmetry Parameterizations of the Angular-Overlap Model of the Ligand-Field. Relation to the Crystal-Field Model. Vol. 14, pp. 69–110.
Scheidt, W. R., Lee, Y. J.: Recent Advances in the Stereochemistry of Metallotetrapyrroles. Vol. 64, pp. 1–70.
Schmid, G.: Developments in Transition Metal Cluster Chemistry. The Way to Large Clusters. Vol. 62, pp. 51–85.
Schmidt, P. C.: Electronic Structure of Intermetallic B 32 Type Zintl Phases. Vol. 65, pp. 91–133.
Schneider, W.: Kinetics and Mechanism of Metalloporphyrin Formation. Vol. 23, pp. 123–166.

Schubert, K.: The Two-Correlations Model, a Valence Model for Metallic Phases. Vol. 33, pp. 139–177.

Schutte, C. J. H.: The Ab-Initio Calculation of Molecular Vibrational Frequencies and Force Constants. Vol. 9, pp. 213–263.

Schweiger, A.: Electron Nuclear Double Resonance of Transition Metal Complexes with Organic Ligands. Vol. 51, pp. 1–122.

Sen, K. D., Böhm, M. C., Schmidt, P. C.: Electronegativity of Atoms and Molecular Fragments. Vol. 66, pp. 99–123.

Shamir, J.: Polyhalogen Cations. Vol. 37, pp. 141–210.

Shannon, R. D., Vincent, H.: Relationship between Covalency, Interatomic Distances, and Magnetic Properties in Halides and Chalcogenides. Vol. 19, pp.1–43.

Shriver, D. F.: The Ambident Nature of Cyanide. Vol. 1, pp. 32–58.

Siegel, F. L.: Calcium-Binding Proteins. Vol. 17, pp. 221–268.

Simon, A.: Structure and Bonding with Alkali Metal Suboxides. Vol. 36, pp. 81–127.

Simon, W., Morf, W. E., Meier, P. Ch.: Specificity for Alkali and Alkaline Earth Cations of Synthetic and Natural Organic Complexing Agents in Membranes. Vol. 16, pp. 113–160.

Simonetta, M., Gavezzotti, A.: Extended Hückel Investigation of Reaction Mechanisms. Vol. 27, pp. 1–43.

Sinha, S. P.: Structure and Bonding in Highly Coordinated Lanthanide Complexes. Vol. 25, pp. 67–147.

Sinha, S. P.: A Systematic Correlation of the Properties of the f-Transition Metal Ions. Vol. 30, pp. 1–64.

Schmidt, W.: Physiological Bluelight Reception. Vol. 41, pp. 1–44.

Smith, D. W.: Ligand Field Splittings in Copper(II) Compounds. Vol. 12, pp. 49–112.

Smith, D. W., Williams, R. J. P.: The Spectra of Ferric Haems and Haemoproteins, Vol. 7, pp. 1–45.

Smith, D. W.: Applications of the Angular Overlap Model. Vol. 35, pp. 87–118.

Solomon, E. I., Penfield, K. W., Wilcox, D. E.: Active Sites in Copper Proteins. An Electric Structure Overview. Vol. 53, pp. 1–56.

Somorjai, G. A., Van Hove, M. A.: Adsorbed Monolayers on Solid Surfaces. Vol. 38, pp. 1–140.

Speakman, J. C.: Acid Salts of Carboxylic Acids, Crystals with some "Very Short" Hydrogen Bonds. Vol. 12, pp. 141–199.

Spiro, G., Saltman, P.: Polynuclear Complexes of Iron and their Biological Implications. Vol. 6, pp. 116–156.

Strohmeier, W.: Problem and Modell der homogenen Katalyse. Vol. 5, pp. 96–117.

Sugiura, Y., Nomoto, K.: Phytosiderophores – Structures and Properties of Mugineic Acids and Their Metal Complexes. Vol. 58, pp. 107–135.

Tam, S.-C., Williams, R. J. P.: Electrostatics and Biological Systems. Vol. 63, pp. 103–151.

Teller, R., Bau, R. G.: Crystallographic Studies of Transition Metal Hydride Complexes. Vol. 44, pp. 1–82.

Thompson, D. W.: Structure and Bonding in Inorganic Derivates of β-Diketones. Vol. 9, pp. 27–47.

Thomson, A. J., Williams, R. J. P., Reslova, S.: The Chemistry of Complexes Related to *cis*-$Pt(NH_3)_2Cl_2$. An Anti Tumor Drug. Vol. 11, pp. 1–46.

Tofield, B. C.: The Study of Covalency by Magnetic Neutron Scattering. Vol. 21, pp. 1–87.

Trautwein, A.: Mössbauer-Spectroscopy on Heme Proteins. Vol. 20, pp. 101–167.

Tressaud, A., Dance, J.-M.: Relationships Between Structure and Low-Dimensional Magnetism in Fluorides. Vol. 52, pp. 87–146.

Tributsch, H.: Photoelectrochemical Energy Conversion Involving Transition Metal d-States and Intercalation of Layer Compounds. Vol. 49, pp. 127–175.

Truter, M. R.: Structures of Organic Complexes with Alkali Metal Ions. Vol. 16, pp. 71–111.

Umezawa, H., Takita, T.: The Bleomycins: Antitumor Copper-Binding Antibiotics. Vol. 40, pp. 73–99.

Vahrenkamp, H.: Recent Results in the Chemistry of Transition Metal Clusters with Organic Ligands. Vol. 32, pp. 1–56.

Valach, F., Koreň, B., Sivý, P., Melník, M.: Crystal Structure Non-Rigidity of Central Atoms for Mn(II), Fe(II), Fe(III), Co(II), Co(III), Ni(II), Cu(II) and Zn(II) Complexes. Vol. 55, pp. 101–151.

Wallace, W. E., Sankar, S. G., Rao, V. U. S.: Field Effects in Rare-Earth Intermetallic Compounds. Vol. 33, pp. 1–55.

Warren, K. D.: Ligand Field Theory of Metal Sandwich Complexes. Vol. 27, pp. 45–159.
Warren, K. D.: Ligand Field Theory of f-Orbital Sandwich Complexes. Vol. 33, pp. 97–137.
Warren, K. D.: Calculations of the Jahn-Teller Coupling Costants for d^x Systems in Octahedral Symmetry via the Angular Overlap Model. Vol. 57, pp. 119–145.
Watson, R. E., Perlman, M. L.: X-Ray Photoelectron Spectroscopy. Application to Metals and Alloys. Vol. 24, pp. 83–132.
Weakley, T. J. R.: Some Aspects of the Heteropolymolybdates and Heteropolytungstates. Vol. 18, pp. 131–176.
Wendin, G.: Breakdown of the One-Electron Pictures in Photoelectron Spectra. Vol. 45, pp. 1–130.
Weissbluth, M.: The Physics of Hemoglobin. Vol. 2, pp. 1–125.
Weser, U.: Chemistry and Structure of some Borate Polyol Compounds. Vol. 2, pp. 160–180.
Weser, U.: Reaction of some Transition Metals with Nucleic Acids and their Constituents. Vol. 5, pp. 41–67.
Weser, U.: Structural Aspects and Biochemical Function of Erythrocuprein. Vol. 17, pp. 1–65.
Weser, U.: Redox Reactions of Sulphur-Containing Amino-Acid Residues in Proteins and Metalloproteins, an XPS-Study. Vol. 61, pp. 145–160.
Willemse, J., Cras, J. A., Steggerda, J. J., Keijzers, C. P.: Dithiocarbamates of Transition Group Elements in "Unusual" Oxidation State. Vol. 28, pp. 83–126.
Williams, R. J. P.: The Chemistry of Lanthanide Ions in Solution and in Biological Systems. Vol. 50, pp. 79–119.
Williams, R. J. P., Hale, J. D.: The Classification of Acceptors and Donors in Inorganic Reactions. Vol. 1, pp. 249–281.
Williams, R. J. P., Hale, J. D.: Professor Sir Ronald Nyholm. Vol. 15, pp. 1 and 2.
Wilson, J. A.: A Generalized Configuration-Dependent Band Model for Lanthanide Compounds and Conditions for Interconfiguration Fluctuations. Vol. 32, pp. 57–91.
Winkler, R.: Kinetics and Mechanism of Alkali Ion Complex Formation in Solution. Vol. 10, pp. 1–24.
Wood, J. M., Brown, D. G.: The Chemistry of Vitamin B_{12}-Enzymes. Vol. 11, pp. 47–105.
Woolley, R. G.: Natural Optical Activity and the Molecular Hypothesis. Vol. 52, pp. 1–35.
Wüthrich, K.: Structural Studies of Hemes and Hemoproteins by Nuclear Magnetic Resonance Spectroscopy. Vol. 8, pp. 53–121.
Xavier, A. V., Moura, J. J. G., Moura, I.: Novel Structures in Iron-Sulfur Proteins. Vol. 43, pp. 187–213.
Zumft, W. G.: The Molecular Basis of Biological Dinitrogen Fixation. Vol. 29, pp. 1–65.